AF595344

Power System Simulation, Control and Optimization

Power System Simulation, Control and Optimization

Editors

José Antonio Domínguez-Navarro
José María Yusta-Loyo

MDPI • Basel • Beijing • Wuhan • Barcelona • Belgrade • Manchester • Tokyo • Cluj • Tianjin

Editors
José Antonio Domínguez-Navarro
University of Zaragoza
Spain

José María Yusta-Loyo
University of Zaragoza
Spain

Editorial Office
MDPI
St. Alban-Anlage 66
4052 Basel, Switzerland

This is a reprint of articles from the Special Issue published online in the open access journal *Energies* (ISSN 1996-1073) (available at: https://www.mdpi.com/journal/energies/special_issues/power_simulation).

For citation purposes, cite each article independently as indicated on the article page online and as indicated below:

LastName, A.A.; LastName, B.B.; LastName, C.C. Article Title. *Journal Name* **Year**, *Volume Number*, Page Range.

ISBN 978-3-0365-0748-4 (Hbk)
ISBN 978-3-0365-0749-1 (PDF)

Contents

About the Editors

José Antonio Domínguez-Navarro has been Associate Professor since 2003 at the Department of Electrical Engineering of University of Zaragoza, Spain, and an IEEE Senior Member. His research was done at the INESCN research center in Oporto (Portugal) in 1993, at the University of Strathclyde in Glasgow (United Kingdom) in 2013, and at the Norwegian University of Science and Technology in Trondheim (Norway) in 2015. He has published 32 indexed articles in JCR and participated in 40 congresses. The lines of research in which he has worked are: Electrical network planning, renewable energy integration, and application of computing techniques (neural networks, fuzzy systems, and heuristic optimization algorithms) in power systems.

José María Yusta-Loyo has been a Lecturer since 1994 and Associate Professor since 2003 at the Department of Electrical Engineering of University of Zaragoza, Spain. He is also an IEEE Senior Member, and was Visiting Scientist at the Joint Research Center of the European Commission in Petten, The Netherlands, unit "Energy security, systems and markets", in 2019. He has 40 scientific articles published in JCR ranked journals and 70 conference contributions. He has been involved in 30 competitive R&D projects and 70 contracts. Vicedean of the School of Engineering—University of Zaragoza (2004–2007). He is a member of the board and secretary of the Association of Professional Engineers (2005–2017). His areas of scientific interest are electricity markets, renewables integration, and critical energy infrastructures.

Article

Sensitivity Analysis of the Impact of the Sub-Hourly Stochastic Unit Commitment on Power System Dynamics

Taulant Kërçi [1], Juan S. Giraldo [2] and Federico Milano [1,*]

[1] School of Electrical and Electronic Engineering, University College Dublin, Belfield, Dublin 4, Ireland; taulant.kerci@ucdconnect.ie

[2] Electrical Energy Systems, Department of Electrical Engineering, Eindhoven University of Technology, 5600 Eindhoven, The Netherlands; jnse@ieee.org

* Correspondence: federico.milano@ucd.ie

Received: 17 February 2020; Accepted: 17 March 2020; Published: 20 March 2020

Abstract: Subhourly modeling of power systems and the use of the stochastic optimization are two relevant solutions proposed in the literature to address the integration of stochastic renewable energy sources. With this aim, this paper deals with the effect of different formulations of the subhourly stochastic unit commitment (SUC) problem on power system dynamics. Different SUC models are presented and embedded into time domain simulations (TDS) through a cosimulation platform. The objective of the paper is to study the combined impact of different frequency control/machine parameters and different SUC formulations on the long-term dynamic behaviour of power systems. The analysis is based on extensive Monte Carlo TDS (MC-TDS) and a variety of scenarios based on the New England 39-bus system.

Keywords: subhourly modeling; stochastic unit commitment; cosimulation method; sensitivity analysis; power system dynamic performance

---

1. Introduction

1.1. Motivation

Subhourly modeling of power systems is seen as a solution to different problems in power systems. For example, ENTSO-E has recommended that in order to tackle the phenomena of deterministic frequency deviation (DFD) in the European power systems the following measures have to be taken: *Introduction of 15-min market schedules and balancing; Introduction of 15-min period imbalance settlement in each balancing area* [1]. Subhourly modeling is thus to be preferred compared to the conventional hourly dispatch. Given its short time scale, it makes sense to embed the subhourly unit commitment (UC) into a time domain simulator as it can overlap with relevant long-term power system dynamics [2], and analyse its impact on power system dynamic behaviour [3]. On the other hand, while it is well-known that inertia of synchronous machines is the parameter on which system stability mostly depends in the first seconds after a major contingency, other frequency control parameters help keeping the frequency within certain limits most of the time (which is another major task of system operators). To study the effect of these parameters on long-term frequency deviations, one would need to use dynamic equations with stochastic terms and perform long-term dynamic simulations [4].

The objective of the paper is to assess the impact of different implementations of stochastic UC problem formulations on power system dynamics and their sensitivity with respect to the parameters of primary and secondary controllers of conventional power plants.

1.2. Literature Review

Subhourly variability of renewable energy sources (RES) is not the only challenge that system operators have to deal with. The inherent uncertainty related to RES forecast, in particular wind and solar power, has also to be considered when scheduling the system [5]. A way to address this problem proposed in the literature is by making use of subhourly stochastic UC formulations (SUC) [6]. For instance, in [7] the authors propose a stochastic real-time market with 15-min dispatch intervals and show that it outperforms the relevant deterministic approach. In [8], a stochastic economic dispatch (15 min dispatch interval) is proposed where the authors conclude that the stochastic approach leads to lower operating costs compared to that of the deterministic approach. Nevertheless, they acknowledge the computational burden of the stochastic approach as the main limitation for practical applications.

There is a concern about the impact of low level of inertia and reduced frequency regulation on the long-term dynamic behaviour of power systems as the RES penetration increases [9]. Regarding UC models, this issue has been tackled so far by proposing some linear frequency constraints [10]. The major limitation of these approaches is that the dynamics of the system are oversimplified. The recently proposed cosimulation framework proposed in [3] solves this issue and is further developed in this work.

There are currently very few works that focus on the effect of frequency control parameters on the frequency deviations of the system. In [11], the authors propose a feedback controller method for frequency control and perform a sensitivity analysis of the effect of varying governor droop parameter (R). This reference shows that a value of R between 3% and 4% leads to a better dynamic behaviour of the system after a disturbance. However, only short-term dynamic simulations are considered. In a recent work [4], the authors propose a method that allows finding the frequency probability density function of power systems, and conclude that the dead-band width of turbine governors (TGs) and the aggregate droop are the only parameters that have great impact on the long-term frequency deviations, while inertia has little impact.

In this work, we use as starting point [12,13]. Reference [12] discusses the impact of a subhourly deterministic UC problem power system dynamics. Whereas the focus of [13] is on the effect of different SUC strategies and different wind power uncertainty and volatility within SUC on power system dynamics. Based on these works, we present here a thorough sensitivity analysis related to different control parameters as well as a comparison on the effect of different SUC models on the long-term dynamic behaviour of power systems.

1.3. Contributions

This paper provides the following original contributions.

- A cosimulation framework that allows to assess the effect that different subhourly SUC models have on the long-term dynamic behaviour of power systems.
- A thorough sensitivity analysis with respect to the interaction between different subhourly SUC models, different frequency control/machine parameters and power system dynamics.
- Show through the aforementioned analysis that synchronous machine inertia has little effect on the standard deviation of the frequency. This result is in accordance with the discussion provided in [4].
- Show that the gain of automatic generation control (AGC) is (most of the time) the main parameter affecting the standard deviation of the frequency.

1.4. Organization

The rest of the paper is organised as follows. Section 2 starts by presenting the long-term dynamic model of power systems. Then, it describes the modeling of wind power based on stochastic differential equations (SDEs). Furthermore, it provides the models of primary and secondary frequency controllers as well as the mathematical formulation of the standard mixed-integer linear programming (MILP),

SUC. Section 2 also presents the formulation of a simplified and an alternative model of subhourly SUC. Lastly, Section 2 also describes the cosimulation platform that combines the solution of the SUC problem and the time integration of the SDAEs and explain the interactions between the two tools. Section 3 discusses the results of the case study based on the New England IEEE 39-bus system. Finally, Section 4 provides the relevant conclusions and future work.

2. Modeling

2.1. Stochastic Long-Term Power System Model

The large-scale deployment of stochastic renewable energy sources, e.g., wind, significantly increases the random perturbations and inevitably affects the safe and stable operation of power systems [14]. As a result, the adoption of SDEs to model this randomness in power systems has gained significant interest in recent years. However, most of the literature focus on the impact of wind power variability (modeled as SDEs) on transient and small-signal stability, respectively [15]. On the other hand, stochastic long-term stability analysis of power systems has received little attention. This is of particular importance considering the expected trends regarding the integration of RES into modern power systems.

The stochastic long-term dynamic model of power systems can be represented as a set of hybrid nonlinear SDAEs [16], as follows:

$$\begin{aligned} \dot{\boldsymbol{x}} &= \boldsymbol{f}(\boldsymbol{x}, \boldsymbol{y}, \boldsymbol{u}, \boldsymbol{z}, \dot{\boldsymbol{\eta}}) \,, \\ \boldsymbol{0} &= \boldsymbol{g}(\boldsymbol{x}, \boldsymbol{y}, \boldsymbol{u}, \boldsymbol{z}, \boldsymbol{\eta}) \,, \\ \dot{\boldsymbol{\eta}} &= \boldsymbol{a}(\boldsymbol{x}, \boldsymbol{y}, \boldsymbol{\eta}) + \boldsymbol{b}(\boldsymbol{x}, \boldsymbol{y}, \boldsymbol{\eta})\, \boldsymbol{\zeta} \,, \end{aligned} \tag{1}$$

where $\boldsymbol{f}$ and $\boldsymbol{g}$ represent the differential and algebraic equations, respectively; $\boldsymbol{x}$ and $\boldsymbol{y}$ represent the state and algebraic variables, such as generator rotor speeds and bus voltage angles, respectively; $\boldsymbol{u}$ represents the inputs, such as the schedules of synchronous generators; $\boldsymbol{z}$ represents discrete variables; $\boldsymbol{\eta}$ represents the stochastic characterization of wind speed; $\boldsymbol{a}$ and $\boldsymbol{b}$ are the *drift* and *diffusion* of the SDEs, respectively; and $\boldsymbol{\zeta}$ is the white noise. It is worth mentioning that (1) is solved using numerical integration techniques. This is possible as the algebraic equations $\boldsymbol{g}$, in (1), do not explicitly depend on white noises $\boldsymbol{\zeta}$, or on $\dot{\boldsymbol{\eta}}$ [16]. In this work, an implicit trapezoidal integration method is used for functions, $\boldsymbol{f}$ and $\boldsymbol{a}$, while the Euler–Maruyama method is used to integrate the stochastic term $\boldsymbol{b}$. Further details related to the numerical integration of the SDAEs can be found in [16].

It is necessary to consider both electromechanical and long-term dynamic models, when one performs stochastic long-term dynamic simulations [17]. With this regard, (1) includes the dynamic models of conventional machines (4th order models) and their primary controllers; AGC; wind power plants (5th order Doubly-Fed Induction Generator) [18]; the model of subhourly SUC.

2.2. Wind Power Modeling

Modeling wind power as a stochastic source is crucial in long-term dynamic studies [19]. With this aim, Equation (1) model wind power variations as a stochastic perturbation with respect to the wind generation forecast. More specifically, we use Ornstein–Uhlenbeck processes (modeled as SDEs) to represent the volatile nature of the wind speed variations v_s, that feeds the wind turbines, as follows:

$$\begin{aligned} v_s(t) &= v_{s0} + \eta_v(t) \,, \\ \dot{\eta}_v(t) &= \alpha_v(\mu_v - \eta_v(t)) + b_v \zeta_v \,, \end{aligned} \tag{2}$$

where v_{s0} represents the initial value of the wind speed; η_v represents the stochastic variable which depends on the drift $\alpha_v(\mu_v - \eta_v)$, and the diffusion term b_v of the SDEs; α_v represents the mean reversion speed, and shows how quickly η_v tends towards its mean μ_v; finally, the white noise is represented by ζ_v.

2.3. Primary and Secondary Frequency Controllers of Conventional Power Plants

The displacement of conventional synchronous generators and their replacement with non-synchronous stochastic sources leads to a decrease in the overall inertia of the system, reduction of secondary frequency regulation and an increase of the aggregated system droop [4]. In general, this can lead to large frequency deviations and therefore it is crucial to study such an impact on the dynamic behaviour of power systems. Motivated by the above, the paper make use of a recent cosimulation platform proposed by the authors to perform a sensitivity analysis where the parameters of primary and secondary controllers are varied and their effect on the long-term dynamic behaviour of power systems is observed. In this work, the frequency of the center of inertia of the system, $\omega_{\rm COI}$, is utilised to monitor the overall long-term dynamic behaviour of power systems. Its expression is [18]:

$$\omega_{\rm COI} = \frac{\sum_{g \in \mathcal{G}} M_g \omega_g}{\sum_{g \in \mathcal{G}} M_g}, \tag{3}$$

where ω_g and M_g are the rotor speed and the starting time (which is twice the inertia constant) of the gth synchronous machine, respectively. M_g is an intrinsic part of the machine and as such deeply impacts on the dynamic behaviour of the machine and of the system. It appears in the electromechanical swing equations of the machine, as follows [17]:

$$\begin{aligned} \frac{d\delta_g}{dt} &= \omega_g - \omega_{\rm COI}, \\ M_g \frac{d\omega_g}{dt} &= p_{{\rm m},g} - p_{{\rm e},g}(\delta_g), \end{aligned} \tag{4}$$

where δ_g, $p_{{\rm m},g}$ and $p_{{\rm e},g}$ are the rotor angle, the mechanical power and the electrical power of the gth synchronous machine, respectively.

A standard control scheme of a primary frequency control of a synchronous power plant is depicted in Figure 1 [18]. The model of the primary frequency control of synchronous machines includes an ensemble of the models of the turbine, the valve and the turbine governor. These regulate the mechanical power of the machine ($p_{{\rm m},g}$) through the measurement of the rotor speed ω_g. The conventional primary frequency regulator consists of a droop (R_g) and a lead-lag transfer function that models regulator and turbine combined dynamics. Finally $p_{{\rm ord},g}$ is the active power order set-point as obtained by the solution of the SUC problem ($p_{g,t,\xi}$)—as thoroughly discussed in Section 2.4—and the signal sent by the secondary frequency control (Δp_g)—discussed at the end of this section. Specifically, $p_{{\rm ord},g}$ is given by:

$$p_{{\rm ord},g} = p_{g,t,\xi} + \Delta p_g\,. \tag{5}$$

The main purpose of the primary frequency control is to restore the power system frequency at a quasi-steady state value following a disturbance into the power system. This control is locally implemented and takes place on time scales of tens of seconds.

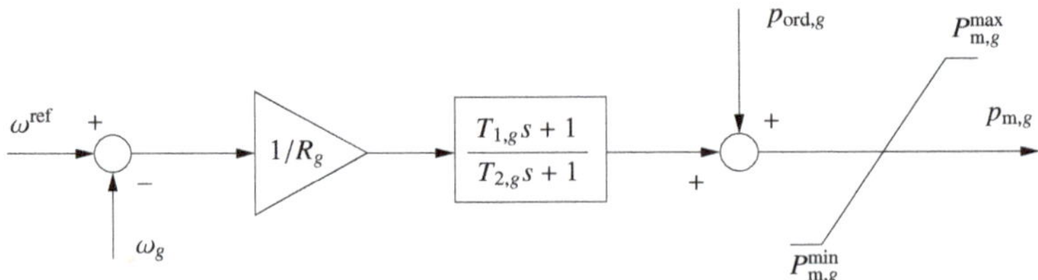

Figure 1. Standard turbine governor control scheme.

A simple control scheme of a secondary frequency control or automatic generation control (AGC) is shown in Figure 2. In its simplest implementation, the AGC consists of an integrator block with gain K_0 that coordinates the TGs of synchronous machines to nullify the frequency steady-state error originated by the primary frequency control by sending the active power corrections set-points Δp_g. These signals are proportional to the capacity of the machines and the TG droops R_g. The AGC is a centralised control and takes place on the time scales of tens of minutes.

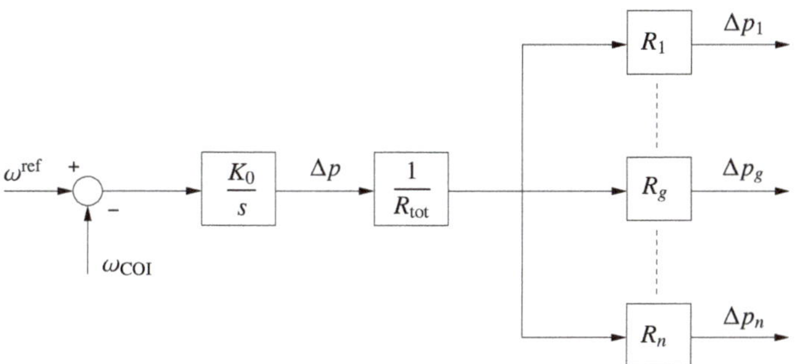

Figure 2. Simple automatic generation control (AGC) control scheme.

An in-depth discussion of power system dynamics and frequency control is beyond the scope of this paper. The interested reader can find further information regarding the relevant relationship between the above variables, namely, inertia M, droop R, and gain K, in dedicated monographs, e.g., [17,18].

2.4. Stochastic Unit Commitment

Over the last decade, there has been a significant interest to better represent uncertainty in the UC models due to the integration of highly variable RES. Traditionally, the uncertainty in these models have been managed by scheduling some specific amount of reserves (e.g., as a percentage of the total demand). However, this may lead to suboptimal solution of the UC problem and thus impact system security. A common solution to this problem is to use stochastic optimization, in particular, two-stage stochastic programming [20]. A two-stage SUC model uses a probabilistic description of uncertain nature of parameters, e.g., wind power generation, in the form of scenarios, [21], as follows:

$$\min_{H,W} \sum_{t\in\mathcal{T}} \sum_{g\in\mathcal{G}} (C_g^F z_{g,t}^F + C_g^{SU} z_{g,t}^{SU} + C_g^{SD} z_{g,t}^{SD}) \tag{6}$$

$$+ \sum_{\xi\in\Xi} \pi_\xi \left[\sum_{t\in\mathcal{T}} \sum_{l\in\mathcal{L}} C_g^V p_{g,t,\xi} + \sum_{t\in\mathcal{T}} \sum_{l\in\mathcal{L}} C^L L_{l,t,\xi}^{SH} \right]$$

such that

$$z_{g,t}^{SU} - z_{g,t}^{SD} = z_{g,t}^F - z_{g,t-1}^F, \quad \forall g \in \mathcal{G}, \forall t \in \{2..., T\} \tag{7}$$

$$z_{g,t}^{SU} - z_{g,t}^{SD} = z_{g,t}^F - IS_g, \quad \forall g, \forall t \in \{1\} \tag{8}$$

$$z_{g,t}^{SU} + z_{g,t}^{SD} \leq 1, \quad \forall g, \forall t \in \{1..., T\} \tag{9}$$

$$z_{g,t}^F = IS_g, L_g^{UP} + L_g^{DW} > 0, \quad \forall g, \forall t \leq L_g^{UP} + L_g^{DW} \tag{10}$$

$$\sum_{\tau=t-UT_g+1}^{t} z_{g,\tau}^{SU} \leq z_{g,t}^F, \quad \forall g, \forall t > L_g^{UP} + L_g^{DW} \tag{11}$$

$$\sum_{\tau=t-DT_g+1}^{t} z_{g,\tau}^{SD} \leq 1 - z_{g,t}^F, \quad \forall g, \forall t > L_g^{UP} + L_g^{DW} \tag{12}$$

$$\sum_{g\in\mathcal{G}_n} p_{g,t,\xi} - \sum_{l\in\mathcal{L}_n} L_{l,t} + \sum_{l\in\mathcal{L}_n} L^{SH}_{l,t,\xi} + \sum_{f\in\mathcal{F}_n} W_{f,t,\xi} \tag{13}$$

$$- \sum_{f\in\mathcal{F}_n} W^{SP}_{f,t,\xi} = \sum_{m\in\mathcal{M}_n} \frac{(\theta_{n,t,\xi} - \theta_{m,t,\xi})}{X_{n,m}}, \qquad \forall n, \forall t, \forall \xi \in \Xi$$

$$p_{g,t,\xi} \leq P^{\max}_g z^F_{g,t}, \qquad \forall g, \forall t, \forall \xi \in \Xi \tag{14}$$

$$p_{g,t,\xi} \geq P^{min}_g z^F_{g,t}, \qquad \forall g, \forall t, \forall \xi \in \Xi \tag{15}$$

$$p_{g,t,\xi} \leq (P^{IS}_g + RU_g) z^F_{g,t}, \qquad \forall g, \forall t \in \{1\}, \forall \xi \in \Xi \tag{16}$$

$$p_{g,t,\xi} \geq (P^{IS}_g - RD_g) z^F_{g,t}, \qquad \forall g, \forall t \in \{1\}, \forall \xi \in \Xi \tag{17}$$

$$p_{g,t,\xi} - p_{g,t-1,\xi} \leq (2 - z^F_{g,t-1} - z^F_{g,t}) P^{SU}_g \tag{18}$$

$$+ (1 + z^F_{g,t-1} - z^F_{g,t}) RU_g), \qquad \forall g, \forall t \in \{2, ..., T\}, \forall \xi \in \Xi$$

$$p_{g,t-1,\xi} - p_{g,t,\xi} \leq (2 - z^F_{g,t-1} - z^F_{g,t}) P^{SD}_g \tag{19}$$

$$+ (1 - z^F_{g,t-1} + z^F_{g,t}) RD_g), \qquad \forall g, \forall t \in \{2, ..., T\}, \forall \xi \in \Xi$$

$$L^{SH}_{l,t,\xi} \leq L_{l,t}, \qquad \forall l, \forall t, \forall \xi \in \Xi \tag{20}$$

$$W^{SP}_{f,t,\xi} \leq W_{f,t,\xi}, \qquad \forall f, \forall t, \forall \xi \in \Xi \tag{21}$$

$$- P^{\max}_{n,m} \leq \frac{(\theta_{n,t,\xi} - \theta_{m,t,\xi})}{X_{n,m}} \leq P^{\max}_{n,m}, \qquad \forall n, m \in M_n, \forall t, \forall \xi \in \Xi \tag{22}$$

$$p_{g,t,\xi}, L^{SH}_{l,t,\xi}, W^{SP}_{f,t,\xi} \geq 0, \qquad \forall g, \forall l, \forall f, \forall t, \forall \xi \in \Xi \tag{23}$$

$$z^F_{g,t}, z^{SU}_{g,t}, z^{SD}_{g,t} \in \{0,1\}, \qquad \forall g, \forall t \tag{24}$$

with initial state conditions:

$$IS_g = \begin{cases} 1 & \text{if } ON_g > 0 \\ 0 & \text{if } ON_g = 0 \end{cases}$$
$$L^{UP}_g = min\{T, (UT_g - ON_g) IS_g\}$$
$$L^{DW}_g = min\{T, (DT_g - OFF_g)(1 - IS_g)\}$$

The SUC problem (6)–(24) is a standard model proposed in the literature. Its objective is to minimise the fixed cost (C^F_g), start-up cost (C^{SU}_g), shut-down cost (C^{SD}_g), and variable cost (C^V_g) of the synchronous generators, as well as the cost of involuntarily demand curtailment, (6). Equations (7)–(9) represent the logical expression between different binary variables (ON/OFF commitment status). Equations (10)–(12) represent the minimum and maximum up- and down-time constraints of the synchronous generators. Equation (13) model the power balance constraint. Equations (14) and (15) model the capacity limits of synchronous generators, while their respective ramping limits are modeled through (16)–(19). The limits of the demand and wind power curtailment are modeled through (20) and (21). Equation (22) represent the transmission capacity limits. Last, Equations (23) and (24) represent variable declarations.

The SUC problem (6)–(24) includes first-stage variables $z^F_{g,t}, z^{SU}_{g,t}, z^{SD}_{g,t}$ that model the status of the machines g in time period t (e.g., start-up/shut-down status) and second-stage variables $p_{g,t,\xi}, L^{SH}_{l,t,\xi}, W^{SP}_{f,t,\xi}, \theta_{n,t,\xi}$ that model the active power of the synchronous machines g, the power curtailment from load l, wind power curtailment from wind production unit f, and voltage angle at node n, in time period t and scenario ξ, respectively. The interested reader can find further details of the SUC in [22] and references therein.

The problem (6)–(24) is based on [22] and is the reference complete SUC formulation considered in the case study. Note that the cosimulation framework is general and allows to assess the impact of any other model of SUC on power system dynamics.

If one considers only one scenario, the set of Equations (6)–(24) reduces to a deterministic UC problem. In the following we use the notation "SUC" to indicate a stochastic formulation, i.e., the set Ξ consists of more than one scenario, and the notation "DUC" for the deterministic case.

2.5. Simplified SUC Formulation

The level of complexity of SUC formulations proposed so far in the literature vary significantly [23]. For example, a well-assessed MILP SUC formulation is provided in [22]. Such a model takes into consideration several technical constraints, e.g., ramping limits of generators and capacity limits of transmission lines, just to mention some. These constraints improve the performance of the schedules but are not crucial in this paper, which focuses on the impact on long-term power system dynamics. Hence, a simplified model of SUC is considered, as follows:

$$\underset{z^F_{g,t},p_{g,t,\xi}}{\text{minimize}} \sum_{t\in\Xi_T}\sum_{g\in\mathcal{G}} C^F_g z^F_{g,t} + \sum_{\xi\in\Xi}\pi_\xi\left[\sum_{t\in\Xi_T}\sum_{g\in\Xi_G} C^V_g p_{g,t,\xi}\right] \tag{25}$$

such that

$$p_{g,t,\xi} \le P^{\max}_g z^F_{g,t}, \qquad \forall g,\forall t,\forall\xi\in\Xi \tag{26}$$

$$p_{g,t,\xi} \ge P^{min}_g z^F_{g,t}, \qquad \forall g,\forall t,\forall\xi\in\Xi \tag{27}$$

$$\sum_{g\in\Xi_{G_n}} p_{g,t,\xi} - \sum_{l\in\Xi_{L_n}} L_{l,t} + \sum_{k\in\Xi_{K_n}} W_{k,t,\xi} = \sum_{m\in\Xi_{M_n}} \frac{(\theta_{n,t,\xi}-\theta_{m,t,\xi})}{X_{n,m}}, \qquad \forall n,\forall t,\forall\xi\in\Xi \tag{28}$$

$$z^F_{g,t}\in\{0,1\}, \qquad \forall g\in\Xi_G,\forall t\in\Xi_T \tag{29}$$

where $z^F_{g,t}$ is a first-stage decision variable that models the status (ON/OFF) of the conventional machines in time period t; C^F_g and C^V_g represents the fixed and variable production cost of generation unit g, respectively; π_ξ is the probability of wind power scenario ξ; $p_{g,t,\xi}$ is a variable that represents the active power of the machine g in scenario ξ and time period t; $P^{\max}_g$ and $P^{\min}_g$ represents the maximum and minimum active power limits of generation unit g, respectively; $L_{l,t}$ represents the demand for load l at time period t; $W_{k,t,\xi}$ is the wind power production of wind generation unit k, $\theta_{n,t,\xi}$ are the voltage angles at node n and at time period t in scenario ξ, respectively; $X_{n,m}$ is the reactance of the transmission line $n-m$; and Ξ_{K_n}, Ξ_{M_n} represents the sets of wind power generation connected at node n, and nodes $m\in N$ connected to node n through a line, respectively. Finally, constraints (25) to (29) model the objective function, maximum and minimum power output of generators, nodal power balance equation (DC power flow), and variable declarations, respectively.

2.6. Alternative SUC Formulation

The literature provides several different formulations of the SUC problem [24]. Therefore it is relevant to compare and study the impact that these models have on the power system dynamic behaviour. In this context, an alternative subhourly SUC model with respect to the one discussed in the previous section has been adapted based on [25], as follows:

$$\underset{z^F_{g,t},p_{g,t},r^U_{g,t,\xi},r^D_{g,t,\xi},W_{k,t}}{\text{minimize}} \sum_{t\in\Xi_T}\sum_{g\in\Xi_G} C^{SU}_{g,t} + C^V_g p_{g,t} \tag{30}$$

$$+\sum_{\xi\in\Xi}\pi_\xi\left[\sum_{t\in\Xi_T}\sum_{g\in\Xi_G} C^V_g(r^U_{g,t,\xi}-r^D_{g,t,\xi})\right]$$

such that

$$C^{SU}_{g,t} \geq C^{SU}_{g}(z^F_{g,t} - z^F_{g,t-1}), \qquad \forall g, \forall t \tag{31}$$

$$C^{SU}_{g,t} \geq 0, \qquad \forall g, \forall t \tag{32}$$

$$z^F_{g,t} P^{min}_{g} \leq p_{g,t} \leq z^F_{g,t} P^{\max}_{g}, \qquad \forall g, \forall t \tag{33}$$

$$\sum_{g \in \Xi_G} p_{g,t} + \sum_{k \in \Xi_K} W_{k,t} = L_t, \qquad \forall t \tag{34}$$

$$z^F_{g,t} P^{min}_{g} \leq (p_{g,t} + r^U_{g,t,\xi} - r^D_{g,t,\xi}) \leq z^F_{g,t} P^{\max}_{g}, \qquad \forall g, \forall t, \forall \xi \tag{35}$$

$$\sum_{g \in \Xi_G} (p_{g,t} + r^U_{g,t,\xi} - r^D_{g,t,\xi}) + \sum_{k \in \Xi_K} W_{k,t,w} = L_t, \qquad \forall t, \forall \xi \tag{36}$$

$$0 \leq r^U_{g,t,\xi} \leq R^{U\max}_{g,t}, \qquad \forall g, \forall t, \forall \xi \tag{37}$$

$$0 \leq r^D_{g,t,\xi} \leq R^{D\max}_{g,t}, \qquad \forall g, \forall t, \forall \xi \tag{38}$$

$$z^F_{g,t} \in \{0,1\}, \qquad \forall g, \forall t \tag{39}$$

where, apart from the variables and parameters already defined above, $C^{SU}_{g,t}$ models the start-up cost of generation unit g incurred at the beginning of time period t; $r^U_{g,t,\xi}$ and $r^D_{g,t,\xi}$ are decision variables and represents the increase and decrease in the active power output of generation unit g at time period t, respectively, say, during real-time operation (e.g. to compensate wind power fluctuations); while all other parameters and variables have analogous meanings as in the previous SUC formulation.

The main difference between problem (30)–(39) and problem (25)–(29) is that in the former the active power of generation units, $p_{g,t}$, is a first-stage variable and does not adapt to the uncertainty realization. Note that model (25)–(29) has a slightly different objective function, namely, it does not include the start-up cost ($C^{SU}_{g,t}$) as compared to that of model (30)–(39).

2.7. Cosimulation Framework

The cosimulation method allows coupling different subdomain models, like, for example, power systems and electricity markets, and is currently one of the most used to study the behaviour of modern power grids [26]. Figure 3 depicts the structure of the cosimulation framework, and shows the interaction between the discrete model of the SUC and dynamic model of SDAEs. The software tool DOME [27] takes care of the entire cosimulation, in particular, the exchange of information from one model to the other. Note that each model is solved independently, i.e., the subhourly SUC in Gurobi [28], and SDAEs in the dynamic software tool DOME. Both models communicate together in every time period t, e.g., 5 min, by exchanging their outputs as indicated in Equation (5).

On the other hand, the subhourly SUC models uses as an input the forecast of wind and loads, which are obtained based on the realizations of the stochastic processes (utilizing a Monte Carlo method). These realizations (output of SDAEs) define the "reality" that has to be updated to solve the next SUC problem (see Figure 3).

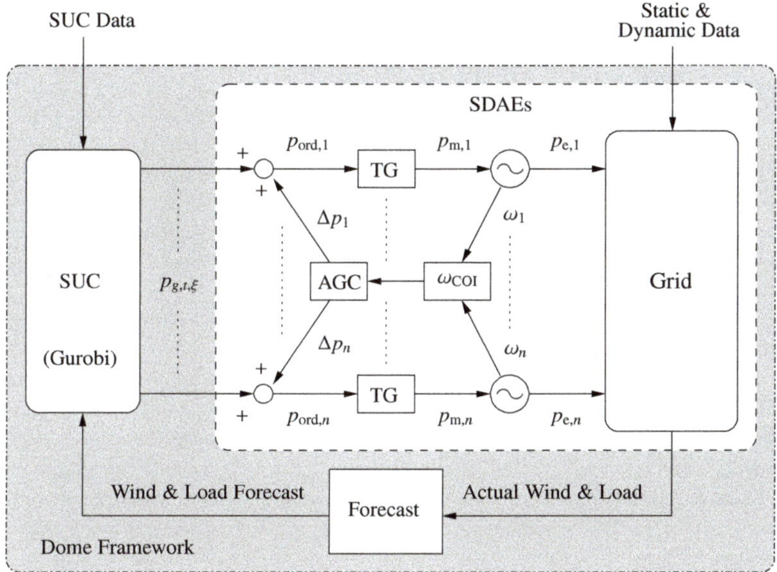

Figure 3. Structure of the interaction between the discrete model of the stochastic unit commitment (SUC) and dynamic model of SDAEs.

3. Case Study

This Section is organised as follows. Section 3.1 provides a sensitivity analysis of the impact of different frequency controllers/machine parameters using a complete SUC, a DUC, a simplified SUC and an alternative SUC model, and different scheduling time periods. This analysis allows comparing and drawing conclusions on the effect of these parameters on the dynamic performance of the system. Section 3.2 compares the impact that different models of SUC have on the dynamic performance of power systems.

A modified New England IEEE 39-bus system [29] is used in all simulations, where the data of the SUC are taken from [30]. In particular, the value of load curtailment (present in the reference model of SUC) is taken equal to $1000/MWh [21]; the cost of wind is assumed zero; the value of the fixed (C_g^F) and variable (C_g^V) cost coefficients is taken equal to the fixed and proportional cost coefficients a ($/h) and b ($/MWh), respectively, in [30]. The focus of this work is on the first 4 h of the planning horizon. Furthermore, the wind profile is modeled as in [13].

Wind power uncertainty, volatility and rolling planning horizon within the SUC are modeled as in [13]. In particular, three wind power scenarios are considered, namely, low, medium, and high. The medium scenario considers a 25% wind penetration level, while the low and high scenario are built according to the maximum variation width. Moreover, three wind power plants are considered and connected at bus 20, 21 and 23, respectively, with a nominal capacity of 300 MW each. It should be noted that this number of scenarios is considered sufficient as it was shown in [13] that increasing the number of scenarios leads to similar effect on the dynamic performance of the system.

The results of the case study are based on a Monte Carlo method, where 50 simulations are considered for each scenario. Moreover, the standard deviation of the frequency of the COI, σ_{COI}, is computed as the average of the standard deviation obtained for each trajectory. The SUC models (6)–(24), (25)–(29) and (30)–(39) are modeled using the Gurobi Python interface [28], whereas all simulations are obtained using a Python-based software tool for power system analysis, called DOME [27].

3.1. Sensitivity Analysis of the Impact of Different Frequency Controllers/Machine Parameters

In this Section, we vary (one at a time) three relevant frequency control/machine parameters and observe their impact on the standard deviation of the frequency.

The following base-case scenario is considered: the value of the gain of the AGC is taken equal to $K_0 = 50$, the value of the droop of the TGs is taken equal to $R = 0.05$, and the value of inertia of the machines is taken equal to the original values. In the scenarios below, the total inertia of the system and the gain of AGC is decreased up to 45% from the base case. Similarly, the aggregated system droop is increased up to 45% from the base case. Finally, as system operators still rely on deterministic UC and different subhourly scheduling time periods [31], this sensitivity is performed using 15- and 5-min time periods and different SUC formulations.

In the following, the subindexes S and D indicate control parameters for stochastic and deterministic scenarios, respectively.

3.1.1. SUC with 15-min Time Period

In this scenario, we use a 15-min scheduling time period and set the SUC probabilities for the low, medium and high wind power scenario equal to 20%, 60%, 20%, respectively. Same probabilities are used when solving the MC-TDS. Next, the relevant frequency control/machine parameters are varied accordingly up to 45% of their base case value. Figure 4 shows the effect of the variation of these parameters on σ_{COI}.

The gain of AGC, K_S, has the highest impact on σ_{COI}. The relationship between the gain K_S, the droop R_S, and σ_{COI} is almost linear within the used range. This indicates larger frequency deviations as synchronous generators are replaced with RES (assuming that RES will not provide frequency regulation). On the other hand, the inertia M_S appears to have a small impact on long-term frequency deviation. This result indicates that while the inertia is the main parameter impacting on the frequency dynamics following a major contingency, this is not the case on its impact on the σ_{COI} (see also [4], which draws a similar conclusion).

3.1.2. DUC with 15-min Time Period

As discussed above, system operators still rely on a DUC formulation when scheduling the system. Therefore it is important to compare the impact of the variation of the relevant control/machine parameters on σ_{COI}, using a SUC and a DUC. With this aim, we set the DUC probabilities for the low, medium and high wind power scenarios equal to 0%, 100%, 0%, respectively. Thus, perfect forecast, which corresponds to the medium scenario, is assumed. However, when solving the MC-TDS, 20%, 60%, 20% probabilities are used to generate the three wind power scenarios. This creates a mismatch between forecast and actual wind variations and allows evaluating the robustness of the SUC and DUC formulations.

Figure 4 shows the impact on σ_{COI} of varying control/machine parameters. A relevant difference with respect to the scenario above is that the droop R_D has the highest impact on σ_{COI} when its value is $\geq$ 30% with respect to the base case. On the other hand, the gain of the AGC, K_D and the inertia M_D have similar impact on σ_{COI} as in the previous scenario. It appears that, the impact of different control parameters on σ_{COI} depends on the UC formulation (deterministic or stochastic). It is thus not obvious which one (i.e., R or K) has the highest impact on σ_{COI}. Figure 4 compares the results of scenario 1 and 2 and shows that using a SUC leads to lower variations of the frequency. Furthermore, it should be noted that the differences between scenarios in Figure 4, for example, R_S and R_D after 28%, are due to the fact that both models, namely, SUC and DUC, produce different schedules for generators.

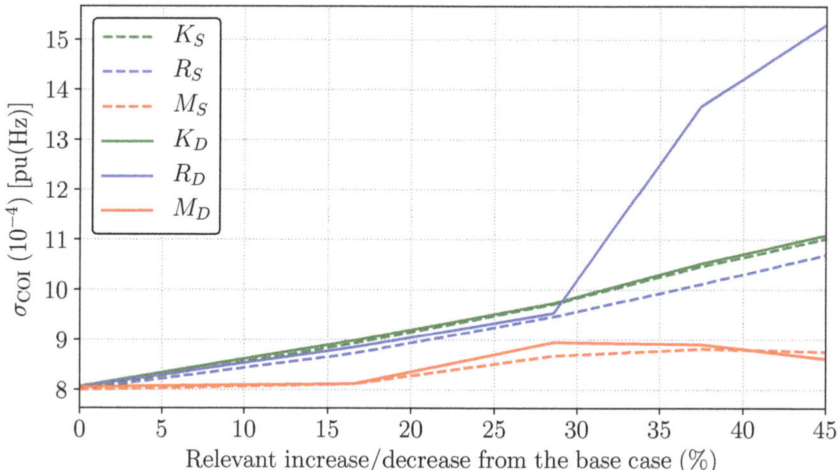

Figure 4. 15-min time period—σ_{COI} as a function of different frequency controllers/machine parameters using the complete SUC and DUC models.

Finally, although the focus of this paper is on the impact of different SUC models on the dynamic behaviour of the system, it is relevant to show the total operating cost for each model. With this aim, Table 1 shows the total operating cost for each subhourly UC model. For the considered case, the complete SUC and DUC produce very similar operating costs, namely, 412,000$ and 411,580$, respectively. While the simplified SUC model provides a lower estimate of the operating costs, namely, 398,000$, as it is less constrained. The alternative SUC model results in lower operating costs compared to others models (complete SUC, simplified SUC and DUC), namely, 339,470$, mainly due to the fact that it has a slightly different objective function, e.g., it does not include the fixed cost.

Table 1. 15-min time period—total operating costs for different UC models.

Model	Total Operation Cost ($)
SUC (Complete)	412,000
SUC (Simplified)	398,000
SUC (Alternative)	339,470
DUC	411,580

3.1.3. Sensitivity Analysis Using the Simplified and Alternative SUC, and 15-min Time Period

Here, we perform the same sensitivity analysis as above, but this time using the simplified and alternative SUC models, respectively. Such an analysis allows comparing and drawing conclusions on the impact of different frequency control/machine parameters on σ_{COI}, using different subhourly SUC models. With this aim, Figure 5 shows the relevant results of the sensitivity analysis. Note that the subindexes *Sim* and *Alt* indicate control parameters for simplified and alternative SUC scenarios, respectively. As expected, results are, in general, very similar to the ones discussed in previous sections. The gain of the AGC, in this case, K_{Sim} and K_{Alt}, has most of the time the highest impact on σ_{COI}, as well as the inertia of the machines appears to have a small impact. Moreover, using an alternative SUC model and varying its relevant parameters leads to lower variations of the frequency. This is due to the fact that the alternative SUC model schedules more generators to be online compare to the other three UC models. Thus, depending on the SUC model, different control parameters (e.g., K or R) can have different impact on σ_{COI}.

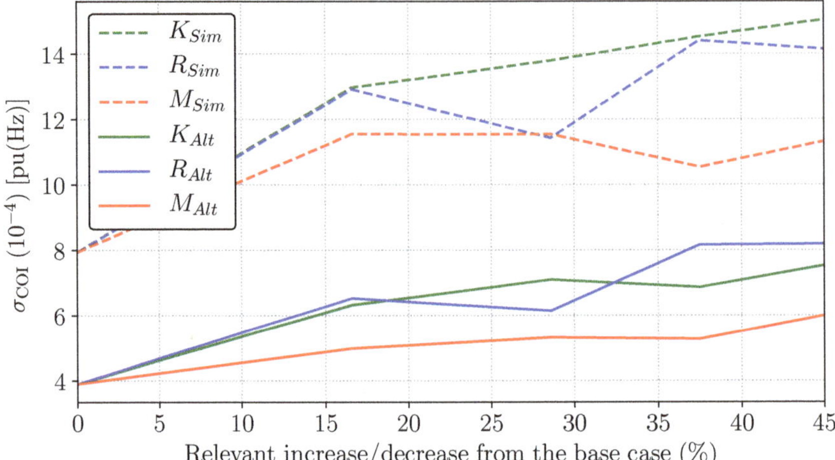

Figure 5. 15-min time period—σ_{COI} as a function of different frequency controllers/machine parameters using the simplified and alternative SUC models.

3.1.4. SUC with 5-min Time Period

This scenario investigates whether a shorter scheduling time period of SUC, namely, 5 min, changes the results and conclusions drawn for the 15-min time period. Due to the shorter time period, the wind power uncertainty level within SUC is lower compared to the previous section. Both SUC and MC-TDS probabilities for the low, medium and high scenarios are set equal to 20%, 60% and 20%, respectively.

Figure 6 shows the effect of the variation of the relevant control parameters on σ_{COI}. The most visible effect of reducing the time period to 5 min is that the value of σ_{COI} decreases in all scenarios due to a lower wind power uncertainty. In this case, the gain of the AGC, K_S, has the highest impact on σ_{COI}. The relationship is linear in the considered range, which indicates the need to increase the frequency regulation to keep long-term frequency deviations within certain limits. Finally, the inertia has little effect on σ_{COI}, supporting the conclusions above.

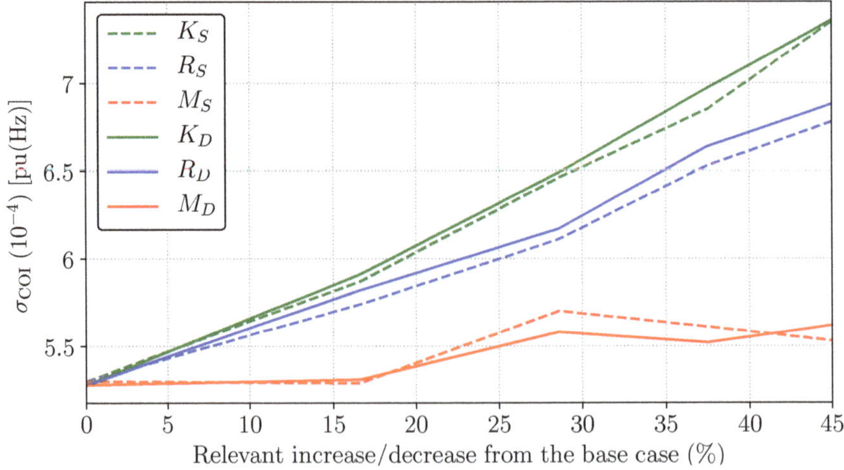

Figure 6. 5-min time period—σ_{COI} as a function of different frequency controllers/machine parameters.

3.1.5. DUC with 5-min Time Period

Figure 6 shows the effect of the variation of control/machine parameters using a 5-min scheduling and DUC. Results are very similar to the 5-min scheduling SUC. To better show the differences, Figure 6 compares the results of both scenarios, where it can be observed that the SUC leads, in general, to lower frequency variations. In both scenarios, the gain of the AGC has the highest impact on σ_{COI}, whereas the inertia has a small impact.

3.2. Comparison of Different SUC Models

Even though system operators are skeptical regarding the use of SUC approaches due to their complexity and transparency, they acknowledge the need to better represent uncertainty when scheduling the system [24].

For this reason, the objective of this section is to compare the impact that these models have on long-term power system dynamics. Such a comparison is made using a 15-min time period.

We first compare the impact on the dynamic response of the New England 39-bus system of the SUC formulations (6)–(24) (complete SUC) and (25)–(29) (simplified SUC).

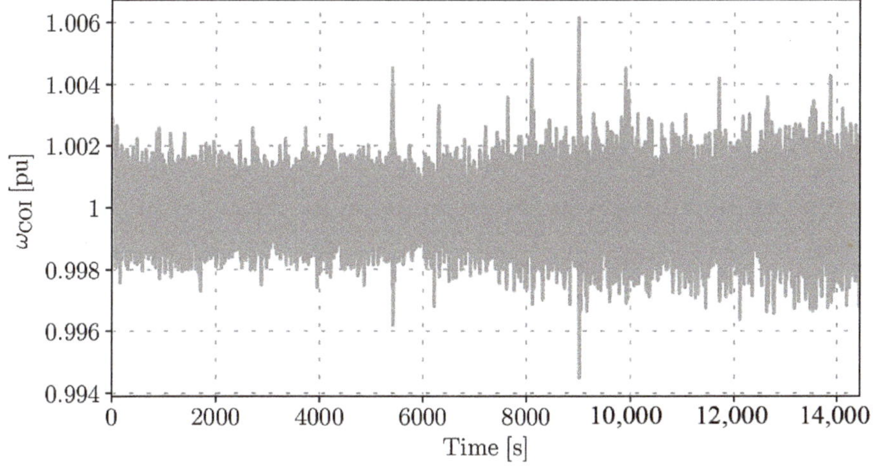

Figure 7. Trajectories of ω_{COI} for 15-min time period and complete SUC.

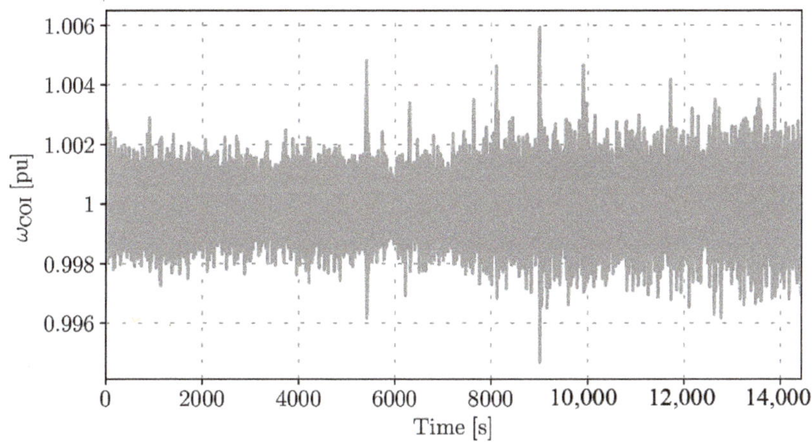

Figure 8. Trajectories of ω_{COI} for 15-min time period and simplified SUC.

Figures 7 and 8 show the ω_{COI}, of the complete and simplified SUC models, respectively. The two formulations returns an almost identical ω_{COI}, namely $\sigma_{\text{COI}} = 0.000800$ pu(Hz) for the complete SUC and, while that of the simplified SUC is $\sigma_{\text{COI}} = 0.000794$. Specifically, the difference is about 1%. Figures 9 and 10 show the mechanical power of the conventional synchronous generators 1, 2 and 4, of the complete and simplified SUC model, respectively. The two models produce similar schedules. In fact, at the beginning of the planning horizon, the simplified SUC model alternates the schedules for generators 2 and 4, and after some time (i.e., after 6000 s) it produces the same schedules as the complete SUC model. It appears, thus, that for this particular system and for normal operation conditions, the differences between using a complete SUC model and a simplified one are negligible. These results suggest thus that an involved UC formulation is not necessarily the best in normal operating conditions of the system.

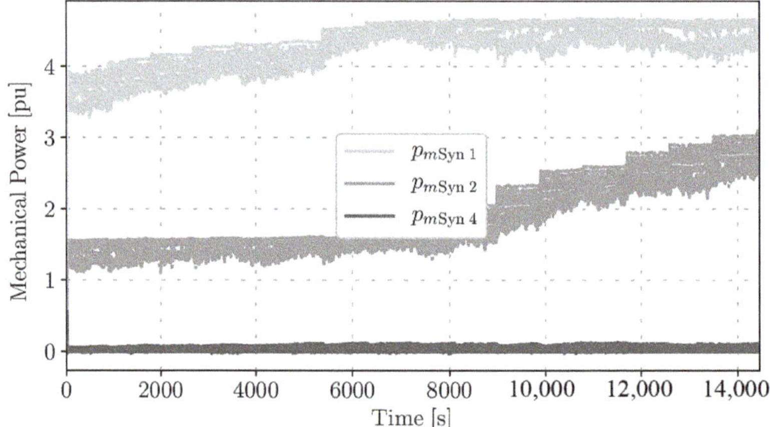

Figure 9. Mechanical power of synchronous generators 1, 2 and 4, for 15-min time period and complete SUC model.

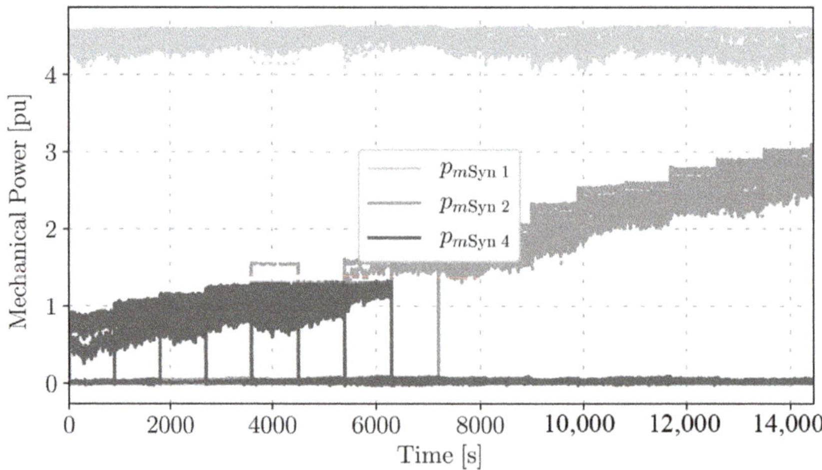

Figure 10. Mechanical power of synchronous generators 1, 2 and 4, for 15-min time period and simplified SUC model.

Next, we compare the impact on system dynamics of the problem (6)–(24) (complete SUC) and the problem (30)–(39) (alternative SUC) described in Section 2.6. With this aim, a MC-TDS per each SUC formulation is carried out.

Figure 11 shows ω_{COI} for the alternative SUC model. Compared to the complete SUC model (see Figure 7), the alternative formulation leads to lower frequency variations, i.e., $\sigma_{\mathrm{COI}} = 0.000390$. In other words, the differences between the models is about 48%. This is due to fact that the two SUC formulations produce slightly different schedule for generators. Specifically, the alternative SUC model schedules a few more generators (Figure 12) compared to the complete SUC model (Figure 9). This, in turn, implies more regulation, which helps better manage wind uncertainty. This can be observed in Figure 11 where, in the time between two scheduling events—i.e., 15 min—frequency variations are lower due to the increased frequency regulation available in the system.

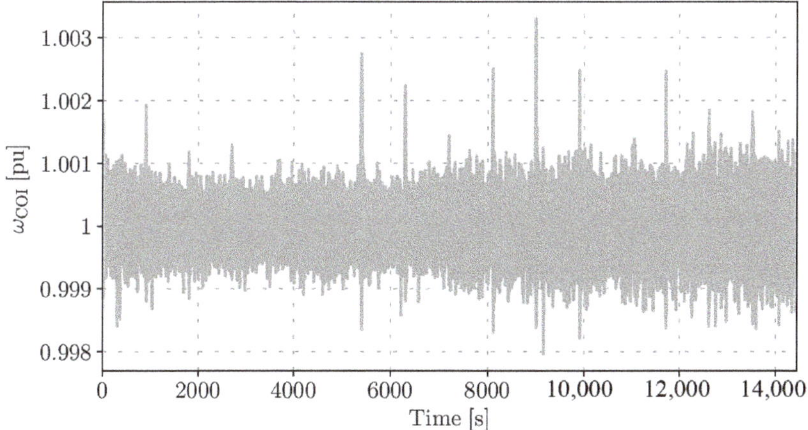

Figure 11. Trajectories of ω_{COI} for 15-min time period and alternative SUC model.

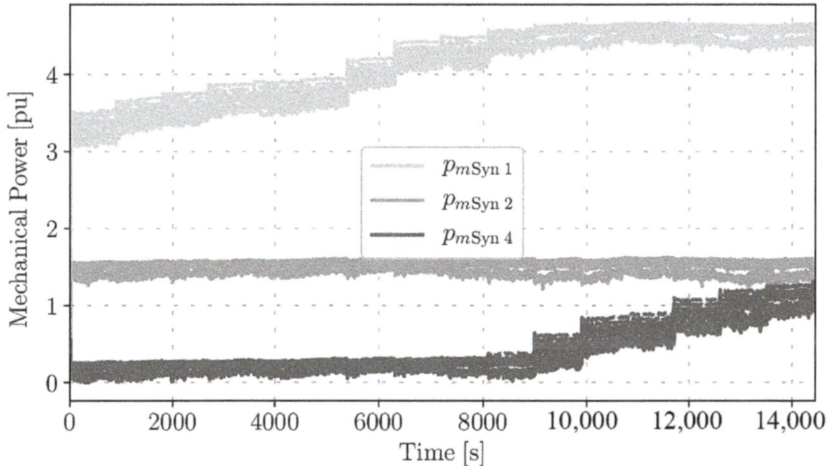

Figure 12. Mechanical power synchronous generators 1, 2 and 4, for 15-min time period and alternative SUC model.

This result had to be expected. From the dynamic point of view, in fact, it is better to schedule more conventional synchronous generators, which provide both primary and secondary frequency regulations, rather than wind generation.

4. Conclusions

In this paper we have performed a thorough sensitivity analysis in order to assess the impact on long-term power system dynamics of different frequency control/machine parameters and different subhourly SUC formulations.

Simulation results show that the gain of the AGC is the main parameter impacting the long-term frequency deviation. On the other hand, synchronous machine inertia has a small effect on the standard deviation of the frequency. Moreover, results suggest that the difference on the dynamic behaviour of the power system are marginal when using a complete or a simplified SUC model. Furthermore, a SUC leads to lower variations of the frequency compared to a DUC.

An interesting result is also that different formulations of the SUC problem can lead to schedule conventional power plants in different locations and/or number. These differences can have a significant impact on long-term power system dynamics and hence the formulation of the SUC has to be carefully chosen.

Regarding future work, we will consider closing the loop of the proposed cosimulation framework, i.e. using relevant output of SDAEs for adjustment of SUC, and perform a similar sensitivity analysis in order to illustrate the effectiveness and impact on electricity market of such a feedback.

Author Contributions: All authors equally contributed to this work. All authors have read and agreed to the published version of the manuscript.

Funding: This work was supported by Science Foundation Ireland, by funding T. Kërçi and F. Milano under project ESIPP, Grant No. SFI/15/SPP/E3125; and F. Milano under project AMPSAS, Grant No. SFI/15/IA/3074.

Conflicts of Interest: The authors declare no conflict of interest.

References

1. ENTSO-E. Continental Europe Significant Frequency Deviations January 2019. Available online: https://www.entsoe.eu (accessed on 17 February 2020).
2. Kundur, P.; Paserba, J.; Ajjarapu, V.; Andersson, G; Bose, A.; Cañizares, C.; Hatziargyriou, N.; Hill, D; Stankovic, A.; Taylor, C.; et al. Definition and classification of power system stability IEEE/CIGRE joint task force on stability terms and definitions. *IEEE Trans. Power Syst.* **2004**, *19*, 1387–1401.
3. Kërçi, T.; Milano, F. A framework to embed the unit commitment problem into time domain simulations. In Proceedings of the 2019 IEEE International Conference on Environment and Electrical Engineering, Genova, Italy, 11–14 June 2019.
4. Vorobev, P.; Greenwood, D.M.; Bell, J.H.; Bialek, J.W.; Taylor, P.C.; Turitsyn, K. Deadbands, droop, and inertia impact on power system frequency distribution. *IEEE Trans. Power Syst.* **2019**, *34*, 3098–3108. [CrossRef]
5. Wang, J.; Wang, J.; Liu, C.; Ruiz, J.P. Stochastic unit commitment with sub-hourly dispatch constraints. *Appl. Energy* **2013**, *105*, 418–422. [CrossRef]
6. Gangammanavar, H.; Sen, S.; Zavala, V.M. Stochastic optimization of sub-hourly economic dispatch with wind energy. *IEEE Trans. Power Syst.* **2016**, *31*, 949–959. [CrossRef]
7. Wang, B.; Hobbs, B.F. Real-time markets for flexiramp: A stochastic unit commitment-based analysis. *IEEE Trans. Power Syst.* **2016**, *31*, 846–860. [CrossRef]
8. Gu, Y.; Xie, L. Stochastic look-ahead economic dispatch with variable generation resources. *IEEE Trans. Power Syst.* **2017**, *32*, 17–29. [CrossRef]
9. Milano, F.; Dörfler, F.; Hug, G.; Hill, D.J.; Verbič, G. Foundations and challenges of low-inertia systems (invited paper). In Proceedings of the 2018 Power Systems Computation Conference (PSCC), Dublin, Ireland, 11–15 June 2018.
10. Ahmadi, H.; Ghasemi, H. Security-Constrained Unit Commitment With Linearized System Frequency Limit Constraints. *IEEE Trans. Power Syst.* **2014**, *29*, 1536–1545. [CrossRef]
11. Singh Parmar, K.P.; Majhi, S; Kothari, D.P. Load frequency control of a realistic power system with multi-source power generation. *Int. J. Electr. Power Energy Syst.* **2012**, *42*, 426–433. [CrossRef]
12. Kërçi, T.; Milano, F. Sensitivity Analysis of the Interaction between Power System Dynamics and Unit Commitment. In Proceedings of the 2019 IEEE Milan PowerTech, Milan, Italy, 23–27 June 2019.

13. Kërçi, T.; Giraldo, J. Milano, F. Analysis of the impact of sub-hourly unit commitment on power system dynamics. *Int. J. Electr. Power Energy Syst.* **2020**, *119*, 105819. [CrossRef]
14. Yuan, B.; Zhou, M.; Li, G.; Zhang, X. Stochastic Small-Signal Stability of Power Systems With Wind Power Generation. *IEEE Trans. Power Syst.* **2015**, *30*, 1680–1689. [CrossRef]
15. Wang, X.; Chiang, H.; Wang, J.; Liu, H.; Wang, T. Long-Term Stability Analysis of Power Systems With Wind Power Based on Stochastic Differential Equations: Model Development and Foundations. *IEEE Trans. Power Syst.* **2015**, *30*, 1534–1542. [CrossRef]
16. Milano, F.; Zárate-Miñano, R. A Systematic Method to Model Power Systems as Stochastic Differential Algebraic Equations. *IEEE Trans. Power Syst.* **2013**, *28*, 4537–4544. [CrossRef]
17. Kundur, P. *Power System Stability and Control*; McGraw-Hill: New York, NY, USA, 1994.
18. Milano, F. *Power System Modeling and Scripting*; Springer: London, UK, 2010.
19. IEA-WIND. Design and Operation of Power Systems with Large Amounts of Wind Power. Available online: https://community.ieawind.org (accessed on 17 February 2020).
20. Morales, J.; Conejo, A.; Madsen, H.; Pinson, P.; Zugno, M. *Integrating Renewables in Electricity Markets: Operational Problems*; Springer: Berlin, Germany, 2013.
21. Conejo, A.; Carrión, M.; Morales, J. *Decision Making Under Uncertainty in Electricity Markets*; Springer: Berlin, Germany, 2010.
22. Blanco, I.; Morales, J.M. An Efficient Robust Solution to the Two-Stage Stochastic Unit Commitment Problem. *IEEE Trans. Power Syst.* **2017**, *32*, 4477–4488. [CrossRef]
23. Håberg, M. Fundamentals and recent developments in stochastic unit commitment. *Int. J. Electr. Power Energy Syst.* **2019**, *109*, 38–48. [CrossRef]
24. Zheng, Q.P.; Wang, J.; Liu, A.L. Stochastic Optimization for Unit Commitment—A Review. *IEEE Trans. Power Syst.* **2015**, *30*, 1913–1924. [CrossRef]
25. Gómez Expósito, A.; Conejo, A.; Cañizares, C. *Electric Energy Systems: Analysis and Operation*; CRC Press: Boca Raton, FL, USA, 2018.
26. Palensky, P.; Van Der Meer, A.A.; Lopez, C.D.; Joseph, A.; Pan, K. Cosimulation of Intelligent Power Systems: Fundamentals, Software Architecture, Numerics, and Coupling. *IEEE Ind. Electron. Mag.* **2017**, *11*, 34–50. [CrossRef]
27. Milano, F. A Python-based software tool for power system analysis. In Proceedings of the IEEE PES General Meeting, Vancouver, BC, Canada, 21–25 July 2013; pp. 1–5.
28. Gurobi Optimization, LLC. *Gurobi Optimizer Reference Manual*; Gurobi Optimization, LLC: Houston, TX, USA, 2018.
29. Illinois Center for a Smarter Electric Grid (ICSEG). IEEE 39-Bus System. Available online: https://icseg.iti.illinois.edu/ieee-39-bus-system/ (accessed on 17 February 2020).
30. Carrión, M.; Arroyo, J.M. A computationally efficient mixed-integer linear formulation for the thermal unit commitment problem. *IEEE Trans. Power Syst.* **2006**, *21*, 1371–1378. [CrossRef]
31. Costley, M.; Feizollahi, M.J.; Ahmed, S.; Grijalva, S. A rolling-horizon unit commitment framework with flexible periodicity. *Int. J. Electr. Power Energy Syst.* **2017**, *90*, 280–291. [CrossRef]

Article

Receding Horizon Control of Cooling Systems for Large-Size Uninterruptible Power Supply Based on a Metal-Air Battery System

Bonhyun Gu [1], Heeyun Lee [1], Changbeom Kang [1], Donghwan Sung [1], Sanghoon Lee [1], Sunghyun Yun [2], Sung Kwan Park [3], Gu-Young Cho [4], Namwook Kim [5,*] and Suk Won Cha [1,*]

1 Department of Mechanical and Aerospace Engineering, Seoul National University, Seoul 08826, Korea; gbh980@snu.ac.kr (B.G.); leeheeyun@snu.ac.kr (H.L.); gajaboom@snu.ac.kr (C.K.); kevin3259@snu.ac.kr (D.S.); ynsc234@snu.ac.kr (S.L.)
2 Department of Electrical and Computer Engineering, Seoul National University, Seoul 08826, Korea; ysh154947@snu.ac.kr
3 Department of Materials Science and Engineering, Seoul National University, Seoul 08826, Korea; sungkwan.psk@gmail.com
4 Department of Mechanical Engineering, Dankook University, Gyeonggi-do 16890, Korea; guyoungcho@dankook.ac.kr
5 Department of Mechanical Engineering, Hanyang University, Ansan 15588, Korea
* Correspondence: nwkim21@gmail.com (N.K.); swcha@snu.ac.kr (S.W.C.)

Received: 10 March 2020; Accepted: 29 March 2020; Published: 1 April 2020

Abstract: As application of electric energy have expanded, the uninterruptible power supply (UPS) concept has attracted considerable attention, and new UPS technologies have been developed. Despite the extensive research on the batteries for UPS, conventional batteries are still being used in large-scale UPS systems. However, lead-acid batteries, which are currently widely adopted in UPS, require frequent maintenance and are relatively expensive as compared with some other kinds of batteries, like metal-air batteries. In previous work, we designed a novel metal-air battery, with low cost and easy maintenance for large-scale UPS applications. An extensive analysis was performed to apply our metal-air battery to the hybrid UPS model. In this study, we focus on including an optimal control system for high battery performance. We developed an algorithm based on receding horizon control (RHC) for each fan of the cooling system. The algorithm reflects the operation properties of the metal-air battery so that it can supply power for a long time. We solved RHC by applying dynamic programming (DP) for a corresponding time. Different variables, such as current density, oxygen concentration, and temperature, were considered for the application of DP. Additionally, a 1.5-dimensional DP, which is used for solving the RHC, was developed using the state variables with high sensitivity and considering the battery characteristics. Because there is no other control variable during operation, only one control variable, the fan flow, was used, and the state variables were divided by section rather than a point. Thus, we not only developed a sub-optimal control strategy for the UPS but also found that fan control can improve the performance of metal-air batteries. The sub-optimal control strategy showed stable and 6–10% of improvement in UPS operating time based on the simulation.

Keywords: cooling system; dynamic programming; metal-air battery; receding horizon control; state variables; uninterruptible power supply

1. Introduction

To date, many electrical storage systems (ESS), such as novel types of batteries and ultracapacitors, have been studied and applied to various systems. In particular, these batteries can improve the size or weight aspects of the storage system making it small-size and easy to move. Large size is required for ESS that optimize economical gain by charging/discharging large amounts of electrical energy and for uninterruptible power supplies (UPS) supporting large electrical systems in emergencies. UPS systems have studied by many researchers because UPS are essential for supporting electrical energy stability. Before being applied to practical commercialization, a UPS is developed and studied through modeling and simulation because of the scale [1–3].

However, UPS are often based on conventional batteries because of their economic convenience and reliability. Among the UPS systems, the hybrid UPS (H-UPS) combines an electrical storage system (ESS) with UPS, using lead-acid batteries as the source of DC voltage [4]. Currently, replacing the batteries with an ultracapacitor is not affordable, because of the price of the latter [5]. Lithium-ion batteries, which are mostly used as a secondary battery, are also costly. Additionally, the secondary batteries must be always activated when connected to the main power source. In this aspect, though reliable, lead-acid batteries require frequent maintenance, and therefore they are not affordable [6].

Therefore, UPS requires the introduction of new battery technologies, especially high-power UPSs (those of more than 10 kW), which are more challenging. Among the commercially available batteries, metal-air batteries have a high energy density and are inexpensive, when select metal electrodes are used. Thus, they are widely studied as primary or secondary batteries, and small-capacity metal-air batteries have already been commercialized as primary batteries [7,8]. However, they cannot be used as rechargeable batteries, because they support only a few recharging cycles. For this reason, metal-air batteries are suitable for high-power UPSs, which require high capacity and no recharge. In these systems, the initial standby time after triggering can be supplemented by hybridization with a small-sized lithium-ion battery. Thus, modeling and simulation of UPSs using metal-air batteries are needed for their commercialization. In a previous study, we developed UPS model using experimental cell data and theoretical analysis [9]. Simulation includes the controls for UPS and lithium-ion battery which is a secondary battery on standby during normal situation. However, after initial operation, when the primary battery, metal air battery, which is the main battery operates in earnest, only basic control using constant fan flow was applied.

Metal-air batteries are not widely used as energy storage devices in the industry. Thus, previous researches on controlling metal-air batteries have focused on determining their parameters or supplying stable energy [10–12]. However, the performance and lifespan of a battery is directly related to its thermal management system that controls the temperature of the battery [13–15]. Unlike other batteries for which the cooling system has already been studied, metal-air batteries have a significant limitations on air cooling because of its their consumption. Furthermore, previous studies have barely covered air-cooling of metal-air batteries because of their unusual properties. Although studies for cooling system control of metal-air batteries are fewer than for other batteries, control strategies for cooling systems of other batteries have been studied [16,17]. Especially, Zhan developed a rule-based control strategy for cooling PEM fuel cells in UPS systems [18].

This study devised a control strategy of air-cooling system for fuel supply using battery cell experimental data. As the optimal control strategy for a cooling system affects not only cooling but also oxygen supply, an overall performance improvement of the metal air battery could be expected. To do this, the detailed thermodynamics of metal-air battery among the UPS module were examined, and state variables were calculated for each cell. Because the studies for the UPS system based on metal-air battery are rare, detailed modeling of UPS system can be the first contribution. Though difficult, we applied the existing discrete dynamic programming (DP) algorithm, optimization theory, considering the characteristics of the metal-air battery. This is the second contribution of this paper and main contents.

The paper is structured in five sections. After this Introduction, Section 2 presents the UPS system modeling using prior research and thermodynamic models. Section 3 provides functions, parameters, and analytical theories to solve the main optimal problem for control modeling and simulation. Section 4 arrange the result of the simulation and compare that with other control strategies. Section 5 presents concluding remarks and the performance of the suggested control strategy.

2. UPS System Modeling

2.1. Prior Research

As previously mentioned, we developed a 64 kW UPS system using small-sized stacks of metal-air batteries in a previous study [9]. The research represents an important experimental trial to develop a UPS system based on metal-air batteries. The major point of the prior research was to determine the entire composition of the system and a hybrid system using lithium-ion battery. Thermodynamic and electromechanical analyses were performed assuming that the properties of the cells in each module of the UPS system are uniform:

$$T_1 = T_2 = \cdots = T_n \tag{1}$$

$$Cp_{module} = Cp_{\text{case}} + Cp_{electrolyte} + \sum_{k=1}^{n} Cp_{cell,k} \tag{2}$$

where T is the temperature, Cp is the heat capacity, and n is the number of cells in a module. The parameters of the previous UPS model were also used in this study. Table 1 shows the parameters of the iterative simulation and mechanical calculation used. A total of 20 fans that supply oxygen and cool the module were located at the 20 inlet, and 20 fans at outlet. The flow rate of these fans was constant value in the previous study, i.e.:

$$Q_1^t = Q_{N_2,1}^t + Q_{O_2,1}^t = Q_2^t = \cdots = Q_m^t = const \tag{3}$$

where Q_m^t is the flow rate at time t on m^{th} fan, $Q_{N_2,m}^t$, $Q_{O_2,m}^t$ is the flow rate of gas except oxygen, and the flow rate of oxygen gas.

Table 1. Values of the parameters in the UPS module.

Parameter	Basis	Value
Stack area	current density ~40 mA/cm^2, load current 330 A	1.28 m^2
Max load power	load current 330 A, boost converter efficiency	370 A
Module quantity	load current 330 A, max current density	0.16 m^2, 8 modules
Stack voltage range	min voltage 0.8 V, max voltage 1.1 V	180–200 V
Cell stack quantity	voltage range, cell voltage	170 cells
Metal weight	simulation result, demanded current 576 A	200 g/cell
Metal thickness	cell surface area, metal weight	0.15 mm/cell
Electrolyte thickness	simulation result, other variables	2 mm
Cathode thickness	simulation result, other variables	0.7 mm
Separator thickness	simulation result, stack weight	0.3 mm
Air tube thickness	cell array structure, airflow, air consumption	16 mm
Module volume	cell, airflow	Length : 2 m, Volume : 1.44 m^3

The parameters were designed to satisfy the required specification of up to 2 h runtime under maximum electrical load with a simple constant flow control. The 2 h runtime was the minimum required specification of the UPS until the main source is connected again. Though the simple constant flow control showed good performance under static load power, it was not optimized for different environments like changing electrical loads. Therefore, it did not show stable control performance in this case, because it did not meet the electrical load and boundary conditions of a metal-air battery.

2.2. Thermodynamic Model

To optimize the previous control strategy for a metal-air battery-based UPS, the thermal effect of each cell was modeled for a system of 180 cells per module. The heat in each cell was modeled using the Gibbs potential energy and the voltage from the cell as:

$$Q_{heat}(t_k) = I_{cell}(t_k)(G_{zinc} - V_{cell}(x(t_k))) \tag{4}$$

where Q_{heat} is the heat, I_{cell} is the current in the cell, G_{zinc} is the Gibbs potential of the zinc oxide reaction, V_{cell} is the voltage of the cell, and t_k is the time at step k.

The cooling effect and performance of the fans in each cell were also divided to simulate along with the units. To calculate the heat coefficient on a plate for natural convection, we used the following equations:

$$Nu = \left\{0.825 + \frac{0.387Ra^{\frac{1}{6}}}{\left[1+\left(\frac{0.492}{Pr}\right)^{9.16}\right]^{\frac{8}{27}}}\right\}^2 \tag{5}$$

where Nu is the Nusselt number, Ra is the Rayleigh number for natural convection, and Pr is the Prandtl number, and:

$$C_{heat} = K_{air} * \frac{Nu}{L^*},\ L^* = L_H \times \frac{L_W}{2(L_H + L_W)} \tag{6}$$

where C_{heat} is the heat coefficient of the plate, K_{air} is the thermal conductivity of the air, and L_H and L_W are the height and width of the plate, respectively.

A diagram of the stack structure and thermal effect is presented in Figure 1.

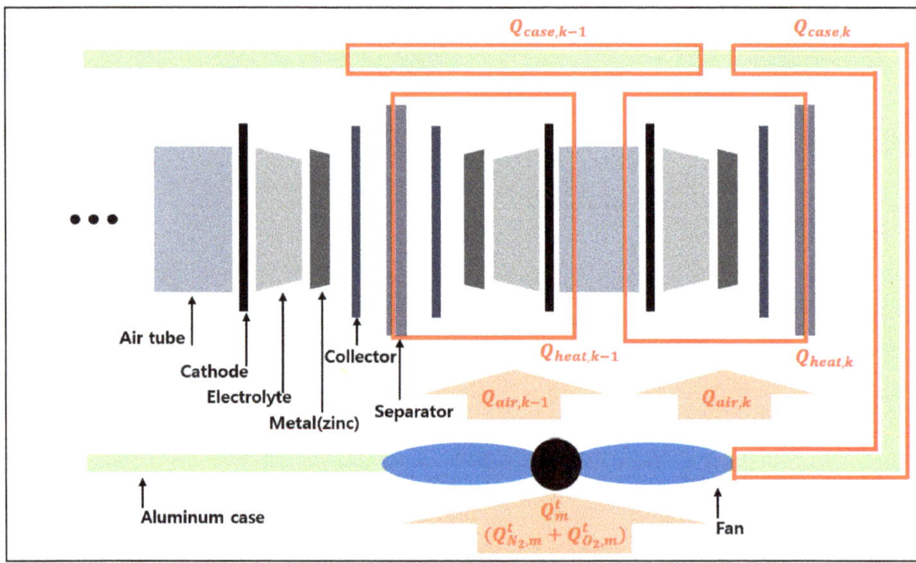

Figure 1. Schematic of stack structure and thermo-dynamical mechanism.

The thermodynamics of each cell in the UPS module described in Figure 1 are modeled by the following equations:

$$Q_{case} = \sum_{k=1}^{end} (C_{ceil,k} \cdot A_{ceil,k} \cdot \Delta T_{ceil,k} + C_{floor,k} \cdot A_{floor,k} \cdot \Delta T_{floor,k}) + C_{side,1} \cdot A_{side,1} \cdot \Delta T_{side,1} + C_{side,end} \cdot A_{side,end} \cdot \Delta T_{side,end} \tag{7}$$

where $C_{x,k}$ and $A_{x,k}$ are the thermal conductivity and surface area of area x in k^{th} cell, respectively; and:

$$T_{k,t} = T_{k,t-1} + \frac{d}{dT}\left(Q_{heat,\ k} - Q_{case,k} - Q_{air,k} - Q_{conduct\ near,\ k}\right) \cdot \Delta T \tag{8}$$

$$Q_{conduct\ near,k} = Q_{conduct,k-1} + Q_{conduct,k+1} \tag{9}$$

where $T_{k,t}$ is the temperature of the kth cell at time t, $Q_{cell,k}$ is the heat of the chemical reaction, $Q_{case,k}$ is the (negative) heat of the natural cooling from the aluminum case, $Q_{air,k}$ is the negative heat forced by air-cooling, and $Q_{conduct\ near,k}$ is the negative heat caused by conduction to the neighboring cell.

2.3. UPS Simulation and Fan Control

Many UPS control studies have been conducted on the control of the circuit-level inverter, input, and output current to stabilize the voltage, e.g., [19–21]. Here, we assumed a stable electrical load and supply from the UPS model to the battery model, because our metal-air battery model focused on the power supply at the stack-level and on the operation time for a given electrical load. The main target of this study is the fan flow, which can be controlled during operation to improve the performance which includes operating time and stability of the UPS model. The system used in this study consists of stacked modules, which were modeled using the experimental data of the cell units. Thus, the system is a nonlinear discrete state-space model that can be defined simply as:

$$x(t+1) = f(x(t), u(t)), y(t) = g(x(t)).\ t = 1,\ 2,\ 3,\ \ldots,\ t_{end} \tag{10}$$

Here, $x(t)$ is the state, $y(t)$ is the output, $u(t)$ is the control input, f is the modeled system function computing the states of next step by the state of prior step, and g is the modeled system function computing the output by the state of this step.

The UPS model consists of a hybrid system that supplies stable electrical power. A lithium-ion battery that operates only initially is always connected to the main power system. A metal-air battery operates as the main source after it starts generating electric energy stably. Here, we developed a strategy to control the module containing the metal-air battery. The metal-air battery module is modeled by electrochemical formulas obtained from experimental data. The formula used to calculate the voltage in the cell is:

$$V_{cell}(t_k) = V_{soc}\left(SOC_{t_k}\right) \times PV_{x_1}(x_1(t_k)) \times PV_{x_2}(x_2(t_k)) \times PV_{x_3}(x_3(t_k)) \tag{11}$$

where V_{cell} is the voltage of the cell, V_{soc} is the voltage output by the state-of-charge, SOC_{t_k} is the SOC on the time step t_k, and PV_{x_k} is the voltage fraction in x_k. Because the variables of the experiment for acquiring the cell data were reduced to three, we set only three state variables to calculate the voltage of the cell.

3. Control Modeling and Simulation

3.1. The Conceptual Framework

As mentioned above, the target of this research is the optimal control strategy of an UPS system based on metal-air battery. For simulating the system, the model is developed using iteration for time step using discrete calculation. The schematic diagram for controlling the UPS module is shown in Figure 2. The modeling of whole system is coded in Matlab, which is a commercial program from Mathworks (Natick, Massachusetts, USA).

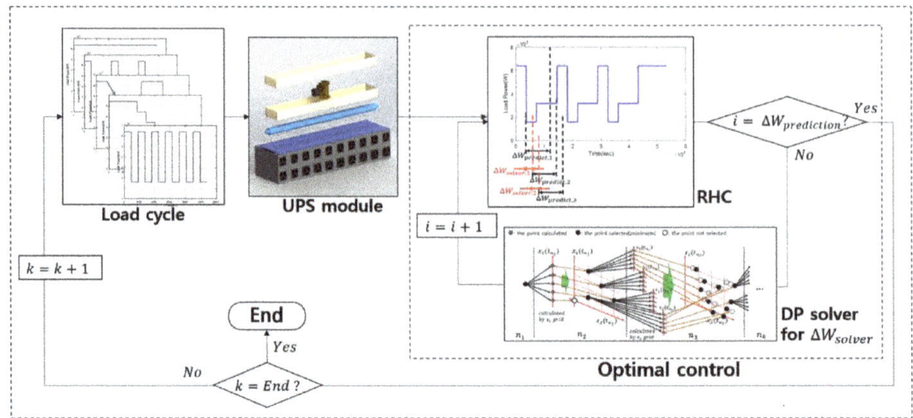

Figure 2. Overall view of the UPS model with the proposed receding horizon control (RHC).

3.2. Receding Horizon Control

The receding horizon control (RHC), also known as moving horizon control, is a feedback control strategy for nonlinear or linear plants with input and state constraints [22]. RHC is suitable for slow nonlinear systems, such as chemical processes, that can be solved sequentially [23]. Many studies have investigated UPS using a model predictive control (MPC), which is another name for RHC [24,25].

The typical electrical load for an UPS can be predicted with a few minutes of advance notice. However, predicting the electrical load continuously is unrealistic, because the conditions of environment of the system and the UPS model can change. Thus, it is common to adjust the system loads supplied by the UPS model according to the pre-scenario or situation of the UPS model, so that the electrical load can be predicted for few minutes from the time of operation. Because of this, RHC is applied to the optimal control proposed in this study. Usually, RHC analyzes the model in the prediction window using linear problem (LP) or a quadratic problem (QP). However, the UPS model of this study is composed of tables of the cell data. Therefore, the key to optimal control is to solve the target function within the prediction window. The target applied to the RHC prediction is the electrical load of the UPS in the system, and the state prediction can be expressed as:

$$x(t_{k+1} + i|t_k) = f(x(t_{k+1}|t_k), u(t_k + i|t_k)) \tag{12}$$

$$y(t_k + i|t_k) = g(x(t_k + i|t_k)) \tag{13}$$

where $x(t_k + i|t_k)$ is the state value of x predicted at time $t_k + i$ using the information available at time t_k, $x(t_k)$ is the state at time t_k, $y(t_k)$ is the cost value at time t_k, and f, g are the functions for state and cost value, respectively.

The range of i (prediction window) at step s is $\Delta W_{predict,s}$. As seen in Figure 3, the range of optimal control in $\Delta W_{predict,s}$ is set to $\Delta W_{solver,s}$ (i.e., the solver window at step s) as:

$$J_{s,s+N_{simul}} = \int_{t_k}^{t_k+\Delta W_{solver}} g(x(t|t_k), u(t|t_k), t|t_k)dt, \ \left(N_{simul} = \frac{\Delta W_{solver}}{\Delta t_{simul}}\right) \tag{14}$$

where $J_{s,s+N_{simul}}$ is the cost function of sth iteration. The solver means one iteration to solve the optimal control values under state variables.

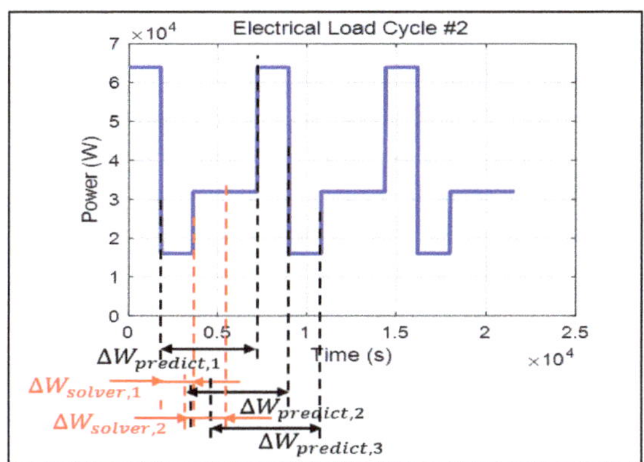

Figure 3. Schematics of the prediction and solver window ($\Delta W_{solver,k}$, $\Delta W_{predict,k}$) in RHC. These parameters are calculated as recursive structure of RHC. The optimal control values are calculated on $\Delta W_{solver,1}$, $\Delta W_{solver,2}$, ... , $\Delta W_{solver,k}$ within the iteration of $\Delta W_{prediction,1}$.

Here, the cost function to be minimized for the metal-air battery-based UPS, $J_{s,\, s+N_{simul}}$, is the metal (zinc) consumption. In state variable of x in Equation (12), the temperature that has a major influence on the performance was included first. The equation to calculate the optimal control in the $\Delta W_{predict,k}$ is:

$$u_k^* = \underset{u}{argmin}\left\{\int_{t_k}^{t_k+\Delta W_{predict,k}} g(x(t|t_k), u(t|t_k), t|t_k)dt\right\},\ (v_{min} \le u \le v_{max}) \tag{15}$$

Here, u_k^* is the optimal control at step k, v_{min} is the minimum fan flow, and v_{max} its maximum.

Although the optimal control should be calculated in the $\Delta W_{predict,k}$, only the optimal control in $\Delta W_{solver,k}$ was applied to the cost function, because in RHC the optimal control should be updated for each ΔW_{solver}. The cost result and optimal cost are calculated as:

$$J^{RHC} = \sum_{k=1}^{N_{total}} J_k^* = \sum_{k=1}^{N_{total}} \int_{t_k}^{t_k+\Delta W_{solver}} g\big(x(t), u_k^*(t), t\big)dt,\ N_{total} = \frac{Total\ time}{\Delta W_{solver}} \tag{16}$$

where J^{RHC} is the RHC cost result of the whole simulation, and J_k^* is the optimal cost at step k. The prediction window covers the electrical system load needed to operate the RHC for a few minutes. After the UPS starts its operation, it is necessary to estimate the load and control the input in the $\Delta W_{predict,k}$.

3.3. Dynamic Programming

The RHC developed in this study cannot be solved mathematically by merely using a discrete variable matrix. In the window of each step of the RHC, J_k^* can be calculated by finding $u_k^*(t)$ using DP, which is a global optimization theory. This control method based on the RHC prediction is similar to the 'Look-ahead DP' [26,27].

Applying the time step of the UPS chemical simulation, Δt_{simul}, to Equation (14), this becomes the discrete equation:

$$J_{s,s+N_{simul}} = \sum_{m=1}^{N_{simul}} g(x(t_m|t_k), u(t_m|t_k), t_m|t_k)\cdot\Delta t_{simul} \tag{17}$$

Although the simulation of the battery differs for each model, it cannot be solved in the grid at each time step. Then, in DP, the state of the battery has to be calculated for each temperature difference

(ΔT), which is less than 0.001 °C/s. Though the range of ΔT varies according to the temperature of the cell, if the total difference is divided in the entire range, computing cost increases exponentially $(\Delta T_{total}/\Delta T \approx 55,000)$. Therefore, the range of each state is calculated for each time step in the prediction window, and state x_k is assigned according to this. Thus, the DP step is set separately from the time step of the UPS chemical simulation, Δt_{simul}, according to Δt_{DP} as:

$$N_{DP} = \frac{\Delta W_{prediction}}{\Delta t_{DP}}, \quad N_{RHC} = \frac{\Delta W_{solver}}{\Delta t_{DP}} \tag{18}$$

Then, Equation (17) can be represented as a recursive relation using Δt_{DP}:

$$J^*_{s,s+N_{simul}}(x(t_k)) = min\{g(x(t_s|t_k), u(t_s|t_k), t_s|t_k)\cdot\Delta t_{DP} + J^*_{s+1,s+N_{simul}}(x(t_{k+1}))\} \tag{19}$$

where $J^*_{s,y}$ is the optimal cost calculated from t_s to t_y, and $s = 1,2,\ldots, N_{DP}$. This recursive relation can conclude minimum cost value using final time t_{end} and backward calculation from $J^*_{s+N_{simul}-1,s+N_{simul}}(x(t_{end}))$.Finally, using Equations (18) and (19), Equation (16) becomes Equation (20), which represents the discrete-type cost result:

$$J^{RHC} = \sum_{k=1}^{N_{total}} J^*_k = \sum_{k=1}^{N_{total}} \sum_{p=1}^{N_{RHC}} g\left(x(t_p), u^*_k(t_p), t\right)\cdot\Delta t_{DP} \tag{20}$$

3.4. Solver Based on 1.5-Dimensional DP

In general, an n-dimensional DP has n states and n control variables. However, in the model developed in this study, there is only one control variable, the fan flow. As shown in the following Equations (21) and (22), there are at least three states for which the data maps are shown in Figure 4:

$$V(t) = h\left(I_J(t), \rho(t), T(t), SOC(t)\right) \tag{21}$$

$$Zn_{consump,t_1} = \int_{t_0}^{t_1} q\left(V(t), I_J(t), \rho(t), T(t), SOC(t)\right)dt \tag{22}$$

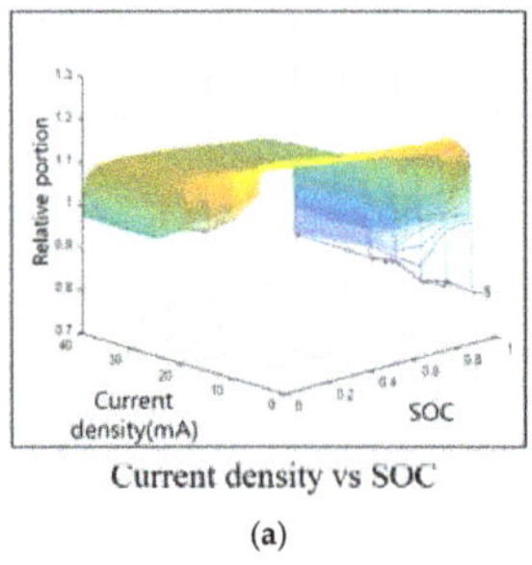

Current density vs SOC

(a)

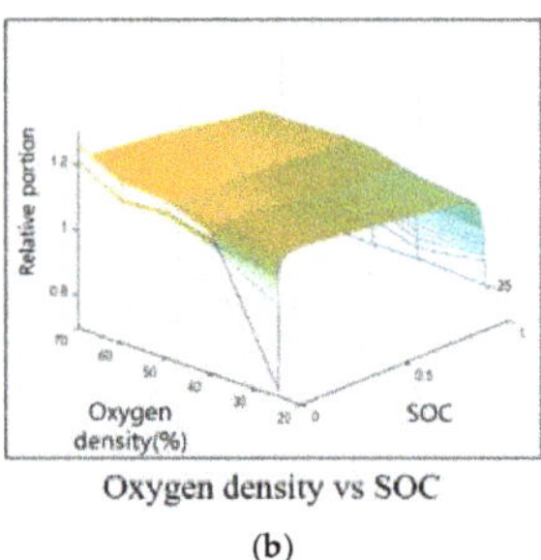

Oxygen density vs SOC

(b)

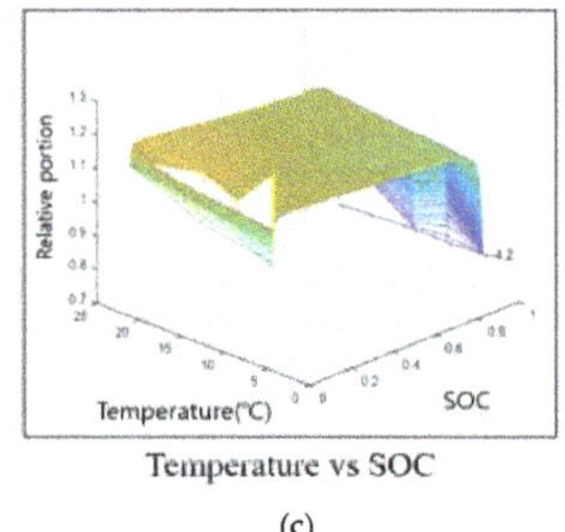

Temperature vs SOC

(c)

Figure 4. Cell output voltage dependence on cell parameters: (**a**) current density and state of charge (SOC); (**b**) oxygen density and SOC; (**c**) temperature and SOC.

Here, $V(t)$ is the output voltage, $I_J(t)$ is the current density, $\rho(t)$ is the oxygen density, $T(t)$ is the temperature, $SOC(t)$ is the state of charge, $Zn_{consump,t_1}$ is the Zn consumption until time t_1, h is the modeled system functions to calculate output voltage using cell experimental data, and q is the modeled functions to calculate metal consumption using cell experimental data.

Because UPS is a chemical model that uses batteries, the variables $I_J(t)$, $\rho(t)$, and $T(t)$ are interrelated states at each moment. Their values depend on each previous state and the control variable.

The sensitivity of the states, i.e., their change in a single step, determines which of the states is the reference for DP.

When the state variables that directly affect the output voltage are simulated for 3600 s, at constant load power and with the minimum and maximum fan flow, the current density $I_J(t)$ changes only slightly, as seen in Figure 5. Therefore, the cell temperature $T(t)$ and oxygen concentration $\rho(t)$ were chosen as the state variables for DP. As discussed, the cell temperature was calculated by distinguishing the time variables Δt_{simul} and Δt_{DP}, and considering that the state of the previous step has a major influence on the next one. However, the oxygen concentration is very sensitive to the states of each step, which depend on the fan flow. Thus, as DP proceeds, the target grid of the oxygen concentration is sparsely split, because this is determined when the cell temperature is applied to the grid by a control input. Therefore, the optimal x_2 was selected among a target region instead of the target point to solve the 1.5-dimensional DP structure:

$$x_{1,t_{m+1}} = f_1(x_{t_m}, u_{t_m}),\ u_m = f'_1\left(x_{1,t_{m+1}}, x_{t_m}\right),\ x_{2,t_{m+1}} = f_2(x_{t_m}, u_{t_m}) \tag{23}$$

$$x_{2,t_{m+1}} = f_2\left(x_{t_m}, f'_1\left(x_{1,t_{m+1}}, x_{t_m}\right)\right),\ x_{t_m} = (x_{1,t_m},\ x_{2,t_m}). \tag{24}$$

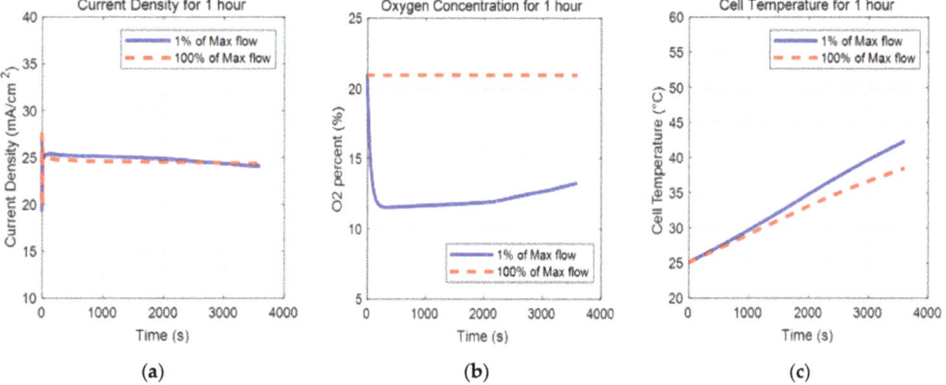

Figure 5. State variables on cell parameter by 1% and 100% of max fan flow control during operation: (**a**) current density; (**b**) oxygen concentration; (**c**) cell temperature.

Here, f'_1 is a reverse function for acquiring the control variable u_m using the 1st state variable at time t_{m+1}, $x_{1,t_{m+1}}$ and the state variables at time t_m, x_{t_m}. The next step, the 2nd state variable at time t_{m+1}, $x_{2,t_{m+1}}$, can be obtained from $x_{1,t_{m+1}}$ (that of the next step) and x_{t_m} (the state variables of the this step). This recursive process make us calculate the optimal control by selecting the minimum points as path in $J^*_{s,s+N_{simul}}$ among each region grid for the next DP step, as shown in Equation (19). This can be expressed as in Figure 6.

The experimental UPS data show that, below 60 °C and within the operational boundary conditions, the zinc consumption is higher for lower temperatures. This means that, when the difference between the lowest and highest temperature in the cells is higher than 10 °C, or the highest temperature is lower than 60 °C, the largest zinc consumption is that of the cell located at the end of the module.

When inverting the control input for the target state T, as in Equation (24), the metal-air battery model cannot calculate backward accurately, because all variables change. Therefore, the target T was calculated after obtaining the minimum and maximum cell temperature in the last step of the solve window, $(k + \Delta W_{prediction})$.

Because of these approximations, the algorithm is not perfectly and globally optimal, though reasonably optimal values are used.

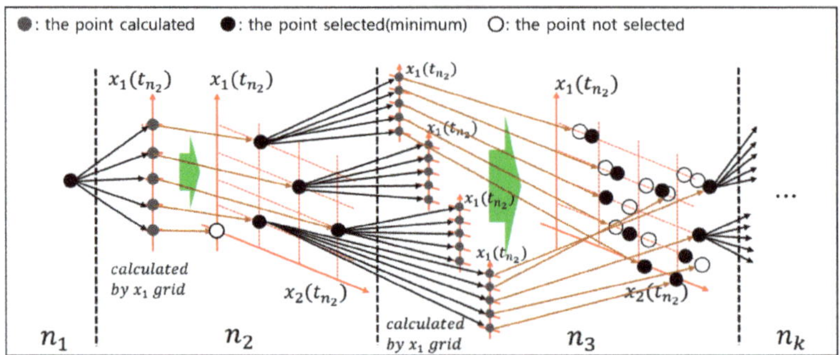

Figure 6. Schematic of structure solving by approximate DP.

3.5. Electrical Load Cycle

A basic electrical load cycle having the performance required by existing UPSs is shown as the first graph in Table 2. The power required was revised by increasing the simulation time from 7200 to 9000 s to evaluate the performance of the UPS control (cycle #1). Additionally, because the load power of the cycle is constant and simple, four more scenarios (cycle #2–#5) of operation time from 2 to 4 h were included in the simulation. These cycles are shown in Table 2.

Table 2. Required electrical load cycles representing different scenarios.

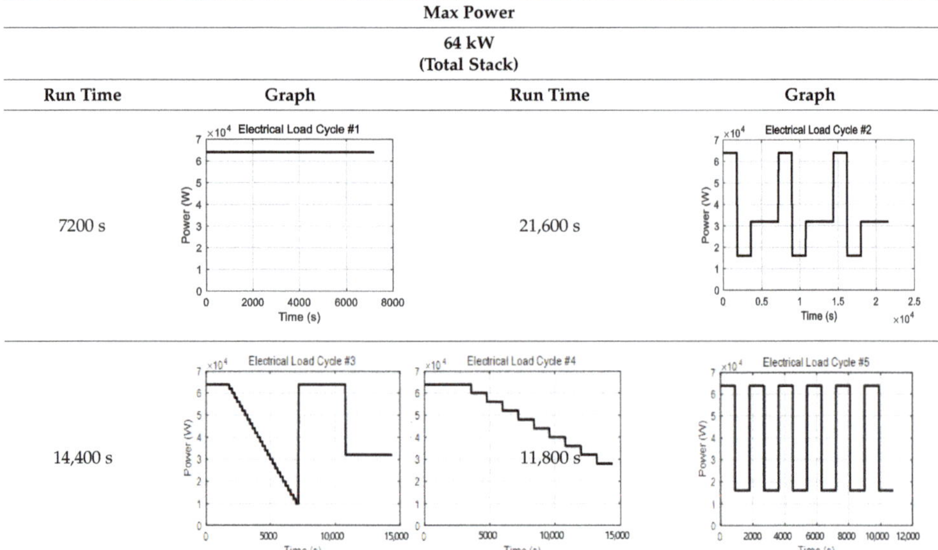

Max Power			
64 kW (Total Stack)			
Run Time	**Graph**	**Run Time**	**Graph**
7200 s	Electrical Load Cycle #1	21,600 s	Electrical Load Cycle #2
14,400 s	Electrical Load Cycle #3	Electrical Load Cycle #4 (11,800 s)	Electrical Load Cycle #5

4. Results

We experimented the proposed RHC algorithm using two-parameter settings. RHC #1 used $\Delta W_{solver} = 10$ and $\Delta W_{prediction} = 200$, and RHC #2 used $\Delta W_{solver} = 20$ and $\Delta W_{prediction} = 100$. We also simulated the UPS model with full DP control which assumes that we know the entire electrical load cycle. This control means globally optimal potential of the UPS model for comparison. In total the simulations were run controlling the fan by RHC, DP control, and constant fan flow. Electrical load cycle #2, which used the DP control, showed a distinct difference, and the resulting cell temperature of the entire module is shown in Figure 7.

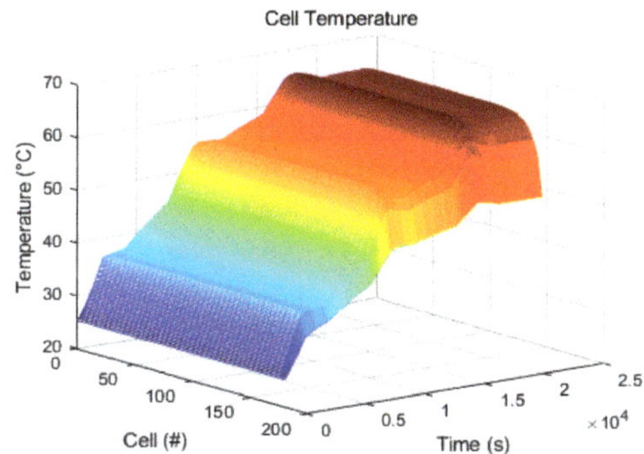

Figure 7. Cell temperature on electrical load cycle #2 by DP control.

As mentioned in Section 3.5, the temperature of the cell at the end of the module is significantly lower than that of the other cells during the entire simulation time. This is caused by the natural air-cooling of the module. Additionally, the temperature difference increases with time. By comparing with electrical load cycle #2 of Figure 6, it can also be seen that the temperature increases rapidly because of the large heat generated when the electrical load increases rapidly.

To compare the different controls, i.e., RHC #1, RHC #2, DP, and constant fan flow (1%, 2%, . . . , 100% of the maximum fan flow), the state variables based on electrical load cycle #2 were used, leading to the result shown in Figure 8.

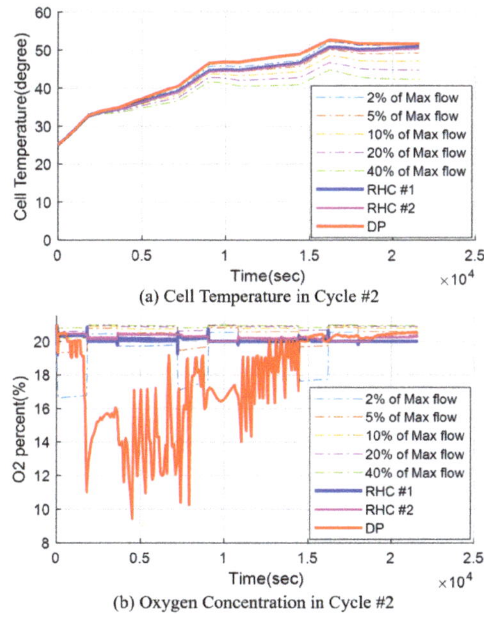

Figure 8. State variables as simulation result in electrical load cycle #2 by different fan control: (**a**) cell temperature; (**b**) oxygen concentration.

The cell temperature reported in Figure 8a indicates that the influence of the fan flow is not significant at first. However, it increases accumulating over time, resulting in a large difference between the controls. In the case of the oxygen concentration of Figure 8b, the effect of electrical load cycle #2 of Figure 6 is critical, because the oxygen concentration by all controls except DP vibrates under electrical load. Particularly in the case of the DP control, the optimal oxygen concentration of the whole cycle drops below 10%. Additionally, the constant fan flow control shows that the oxygen concentration increases gradually as the oxygen consumption decreases with increasing temperature. However, in the case of RHC, the initial oxygen concentration can be maintained through the cycle despite the temperature difference. Because the oxygen concentration hardly drops within a single prediction window, the locally optimal path of the control input is near the high oxygen concentration. The controls which maintained oxygen concentration as high unnecessarily using much electric energy. These trends of the variables by RHC will be similar in different situation because RHC has characteristic to find local optimal regardless of condition of the system and circumstances.

Looking at the fan flow for each control of Figure 9, it is seen that, in RHC, it is different from the DP control, though it gradually follows DP one with a time because of the optimal algorithm. Additionally, the fan flow of the RHC #1 control tends to be lower than that of RHC #2. When $\Delta W_{predict}$ and ΔW_{solver} were respectively increased and decreased (RHC #2 → RHC #1), we found that this control input is better than the DP one. This is usually attributed to low flow in the boundary condition.

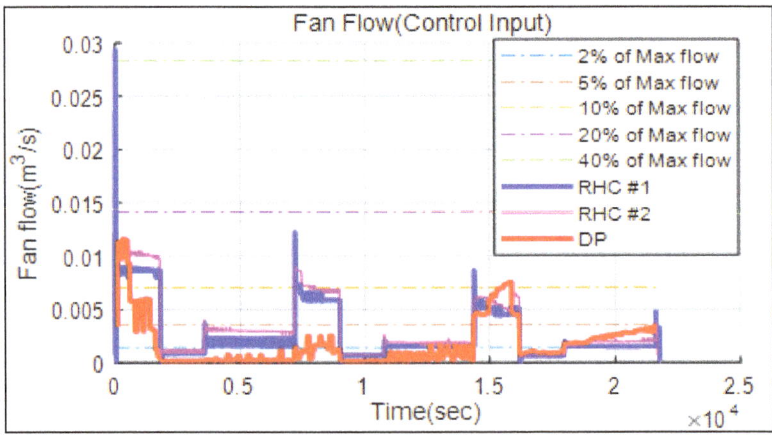

Figure 9. Fan flow control for the different algorithms.

The zinc consumption in all electrical load cycles is shown in Figure 10, using consumption of DP as reference (100%). Here, a grey bar indicates that the simulation stopped because a boundary condition, such as the state-of-charge of the battery, temperature, and oxygen concentration, was met before the total simulation time had passed. The RHC proposed in this study is represented in purple. The results confirm that the RHC control is effective in reducing the zinc consumption of metal-air battery-based UPSs increasing its operation time. In particular, RHC #1 showed stable and excellent performance, in contrast to other constant fan flow controls that fail or consume excessive zinc. Compared to the low-speed flow (that of less than 1–10% of the maximum fan flow), the zinc consumption of RHC #1 is not lower. However, in the case of the low-speed flows, the battery failed because of a poor cooling forced by the fans. Instead, compared with the other controls, RHC #1 displayed a maximum 6–10% difference in zinc consumption.

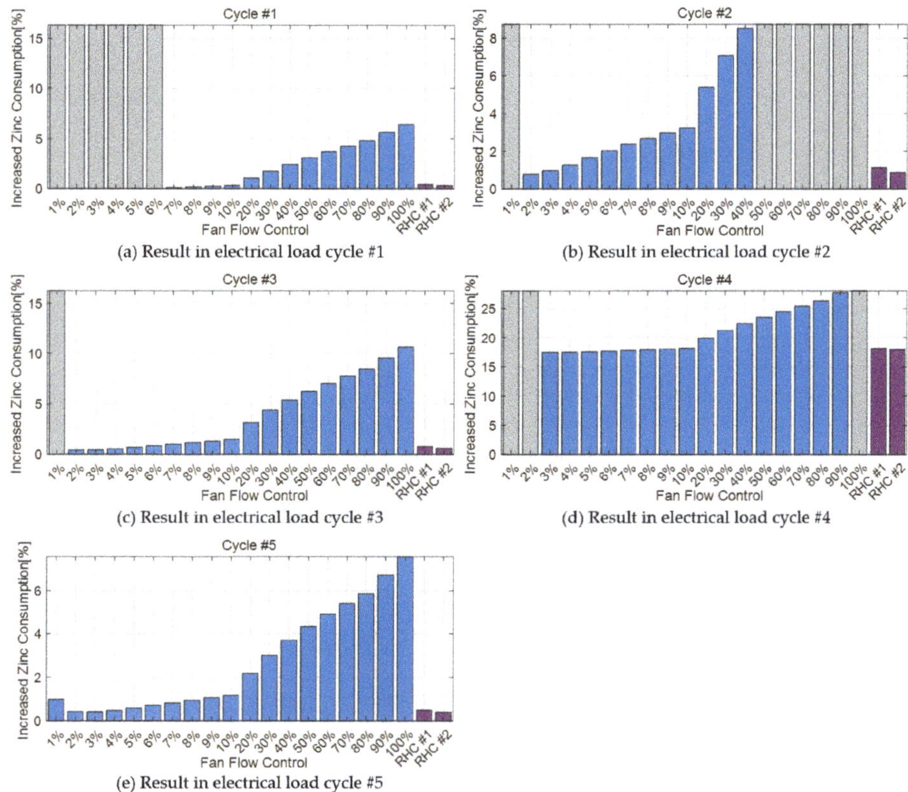

(a) Result in electrical load cycle #1

(b) Result in electrical load cycle #2

(c) Result in electrical load cycle #3

(d) Result in electrical load cycle #4

(e) Result in electrical load cycle #5

Figure 10. A comparison of increased zinc consumption by fan controls according to: (**a**) the electrical load cycles #1; (**b**) the electrical load cycles #2; (**c**) the electrical load cycles #3; (**d**) the electrical load cycles #4; (**e**) the electrical load cycles #5.

In most cycles, the zinc consumption is lower for weaker fan flow, though it is seldom very high in these cases either. However, because the simulation time was determined from the available operation time, all of the zinc metal was consumed, as indicated by the gray bars.

RHC #2 showed a slightly lower performance than RHC #1. However, it displayed a relatively low computing cost, approximately 1/5 of the original cost of RHC #1. As a matter of fact, RHC #1 cannot be used for real-time control because its simulation time is 4–5 times higher in a generic PC (i7-4790 CPU, 16GB RAM). In contrast, RHC #2, which needs a longer prediction range, needs a simulation time similar to the actual UPS running time. Thus, in principle, applying RHC #2 as real-time control in UPS is possible. If the current UPS model is simplified through data mapping, or the RHC control algorithm is optimized further, real-time control can be applied in industry. Additionally, the scenario of the facilities connected to the UPS in the event of a main power outage can be determined. In this case, the DP algorithm can be applied to increase the performance of the UPS as the potential with the predicted overall electrical load cycle.

5. Conclusions

In this work, a study to replace the existing lead-acid battery UPS by applying a metal-air battery was extended. The thermodynamic model is refined by simulating each cell in the stack module. Based on this, we proposed an optimizing algorithm that controls the flow of the fan that blows air to each cell in the module, decreasing its temperature. Because metal-air batteries use oxygen from the air as a fuel, it is difficult to study the correlation between the flow and the battery performance. Additionally, because the battery model is based on a chemical model, it is challenging to apply DP, which is a theory calculated with discrete points. To solve this problem, we developed a 1.5-dimensional DP suitable for zinc-air batteries.

An RHC model using DP is proposed to find the optimal cost function in the $\Delta W_{prediction}$ window, which was estimated in the range of several tens of seconds of electrical load. The load power of the proposed model is more stable than that of the existing control strategies constantly controlling the fans, and it shows excellent performance in terms of overall zinc consumption. Although not globally optimal, appropriate control inputs can be calculated and applied to the oxygen concentration and temperature of each step.

Author Contributions: "Conceptualization, B.G. and G.-Y.C.; methodology, B.G.; software, H.L., D.S.; validation, S.L., S.Y. and S.K.P.; formal analysis, H.L.; investigation, C.K.; resources, S.Y.; data curation, S.K.P.; writing—original draft preparation, B.G.; writing—review and editing, N.K.; supervision, S.W.C. All authors have read and agreed to the published version of the manuscript."

Funding: "This research received no external funding."

Acknowledgments: This work was supported by the Technology Innovation Program (20002762, Development of RDE DB and Application Source Technology for Improvement of Real Road CO2 and Particulate Matter) funded by the Ministry of Trade, Industry and Energy (MOTIE, Korea), and partially supported by the National Research Foundation of Korea (NRF) grant funded by the Korea government (Ministry of Science and ICT) (No. NRF-2019R1A4A1025848, No. NRF-2019R1G1A1100393).

Conflicts of Interest: "The authors declare no conflict of interest."

References

1. Chew, C.; Rama, S.; Kondapalli, R. Optimization (Genetic Algorithm) of dc-dc converter for uninterruptible power supply applications | S S r-. *Int. J. Pure Appl. Math.* **2018**, *118*, 1530–1535.
2. Narvaez, D.I.; Villalva, M.G. Modeling and control strategy of a single-phase uninterruptible power supply (UPS). In Proceedings of the 2015 IEEE PES Innovative Smart Grid Technologies Latin America (ISGT LATAM), Montevideo, Uruguay, 5–7 October 2015; pp. 355–360.
3. Lahyani, A.; Venet, P.; Guermazi, A.; Troudi, A. Battery/Supercapacitors combination in uninterruptible power supply (UPS). *IEEE Trans. Power Electron.* **2013**, *28*, 1509–1522. [CrossRef]
4. Chlodnicki, Z.; Koczara, W.; Al-Khayat, N. Hybrid UPS based on supercapacitor energy storage and adjustable speed generator. In Proceedings of the 2007 Compatibility in Power Electronics, Gdansk, Poland, 29 May–1 June 2007; Volume XIV, pp. 13–24.
5. Mallika, S.; Kumar, R.S. Review on ultracapacitor- battery interface for energy management system. *Int. J. Eng. Technol.* **2011**, *3*, 37–43.
6. Han, X.; Ji, T.; Zhao, Z.; Zhang, H. Economic evaluation of batteries planning in energy storage power stations for load shifting. *Renew. Energy* **2015**, *78*, 643–647. [CrossRef]
7. Linden, D.; Reddy, T.B. *Handbook of Batteries*, 3rd ed.; McGraw-Hill Professional: New York, NY, USA, 2002.
8. Lee, C.W.; Sathiyanarayanan, K.; Eom, S.W.; Kim, H.S.; Yun, M.S. Effect of additives on the electrochemical behaviour of zinc anodes for zinc/air fuel cells. *J. Power Sources* **2006**, *160*, 161–164. [CrossRef]
9. Gu, B.; Yoon, S.H.; Park, S.K.; Byun, S.; Cha, S.W. A study on the application of metal–air battery to large size uninterruptible power supply with a hybrid system. *JMST Adv.* **2019**, *1*, 181–190. [CrossRef]
10. Lee, J.; Yim, C.; Lee, D.W.; Park, S.S. Manufacturing and characterization of physically modified aluminum anodes based air battery with electrolyte circulation. *Int. J. Precis. Eng. Manuf. Green Technol.* **2017**, *4*, 53–57. [CrossRef]
11. Wang, K.; Pei, P.; Wang, Y.; Liao, C.; Wang, W.; Huang, S. Advanced rechargeable zinc-air battery with parameter optimization. *Appl. Energy* **2018**, *225*, 848–856. [CrossRef]
12. Xiao, Y.; Jing, R.; Song, H.; Zhu, N.; Yang, H. Design and experiment on zinc-air battery continuous power controller based on microcontroller. *Chem. Eng. Trans.* **2016**, *51*, 91–96.
13. Li, Z.Z.; Cheng, T.H.; Xuan, D.J.; Ren, M.; Shen, G.Y.; de Shen, Y. Optimal design for cooling system of batteries using DOE and RSM. *Int. J. Precis. Eng. Manuf.* **2012**, *13*, 1641–1645. [CrossRef]

14. Sabbah, R.; Kizilel, R.; Selman, J.R.; Al-Hallaj, S. Active (air-cooled) vs. passive (phase change material) thermal management of high power lithium-ion packs: Limitation of temperature rise and uniformity of temperature distribution. *J. Power Sources* **2008**, *182*, 630–638. [CrossRef]
15. Ling, Z.; Wang, F.; Fang, X.; Gao, X.; Zhang, Z. A hybrid thermal management system for lithium ion batteries combining phase change materials with forced-air cooling. *Appl. Energy* **2015**, *148*, 403–409. [CrossRef]
16. Xie, J.; Ge, Z.; Zang, M.; Wang, S. Structural optimization of lithium-ion battery pack with forced air cooling system. *Appl. Therm. Eng.* **2017**, *126*, 583–593. [CrossRef]
17. Wang, Y.X.; Qin, F.F.; Ou, K.; Kim, Y.B. Temperature control for a polymer electrolyte membrane fuel cell by using fuzzy rule. *IEEE Trans. Energy Convers.* **2016**, *31*, 667–675. [CrossRef]
18. Zhan, Y.; Wang, H.; Zhu, J. Modelling and control of hybrid UPS system with backup PEM fuel cell/battery. *Int. J. Electr. Power Energy Syst.* **2012**, *43*, 1322–1331. [CrossRef]
19. Guerrero, J.M.; Alcala, J.M.; Miret, J.; de Vicuña, L.G.; Matas, J.; Castilla, M. Output impedance design of parallel-connected ups inverters with wireless load-sharing control control and management of single phase microgrids view project converter modelling and digital control view project output impedance design of parallel-connecte. *IEEE Trans. Ind. Electr.* **2005**, *52*, 1126–1135. [CrossRef]
20. Pei, Y.; Jiang, G.; Yang, X.; Wang, Z. Auto-Master-Slave control technique of parallel inverters in distributed AC power systems and UPS. In Proceedings of the 2004 IEEE 35th Annual Power Electronics Specialists Conference (IEEE Cat. No.04CH37551), Aachen, Germany, 20–25 June 2004; Volume 3, pp. 2050–2053.
21. Guerrero, J.M.; Vásquez, J.C.; Matas, J.; Castilla, M.; de Vicuna, L.G. Control strategy for flexible microgrid based on parallel line-interactive UPS systems. *IEEE Trans. Ind. Electron.* **2009**, *56*, 726–736. [CrossRef]
22. Rawlings, J.B.; Muske, K.R. The stability of constrained receding horizon control. *IEEE Trans. Autom. Control* **1993**, *38*, 1512–1516. [CrossRef]
23. Mayne, D. Receding Horzion control of nonlienar system.pdf. *IEEE Trans. Autom. Control* **1990**, *35*, 814–824. [CrossRef]
24. Low, K.S.; Cao, R. Model predictive control of parallel-connected inverters for uninterruptible power supplies. *IEEE Trans. Ind. Electron.* **2008**, *55*, 2884–2893.
25. Kim, S.K.; Park, C.R.; Yoon, T.W.; Lee, Y.I. Disturbance-observer-based model predictive control for output voltage regulation of three-phase inverter for uninterruptible-power-supply applications. *Eur. J. Control* **2015**, *23*, 71–83. [CrossRef]
26. Johannesson, L.; Nilsson, M.; Murgovski, N. Look-ahead vehicle energy management with traffic predictions. *IFAC-PapersOnLine* **2015**, *28*, 244–251. [CrossRef]
27. Hellström, E.; Ivarsson, M.; Åslund, J.; Nielsen, L. Look-ahead control for heavy trucks to minimize trip time and fuel consumption. *Control Eng. Pract.* **2009**, *17*, 245–254. [CrossRef]

Review

Parallel Power Flow Computation Trends and Applications: A Review Focusing on GPU

Dong-Hee Yoon [1] and Youngsun Han [2,*]

1 Department of Railway, Kyungil University, Gyeongsan 38428, Korea; dhyoon@kiu.kr
2 Department of Computer Engineering, Pukyong National University, Pusan 48513, Korea
* Correspondence: youngsun@pknu.ac.kr; Tel.: +82-51-629-6250

Received: 30 March 2020; Accepted: 21 April 2020; Published: 1 May 2020

Abstract: A power flow study aims to analyze a power system by obtaining the voltage and phase angle of buses inside the power system. Power flow computation basically uses a numerical method to solve a nonlinear system, which takes a certain amount of time because it may take many iterations to find the final solution. In addition, as the size and complexity of power systems increase, further computational power is required for power system study. Therefore, there have been many attempts to conduct power flow computation with large amounts of data using parallel computing to reduce the computation time. Furthermore, with recent system developments, attempts have been made to increase the speed of parallel computing using graphics processing units (GPU). In this review paper, we summarize issues related to parallel processing in power flow studies and analyze research into the performance of fast power flow computations using parallel computing methods with GPU.

Keywords: power flow computation; high performance computing (HPC); parallelism; parallel computation; LU decomposition

---

1. Introduction

Power flow (PF) analysis is popularly employed to analyze electrical power systems by estimating the voltage and phase angle of buses inside the power system. Power flow computation is designed to repeatedly apply several numerical methods for solving nonlinear equations of such electric power systems; the calculations thus require a considerable amount of computational execution time. Furthermore, due to a general increase in the demand for electric power, the introduction of emerging renewable energy sources, and the spread of electric vehicles (EVs), modern power systems are continually expanding and display rapidly increasing complexity. The adoption of smart grids and advanced electronic devices creates a need for more electrical power of higher quality, and the spread of EVs is expected to lead to increased power consumption in the near future. Hence, recently, there has been greater demand for more accurate power flow analysis. In addition, because power flow analysis involves a large amount of contingency analysis for all probable accident conditions in power systems, the computational resources and time required for such analyses have increased dramatically.

Numerous studies have been undertaken to significantly reduce the computation time of complex numerical analysis by applying parallel and distributed processing with massive computing resources. In particular, the use of high-performance computing (HPC) machines to reduce computational times in power flow computation has long been studied [1–5]. After D. M. Falcao et al. [1] and V. Ramesh et al. [2] introduced the application of HPC to power system problems, several studies have been conducted in earnest into the use of HPC for accelerating power system analysis. R. Baldick et al. [3] and F. Li et al. [4] proposed rapid power flow computation using distributed computing, which uses multiple computers connected via an Ethernet network. Currently, because HPC technologies such as multi-core processors, clusters, and GPUs have made significant strides, many more studies have been undertaken into accelerating power system analysis by using parallel and cloud computing and adopting HPC technologies [5].

Attempts to reduce the analysis time of power systems by using GPU-based parallel computing have recently increased. Of course, the use of GPU is not confined to power flow studies. One example is the visualization part of power systems. Some studies into the visualization of power systems have been undertaken [6]. Real-time visualization of power systems is necessary for the simulation, state estimation, and prediction of power systems, and many studies have been performed investigating the speedup using HPC [7]. The parallel processing is also valuable in the electric railway system because it requires a lot of computation and data processing in various fields such as scheduling optimizations, signal processing, finite element analysis, train dynamics, and power network simulation [8]. A variety of parallel processing techniques can be applied to the electric railways, and some studies using GPU have been conducted in this area [9–11]

The rest of the paper is organized as follows: In Section 2, we describe the background of our review, such as GPU, power flow computation, and matrix handling methods, briefly. In Section 3, we present GPU application trends in power flow computation. For the sake of understanding, we introduce the previous PF studies with parallel processing and describe the recent trend of GPU-based power flow studies. In Section 4, we provide a detailed description of some representative studies that apply GPU-based parallel computing to power flow analysis and summarize their performance using a comparative analysis table. Finally, the conclusion is made in Section 5.

2. Background

2.1. GPU

GPUs have been widely adopted to accelerate graphical processing algorithms for image processing and computer graphics that handle large amounts of data using their highly parallel architecture [12–16]. In particular, modern GPUs have been developed in the form of general-purpose computing on graphics processing units (GPGPUs) as a substitute for CPUs in HPC for various scientific algorithms, such as deep learning, genome mapping, and power flow analysis. This is because GPUs significantly outperform CPUs in terms of the cost-effectiveness of floating-point computational throughput.

Figure 1 briefly outlines the overall hardware architecture of a GPU and its programming model. The GPU is specialized to dramatically accelerate computation-intensive tasks due to its massive parallel architecture, which employs a large number of streaming multiprocessors (SMs) consisting of 32 scalar processors (SPs) that operate in a lockstep manner. The GPU supports parallel execution of a computational kernel consisting of multiple thread blocks, and each thread block is divided into warps, i.e., groups of 32 threads. All threads in each warp are executed simultaneously on a single SM. A warp scheduler selects one eligible warp at a time and concurrently executes all threads of the warp on the processing cores, i.e., the SPs, of the SM using lockstep synchronization. Therefore, we can exploit extremely high thread-level parallelism (TLP) by interleaving a large number of threads to the processing cores. Programmers can implement parallel programming models on GPUs using open computing language (OpenCL) and compute unified device architecture (CUDA).

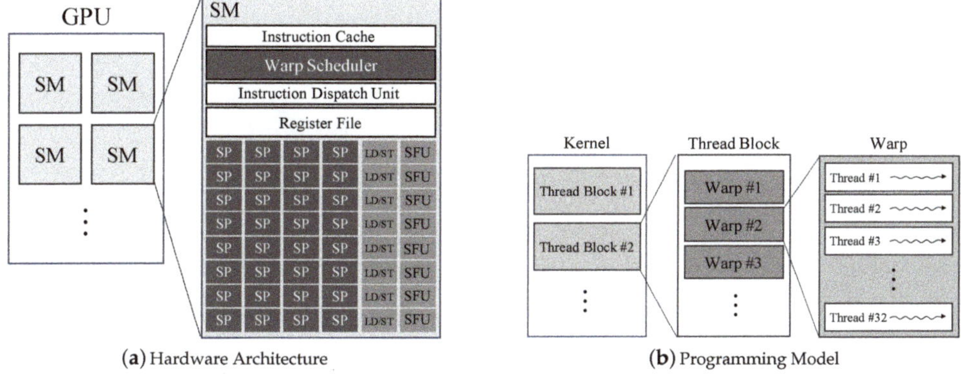

(**a**) Hardware Architecture (**b**) Programming Model

Figure 1. GPU hardware architecture and programming model.

2.2. Numerical Analysis of Power Systems

2.2.1. Power Flow Analysis

A power flow analysis is a representative numerical method that estimates the electric power flow of an interconnected power system. It is designed to repeatedly perform power flow computations to obtain the voltage magnitudes and angles of all the buses in a power system and calculate the real and reactive powers of peripheral equipment connected to the buses. Because the power flow computation mainly consists of nonlinear algebraic equations, various approaches, including Gauss–Seidel (GS), Newton–Raphson (NR) [17], and fast-decoupled power flow (FDPF) [18] methods, have been developed to accelerate the solution of the equations.

The GS method is often used because it is the first practical approach proposed for estimating power flow in large-scale power systems. Because an admittance matrix of a power system contains many zero coefficients, power flow analysis is faster using the GS method than using the NR method for a single iteration. However, the GS method cannot easily be widely adopted in power flow analysis due to its poor convergence characteristics. Therefore, though the GS method is popular in power flow analysis, it is mentioned here only briefly because only a few components can benefit from parallel processing with GPUs.

The NR method is currently the most prevalent for power flow analysis. A number of studies have discussed acceleration of the NR method by applying GPU-based parallel processing techniques. In general, we can write a nodal equation of a power system network as follows:

$$I = Y_{bus} \cdot V \tag{1}$$

where Y_{bus}, I, and V denote an admittance matrix of the network, a current vector, and a voltage vector, respectively. I_i indicates the total power flow at the i-th bus as in Equation (2).

$$I_i = \sum_{k=1}^{n} Y_{ik} V_k \tag{2}$$

A complex power equation can be expressed separately with real and imaginary parts, P and Q, as follows:

$$S_i = V_i (\sum_{k=1}^{n} Y_{ik} V_k)^* = V_i \sum_{k=1}^{n} Y_{ik}^* V_k^* = P_i + jQ_i \tag{3}$$

$$P_i = \sum_{k=1}^{n} |V_i| \cdot |V_k| \cdot (G_{ik} \cos\theta_{ik} + B_{ik} \sin\theta_{ik}) \tag{4}$$

$$Q_i = \sum_{k=1}^{n} |V_i| \cdot |V_k| \cdot (G_{ik}\sin\theta_{ik} - B_{ik}\cos\theta_{ik}) \quad (5)$$

Equations (4) and (5) show the power balance equations of the i-th bus in polar form. The NR method employs an iterative technique to solve the two nonlinear power balance equations. The set of resulting linear equations can be formulated in a matrix form as follows:

$$\Delta f = J \cdot \Delta x \quad (6)$$

where J is a Jacobian coefficient matrix. Equation (7) shows the complete formulation to obtain the power mismatch, i.e., ΔP and ΔQ, through Equations (4) and (5) using the NR method.

$$\begin{bmatrix} \Delta P \\ \Delta Q \end{bmatrix} = \underbrace{\begin{bmatrix} \frac{\partial P}{\partial \theta} & \frac{\partial P}{\partial V} \\ \frac{\partial Q}{\partial \theta} & \frac{\partial Q}{\partial V} \end{bmatrix}}_{\text{Jacobian}} \begin{bmatrix} \Delta \theta \\ \Delta V \end{bmatrix} \quad (7)$$

For the sake of clarity, Figure 2 provides a succinct illustration of the iterative process of calculating a power flow solution using the NR method. The method starts with arbitrary initial values of V and θ, and then calculates the power mismatch using Equation (7). Also, the method determines whether the mismatch is converged to complete the iteration. If it is not converged, the method computes a Jacobian matrix of the power flow system with updated P and Q first, and derives the mismatches of voltage and phase angles, i.e., ΔV and $\Delta\theta$, from Equation (7). Then, the method updates the voltage and phase angles with the mismatches and performs the overall procedures iteratively.

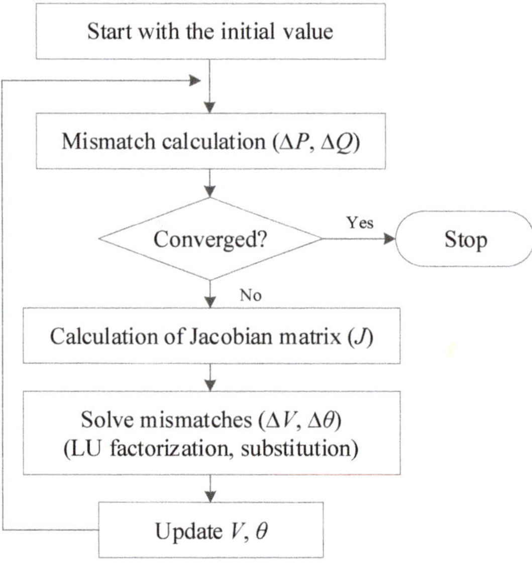

Figure 2. Brief algorithm of the Newton–Raphson (NR) method.

As previously mentioned, we need to carry out heavy matrix calculations, such as an LU decomposition for updating both the Jacobian coefficients and the mismatch of voltage and phase angles. Nevertheless, the NR method performs the power flow computation much faster than the GS method by converging to the solution with fewer iterations; thus, the method is more suitable for large-scale power systems [19]. Also, the Jacobian matrix produced by the NR method can provide an index for sensitivity analysis or some other control problems. Hence, in this paper, we discuss several previous studies focused on accelerating the NR method using GPU-based parallel computing in detail.

Finally, the FDPF is a method of approximating a solution based on the NR method, which required significantly less computational effort [17]. Although the NR method is the most popular method for PF studies, calculation is slow because it requires calculation of many inversions of the Jacobian matrix for each iteration. To resolve this drawback, some methods, including FDPF, have been proposed to apply approximation approaches instead of using the Jacobian coefficient matrix in its full form. In addition, the FDPF method exploits P-Q decoupling, which means the active and reactive powers are not significantly affected by the magnitude and phase angle change of the bus voltage, respectively. The method does not update its Jacobian coefficient matrix during the convergence iterations for reducing the overhead of numerical computations, so the Jacobian matrix is inverted only once during calculation of the solution. As a result, the FDPF method can formulate the NR method more simply; however, this comes with the disadvantage that more iterations are required to reach the solution and the method more often fails to eventually converge. Hence, the method is usually adopted in specific situations where fast calculation is compulsory for the power flow calculation.

2.2.2. LU Decomposition

For computing the power flow calculation by utilizing either the NR or FDPF method, the solution of a linear system of matrix equations is required, as in Equation (1). Notably, the Y_{bus} matrix reflects the admittance information of a power system, so it presents a sparse quality in containing many zeros. As shown in Figure 2, we need to update V and θ with ΔV and $\Delta\theta$ obtained from the power mismatch, i.e., ΔP and ΔQ, using Equation (7). Hence, we apply a mathematical technique such as sparsity oriented LU decomposition to calculate the inverse of the Jacobian coefficient matrix based on the admittance information [17].

$$A \cdot X = (L \cdot U) \cdot X = L \cdot (U \cdot X) = b \tag{8}$$

$$L \cdot y = b \tag{9}$$

$$y = U \cdot X \tag{10}$$

LU decomposition is also known as triangular factorization or LU factorization. To obtain X by solving a linear equation, i.e., $A \cdot X = b$, the LU decomposition first divides the coefficient matrix A into two lower and upper triangular matrices, L and U, as in Equation (8). Then, y is calculated by performing a forward substitution in Equation (9), and X is finally obtained by a backward substitution in Equation (10).

Numerical computation techniques based on such sparsity are popularly employed in procedures to solve the network equations of large power systems [20]. The time required for the matrix calculations requires a high proportion of the total computation time; for instance, it was found to take up to about 80% of the total computation time in [21]. Hence, many related studies have already been performed [22–26], as we can significantly reduce the total execution time of PF analysis by applying parallel processing to the matrix calculations.

2.2.3. QR Decomposition

LU decomposition is more commonly used to solve power flow problems, but it is also possible to adopt QR decomposition as an alternative for leveraging GPU-based parallel processing. This is because LU decomposition without pivoting may suffer from significant numerical instability, as described in [27]. Although QR decomposition has greater computational complexity compared to

LU decomposition, some studies have found that it is more suitable for GPU-based parallel computing due to its high numerical stability even without pivoting [28].

In linear algebra, a QR decomposition of a complex $m \times n$ matrix A is represented as follows:

$$A = Q \cdot R \tag{11}$$

where Q and R indicate an $m \times m$ unitary matrix and an $m \times n$ upper triangular matrix, respectively. Several methods for implementing the QR decomposition are available, such as the Gram–Schmidt process, Householder transformations, and Givens rotations [29].

2.2.4. Conjugate Gradient Method

The conjugate gradient (CG) method is designed to find numerical solutions of linear system equations in which the matrix is symmetric and positive definite. The method is typically implemented as an iterative algorithm for analyzing large sparse power systems that are too large to be handled by direct methods, such as Cholesky decomposition.

The CG method can be employed instead of LU decomposition in a PF study. The use of the CG method in PF studies has been investigated extensively. In [30], a PF study based on the CG method was introduced. In [31], a PF study using the CG method was performed based on GPUs. In [32], a GPU-based PF study was conducted using a combination of preconditioners and a CG solver. In general, preconditioners are widely used for matrix handling. Since it is difficult to secure parallelism in a general preconditioner, a Chebyshev preconditioner is used to secure the parallelism of the preconditioner in [32].

3. GPU Application Trend in Power Flow Computation

3.1. Conventional Parallel Processing in PF Studies

Recent studies applying GPUs to PF studies have usually been conducted based on existing parallel processing research. Therefore, in terms of PF studies, a review of current parallel processing research is necessary to understand the application of GPUs. As mentioned in Section 2.2.1, PF computation using the NR method performs matrix handling during computation. In this process, it is usually necessary to solve a sparse linear system for power system analysis such as static security and transient stability analysis. Most of the power system applications have conventionally employed direct LU decomposition methods to solve such linear equations. Since this decomposition process takes large amounts of time, numerous studies have examined how to perform this operation more quickly. Considerable computational resources are required for fast computation, and parallel computing is an excellent way to secure such resources. Hence, many attempts to conduct large amounts of power flow computation quickly have been made using parallel computing. Parallel computing requires resources for computation, and some attempts have been made to perform power flow computation by connecting multiple computational resources, such as computers with the goal of fast PF computation. Theoretical attempts at parallel processing in PF studies are described in [33,34]. In the 1990s, researches into the parallel processing of PF computation began in earnest. The fundamental concepts of applying the parallel processing to PF computations are precisely presented in [22,26], and a number of further studies have been performed since; most of the studies focused on matrix handling to divide computation for each computational resource [23–25,35,36]. G. A. Ezhilarasi et al. [37] presented an effective method to perform a parallel contingency analysis using high-performance computing. Also, Z. Huang [38] proposed a dynamic computational load balancing method that utilizes multiple machines of high-performance computing for delivering a large number of contingency computations.

It is also essential to employ a massively large number of cores for accelerating the computation. There have been many endeavors to enhance the performance of parallel computing by adopting multiple processing machines in PF computation. In [39], a commercial solver was utilized in a massively parallel computing environment to carry out the contingency analysis of a power system.

To decrease the calculation time significantly, 4127 of the contingencies were spread to 4127 cores and computed simultaneously. I. Konstantelos et al. [40] presented a computational platform, as a part of the iTesla project, for evaluating the dynamic security of an extensive power system using a commercial analysis software on 10,000 cores of Intel Xeon E5-2680. This project utilizes a large number of CPUs, i.e., 80,640 of cores, which are mutually interconnected, but the maximum number of CPU cores used simultaneously is up to 10,000.

3.2. Advantages of GPU-Based PF Computation

Analysis of a power system requires a large number of PF computations, and one method for reducing the computational time is parallel processing. Although the purpose is to calculate quickly by using many computational resources, this approach has obvious disadvantages in terms of time and cost in establishing a parallel processing environment by connection of a large number of machines. Therefore, the use of GPUs that can process many computations at once is an attractive proposal for PF computation, as it overcomes the shortcomings related to physical connections between machines [41].

The LU decomposition mentioned in Section 2.2.2 is a very time-consuming task in PF studies; thus, it is a good candidate for the application of parallel processing with GPUs. LU decomposition with GPUs has been studied outside the field of PF studies, as it is used in various fields that require a large amount of computation. For example, a parallel process for solving the sparse matrix for circuit simulation has been studied in [42–44], and L. Ren et al. [45] presented results that could overcome the difficulty of super-node formation for extremely sparse matrix objects.

Many studies have recently been conducted to reduce the computation time required for LU decomposition through the use of GPUs. Basically, while matrix handling related operations have been physically assigned to multiple machines in the past, recent studies have focused on how to assign the operations to GPUs for efficient calculation. Compared to using physically separated computational sources, the use of a GPU inside the same machine as a computational resource can yield very high-speed delivery of information, which has a significant advantage in terms of reducing computational time compared to past parallel processing studies. The following chapters summarize the trend of GPU-applied PF studies.

3.3. GPU Applications Trends in PF Studies

GPU-based acceleration techniques are highly productive methods to enhance the performance of power flow computations significantly. We can dramatically diminish the computational time of a sparse linear system (SLS) solution by utilizing GPUs. Researches broadly range from investigating conventional direct LU decomposition to using computational solvers other than LU. In [31], the basics of GPU-based parallel computing were explicitly presented, and an efficient method of using GPUs was proposed in [32]. There are some other attempts to GPU-accelerated SLS solution using LU decomposition in [46–48]. Also, in [49,50], the implementation of parallel PF computation being conducted by adopting GPUs were described in detail. Although the methods of using GPUs in the PF study are different, there is a commonality that they tried to use the GPU for a large amount of computation for the analysis of large-scale systems [51–53]. S. Hung et al. [54] showed a typical use of GPU to solve the PF problem. They described a GPU-accelerated FDPF method and a direct linear solver using LU decomposition in detail, and simulation results using Matlab were summarized. G. Zhou et al. [55] described an LU decomposition solver that packaged a large number of LU decomposition tasks. The speed of their proposed solver was up to 76 times greater compared to the KLU library. KLU library is a solver for a sparse linear system [56].

In [32], a GPU-based Chebyshev preconditioner integrated with a GPU-based conjugate gradient solver was proposed to solve SLS for speedup. Also, X. Li et al. [57] combines a GPU with a multifrontal method to solve sparse linear equations in power system analysis, and the multifrontal method requires an inversion of the factorization of a sparse matrix to a series of a dense matrix. Since the inversion process requires significant computation, a reduction of the computation time using GPUs was proposed.

In addition to the typical approach based on the NR method, which requires LU decomposition, in some cases, GPUs have been combined with other methods. In [50], PF computation was studied by combining GPUs and the FDPF method.

In the paper, the authors proposed a new approach using Matpower, i.e., one of Matlab's toolbox, to solve the FDPF problem by combining an iterative linear solver of a preconditioned conjugate gradient method and an Inexact Newton method, alternatively of typical LU decomposition [58,59].

The proposed GPU-based FDPF method in [50] exhibited a speedup performance improvement of up to 2.86 times over 10,000 buses compared to the CPU-based FDPF method. S. Hung et al. [60] proposed a fast batched solution for real-time optimal power flow (RTOPF) considering renewable energy penetration. A primal-dual interior point method including three kernels was used, and several test power systems were simulated on four platforms (regular CPU, parallel CPU, regular GPU, batched GPU). The batched GPU method obtained good results, with a speedup of up to 56.23 compared to a regular CPU. Moreover, an unconventional study used GPUs for computation of probabilistic power flow (PPF) rather than general PF [61]. PPF can be obtained by using Monte Carlo simulation with simple random sampling (MCS-SRS). In [61], a GPU-accelerated algorithm was proposed that could handle largescale MCS-SRS for PFF computation. There are some studies focusing on the CPU-GPU hybrid system [62,63]. In [62], an advanced PF computation technique was proposed. This technique is based on the CPU-GPU hybrid system, vectorization parallelization, and sparse techniques. In [63], the method for performing simultaneous parallel PF computation using a hybrid CPU-GPU system was proposed.

4. Typical GPU-Based PF Studies

4.1. A Study Based on LU Decomposition

In this section, we describe some representative studies that apply GPU-based parallel computing to PF analysis. The most time-consuming part of PF computation using the NR method is the LU decomposition component for matrix handling. Therefore, many studies have been conducted on fast parallel processing of LU decomposition parts using GPUs; this chapter summarizes the LU decomposition parallel processing elements of several related papers.

A GPU-based sparse solver that does not use dense kernels was proposed in [64]. The primary purpose of this study was the application of the parallel processing approach for circuit simulation, so it was not closely related to PF research. However, we mention this study here to explain the parallel processing of LU decomposition using a GPU.

Basically, LU decomposition is a popular method using pivoting. There is partial pivoting in the LU decomposition with proper permutation. In the case of partial pivoting, one of the rows and columns is permutated, and row permutation is generally performed. This can be expressed using the following formula.

$$P \cdot A = L \cdot U \tag{12}$$

In the case of full pivoting, both row and column pivoting are performed, as expressed by the following formula.

$$P \cdot A \cdot Q = L \cdot U \tag{13}$$

where P and Q represent permutation matrices, and L and U mean the lower and upper triangular matrices, respectively. Also, A is an input matrix to be decomposed.

However, this study found that LU decomposition without pivoting is more favorable for parallel processing with GPUs. Hence, a GPU-based sparse solver based on LU decomposition without pivoting was proposed. The proposed GPU solver does not use dense kernels, unlike conventional direct solvers, so it can be used for fast matrix calculation. The framework of the proposed sparse solver is shown in Figure 3. The proposed method is designed to distribute tasks appropriately to CPUs and GPUs, and the row/column reordering performed for the pivoting is replaced by the execution of one pre-processing step on a CPU to reduce the computation time. Most of the iterations for actual factorization take place on a GPU, which receives the CPU's initial calculations. A GPU-based left-looking algorithm is used to factorize a matrix in this method, which is summarized in Algorithm 1. The result obtained through iteration on the GPU is read back from the CPU, and triangular equations are executed. The proposed approach achieves, on average, 7.02 × speedup compared with sequential PARDISO. The simulation results were obtained using a total of 16 cores of Intel E5-2690 CPU and a single NVIDIA GTX580 GPU, respectively.

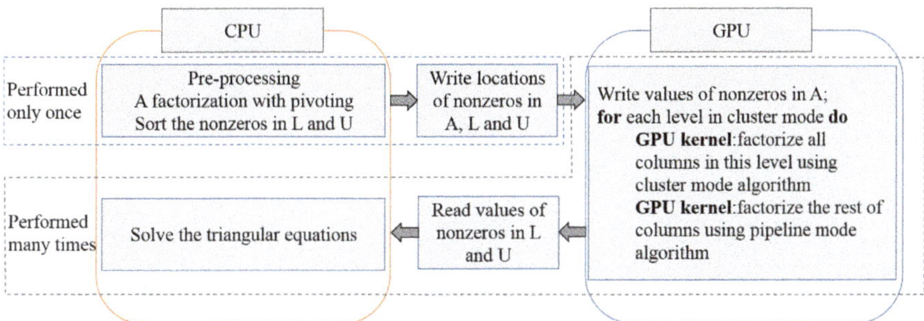

Figure 3. Framework of the sparse solver in [64].

Algorithm 1 left-looking LU factorization (GPU-based) [45,55,64].

INPUT A: an $n \times n$ input matrix to be LU factorized

OUTPUT L, U: lower and upper triangular matrices factorized from A

while there still exist executable columns **do in parallel**
  // e.g., the GPU thread j of the j-th column.
  $\&c = A(:, j)$;
  // Renew the j-th column with all its dependent columns on the left-hand side.
  for $i = 1 : j - 1$ **where** $U(i, j)$ is not equal to zero **do**
    $c(i+1:n) -= c(i) \times L(i+1:n, i)$;
  end for
  Normalize the j-th column;
end while

In [49], a GPU-accelerated GS and NR methods for PF study were analyzed. As discussed earlier, the primary purpose of this review paper is a focus on the acceleration of matrix handling with GPU, so the use of the GS method is beyond the scope of this work. However, this is valuable because it shows the time required for each computational step of the NR method. Hence, the LU decomposition of the NR method is briefly summarized. Also, some preliminary works were presented at the level of data, software composition, and matrix representation for parallel processing using GPU; however, this is not discussed in detail here.

Table 1 summarizes the four computational steps that consume the most execution time in NR implementation. Both building an admittance matrix and obtaining apparent power and line losses are unrelated to matrix handling, and the use of the GS method does not introduce any differences here. The part to be examined carefully is the solution of a sparse linear system that produces and solves the Jacobian matrix. This study demonstrated how some of the iterative solvers provided by CUDA are beneficial for single calculations but have no significant advantages in parallel processing for large scale PF studies. Hence, the Eigen library, an open-source C++ library, is proposed as a solution [56]. In the proposed method, each linear system is solved sequentially, but parallel processing is performed in which multiple systems are distributed to multiple OpenMP threads.

Table 1. Implementation of the NR method with average time and speedup in [49].

Major Procedures (>30 ms)	Time (ms)		Speedup
	CPU	GPU	
Build admittance matrix	154.285	2.820	54.7×
Build Jacobian matrix	61.308	2.925	21.0×
Solve linear system	972.043	55.104	17.6×
Compute S and losses on all lines	69.094	0.971	71.2×

G. Zhou et al. [55] proposed a batch-LU solution method to calculate multiple SLS in the massive-PFs problem (MPFP). The crux of the proposed method is a simultaneous solution of multiple linear systems consisting of $\mathbf{A}_i \cdot x_i = b_i$ (i = 1, 2, ..., N). This method is described in Algorithm 2 and its main features are as follows.

Algorithm 2 GPU-based batch LU factorization [55].

INPUT A: an $n \times n$ input matrix to be LU factorized

OUTPUT L, U: lower and upper triangular matrices factorized from A

1: **while** there still exist executable columns **do in parallel**
    // Thread block j for the j-th column of batch tasks of LU factorization
2:     $j \leftarrow$ block ID, $t \leftarrow$ thread ID in the thread block;
3:     $\&c_t = A_t(:, j)$;
    // Renew the j-th column of all the other subtasks of LU factorization.
4:     **for** $i = 1 : j-1$ **where** $U_t(i, j)$ is not equal to zero **do**
5:         $c_t(i+1 : n) -= c_t(i) \times L_t(i+1 : n, i)$;
6:     **end for**
7:     Normalize the j-th column of A_t ;
8: **end while**

- Unifying sparsity pattern: All SLSs must have a uniform sparsity pattern.
- Performing reordering and symbolic analysis only once: $\mathbf{A}_i = \mathbf{L}_i\mathbf{U}_i$ can be performed only once.
- Achieving extra inter-SLS parallelism: We have achieved N (N = batch size) times parallelism automatically by packaging a large number of LU decomposition subtasks into a new large-scale batch computation task. First, the decompositions on the same indexed columns, which belong to different subtasks of LU decomposition, are designated to a single thread block. In the algorithm, A_t, L_t, and U_t mean the input matrix, output lower and upper triangular matrices, respectively, that are accessed by the t-th thread block. Second, the data of the same indexed columns are located in the adjacent address of a GPU device memory.
- Preventing thread divergence: Lines four to six in Algorithm 2 (updated warp thread operation) will execute the same branch and it does not diverge due to the identical sparsity pattern and aforementioned thread-allocation type.

This method can be used for both forward and backward substitutions of LU decomposition. In the case study using the proposed method, the maximum improvement is a 76-fold speedup compared with the KLU library.

4.2. A Study Based on QR Decomposition

In the case of a system of N elements, an $N-1$ static security analysis (SSA) is required to perform N alternating-current power flows (ACPF), so that usually incurs a tremendously high computational expense for a large-scale power system. Hence, in [28], a GPU-based batch-ACPF method using a QR solver was proposed to reduce the computation time for solving a large number of PFs of $N-1$ SSA. For the sake of clarity, the basic concept of QR decomposition is described in the background Section 2.2.3, and the overall framework of the proposed batch-ACPF solution is shown in Figure 4.

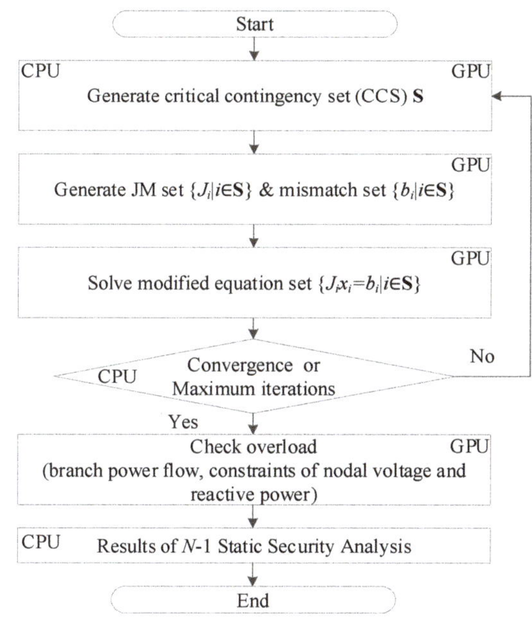

Figure 4. Framework of the proposed method in [28].

Reference [28] proposed a GPU-accelerated contingency screening algorithm to determine the critical contingency set (CCS) required for analysis. The proposed algorithm, which is described in detail in chapter VI of [28], includes dense vector operation in the execution of the algorithm.

Two GPU-based algorithms are used to solve the CCS **S** generated by the screening algorithm. First, the batch Jacobian-Matrix (batch JM) generation is employed to produce the JM set $\{J_i|i \in \mathbf{S}\}$ and power mismatch set $\{b_i|i \in \mathbf{S}\}$. The JM generation can be identified from the modified equations of ACPF analysis in Equation (14).

$$\begin{bmatrix} \Delta P \\ \Delta Q \end{bmatrix} = -J \begin{bmatrix} \Delta\theta \\ \Delta U/U \end{bmatrix} = -\begin{bmatrix} HN \\ ML \end{bmatrix} \begin{bmatrix} \Delta\theta \\ \Delta U/U \end{bmatrix} \tag{14}$$

where ΔP denotes a mismatch vector of nodal active power injection, ΔQ represents a mismatch vector of nodal reactive power injection, $\Delta\theta$ and ΔU are modification vectors of voltage angle and magnitude, respectively. Also, U means a vector of nodal voltage magnitude, H, N, M, and L indicate the sub-matrices of JM J. JM generation is necessary to calculate four sub-matrices. Each submatrix has independence by node, so the larger the power system is, the greater the parallelism.

Second, the batch-QR solver is necessary to accelerate the solution of the modified equation set $\{J_i x_i = b_{out}|i \in \mathbf{S}\}$. Reference [28] showed that a conventional QR solver for solving single SLS was not suitable for the GPU-accelerated algorithm, so a batch-QR solver for a massive number of SLSs is proposed to overcome the disadvantages. The purpose of the batch-QR solver is to simultaneously solve the linear system $A_i x_i = b_i$ (i = 1, 2, ..., N). The proposed solver splits the entire calculation into a single QR subtask, which takes the form of a parallel computation on a GPU to achieve a higher degree of parallelism. This includes the parts for coalesced memory accesses and reordering algorithms for computation.

There are two stop criteria for batch-ACPF. One is that 80% of ACPFs have converged. The other is that the batch-ACPF solver has reached the maximum number of iterations. In [28], the maximum number of iterations was set at 10. Here, the GPU kernels in double-precision format are used to check the overload. One is used to check the branch power flow, and the other is used to check the nodal voltage and reactive power. In Table 2, we summarize the simulation results for a specific case. The first solution was designated as a baseline for performance comparison, and the speedup result for each SSA solution method is indicated to prove the validity of the proposed method.

Table 2. Performance comparison in [28]. All the other speedups were calculated against the analysis time of No. 1.

No.	SSA Solution	Analysis Time (s)	Speedup
1	CPU SSA with a single-threaded UMFPACK [65]	144.8	-
2	CPU SSA with 12-threaded PARDISO [66]	33.6	4.3×
3	CPU SSA with 12-threaded KLU [56]	9.9	14.6×
4	GPU SSA with Batch-ACPF solver [28]	2.5	57.6×

4.3. Computation Time Reduction of GPU-Based Parallel Processing

Research on GPUs to PF study can be categorized into studies aimed at reducing the execution time of matrix handling in a single computation, and studies focused on parallel processing of a large number of computations simultaneously. Many studies focused on parallel processing have suggested methods for efficiently separating matrix operations. Since most of the operations are concentrated in this area, efficiency can be increased if the time required for matrix handling can be reduced significantly [21]. The previous studies introduced in Sections 4.1 and 4.2 are typical matrix handling studies. On the other hand, some studies that attempt parallel processing for a large number of computations suggest a method of reducing the execution time by focusing on parallel processing beyond the division of matrix handling. In [63], parallel power flow using a hybrid CPU-GPU approach was studied. The NR method was used here for PF computation, and some libraries were used for CUDA implementation. This study considered the time needed for the allocation of memory or disadvantageous aspects of GPUs related to CPUs, as well as the PF computation time. The hybrid CPU-GPU approach for the proposed simultaneous parallel power flow computation is shown in Figure 5.

In [63], several systems with 14, 30, 57, 118, 300, 1354, 2869, and 9241 buses were used to obtain results under various conditions, respectively. Single and parallel calculations of 10, 50, 100, and 200 PF computations were performed simultaneously for each system, and the speedup result of simulations with the proposed method was recorded. The results show that if the power system is small, the overhead of memory allocation to the GPU is relatively high for multiple simultaneous PF computations. In contrast, when a large number of PF computations are performed, the memory allocation becomes relatively negligible as the power system grows. Furthermore, focusing the allocation on the beginning of the program reduces the bottleneck effect caused by the nature of GPU behavior. The overall results of the study show that the use of GPUs is efficient in PF studies and that this can be applied to real-time operational planning or control actions in control centers that require large amounts of PF computation.

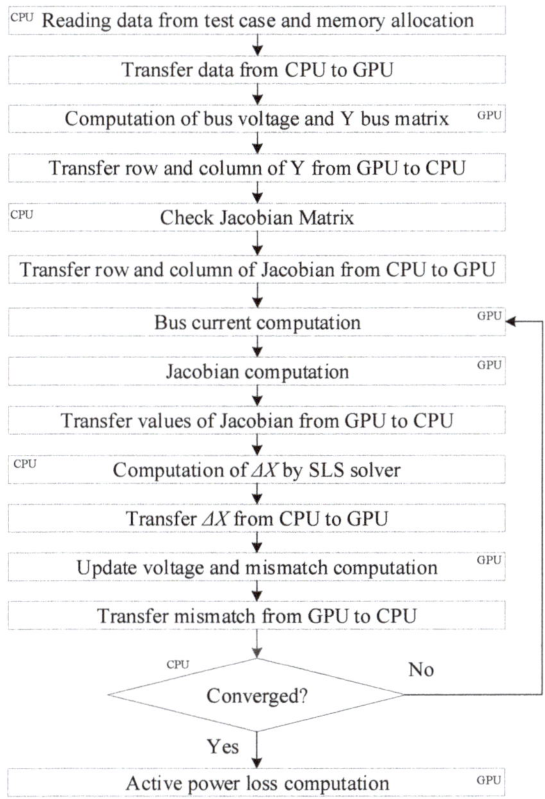

Figure 5. Hybrid CPU-GPU approach in [63].

4.4. Results and Comparison

Studies performing parallel processing using GPUs for PF are summarized in Table 3. In addition to the papers described in detail above, a number of additional meaningful studies are included.

Table 3 shows the solvers or the specific methods used for each study and the speedup results. The NR method is the most popular, and the time-consuming part of this method is matrix handling. Therefore, parallel processing studies using GPUs are also associated with the matrix handling part for speedup of computation. Many studies have been conducted on parallel processing of LU decomposition, which is frequently used in the NR method, and there have been some studies using QR decomposition associated with matrix handling. Table 3 summarizes the speedup results presented in each paper, including what reference they were compared with. The table also shows the hardware specifications of CPU and GPU, which are employed for the speedup evaluation. In some cases, the speedup is measured in comparison to the existing commercial library, and in other cases, the performance is evaluated compared to the CPU alone. Since the speedup reference is differently represented in each paper, it is difficult to understand the inherent speedup value in each case; however, it is clear that GPUs aid speedup when performing PF computations.

Table 3. Comparison of main features and speedup.

References	Main Features (Solver or Method)	Speedup (Comparison Target)	Year	Hardware Specification
[46]	LU decomposition	10.4× (in-house code)	2012	CPU: Phenom 9850 (4 cores) GPU: Tesla S1070
[45]		7.90× (1-core CPU) 1.49× (8-core CPU)	2012	CPU: Xeon X5680 (24 cores) GPU: GTX 580
[54]		4.16× (Matlab counterpart)	2017	CPU: Xeon E5-2620 (6 cores) GPU: GeForce Titan Black
[55]		76.0× (KLU library)	2017	CPU : Xeon E5-2620 (24 cores) GPU : Tesla K40
[63]	LU decomposition, hybrid CPU-GPU	Only graphs (MATPOWER)	2019	CPU : Xeon E5-2620 (6 cores) GPU : Tesla K20c
[32]	CG method	4.7× (Matlab CG)	2014	CPU: Xeon E5607 (8 cores) GPU: Tesla M2070
[28]	Batch-QR	64× (UMFPACK)	2014	CPU: Xeon E5-2620 (24 cores) GPU: Tesla K20c
[50]	GPU-based FDPF with Inexact Newton method	2.86× (traditional FDPF)	2017	CPU: Xeon E5607 (8 cores) GPU: Tesla M2070
[49]	Parallel GS, NR	GS : 45.2×, NR : 17.8× (MATPOWER)	2017	CPU: Xeon E5-2650 (32 cores) GPU: Telsa K20c
[67]	Continuous NR	11.13× (CPU)	2017	CPU: Xeon E5-2620 (6 cores) GPU: Tesla K20Xm

5. Conclusions

In this paper, we present an overview of the current state of knowledge regarding parallel power flow computation using GPU technology. The utilization of GPUs in the PF computation contributes to computational performance improvement and confirms its suitability for parallel processing. In summary, the technical trends and applications of GPUs for the PF computation are reviewed as follows:

- The basic concepts of PF computation methods and matrix handling are described.
- Conventional approaches of parallel processing technology in PF studies are summarized.
- Representative studies are reviewed to describe how to achieve parallel processing using GPUs in PF studies.
- The main features and speedup results are summarized for each study as a table form.

The parallel processing using GPUs is expected to be accelerated significantly further as their hardware performance improves more and more. We would be highly recommended to consider the utilization of GPUs in the near future for electrical power system analysis with a large amount of computations. In particular, it is expected that the GPU-based parallel processing will be very useful in the case of on-line operational planning that reflects the real time changes of the system, or performing simulations considering massive contingencies. However, the studies of performing PF analysis using the GPU techniques are still in a developing phase. Although significant speedup achievements were sufficiently convinced in the studies, but a standard technique has not been yet established completely at this moment. Also, we can expect the employment of cloud computing in the power flow analysis since the power systems are scaling up incrementally as the use of renewable energy is getting increased [68]. By exploiting the coarse and fine-grained parallelism of the PF study with the hybrid of cloud computing and GPU technologies, we can resolve the extremely massive numerical computations to evaluate the future power system properly. We have a plan to study the hybrid further soon.

Author Contributions: All authors contributed to this work. Writing—original draft preparation, D.-H.Y.; writing—review and editing, D.-H.Y. and Y.H.; supervision, Y.H. All authors have read and agreed to the published version of the manuscript.

Funding: This research was supported by the Korea Electric Power Corporation (Grant number: R17XA05-48) and the National Research Foundation of Korea (NRF) grant funded by the Korea government (MSIT) (Grant number: 2019R1C1C1008789).

Conflicts of Interest: The authors declare no conflict of interest.

Nomenclature

PF	Power flow
NR	Newton–Raphson
GS	Gauss–Seidel
EV	Electric vehicle
FDPF	Fast decoupled power flow
CG	Conjugate gradient
RTOPF	Real-time optimal power flow
SLS	Sparse linear system
CUDA	Compute Unified Device Architecture
OpenCL	Open Computing Language
HPC	High performance computing
CPU	Central processing unit
GPU	Graphics Processing Units
GPGPU	General-Purpose computing on Graphics Processing Units
SM	Streaming multiprocessor
TLP	Thread-level parallelism
PPF	Probabilistic power flow

References

1. Falcão, D.M. High performance computing in power system applications. In *International Conference on Vector and Parallel Processing*; Springer: Berlin, Germany, 1996; pp. 1–23.
2. Ramesh, V. On distributed computing for on-line power system applications. *Int. J. Electr. Power Energy Syst.* **1996**, *18*, 527–533. [CrossRef]
3. Baldick, R.; Kim, B.H.; Chase, C.; Luo, Y. A fast distributed implementation of optimal power flow. *IEEE Trans. Power Syst.* **1999**, *14*, 858–864. [CrossRef]
4. Li, F.; Broadwater, R.P. Distributed algorithms with theoretic scalability analysis of radial and looped load flows for power distribution systems. *Electr. Power Syst. Res.* **2003**, *65*, 169–177. [CrossRef]
5. Green, R.C.; Wang, L.; Alam, M. High performance computing for electric power systems: Applications and trends. In Proceedings of the 2011 IEEE Power and Energy Society General Meeting, Detroit, MI, USA, 24–28 July 2011; pp. 1–8.
6. Tournier, J.C.; Donde, V.; Li, Z. Potential of general purpose graphic processing unit for energy management system. In Proceedings of the 2011 Sixth International Symposium on Parallel Computing in Electrical Engineering, Luton, UK, 3–7 April 2011; pp. 50–55.
7. Huang, Z.; Nieplocha, J. Transforming power grid operations via high performance computing. In Proceedings of the 2008 IEEE Power and Energy Society General Meeting-Conversion and Delivery of Electrical Energy in the 21st Century, Pittsburgh, PA, USA, 20–24 July 2008; pp. 1–8.
8. Wu, Q.; Spiryagin, M.; Cole, C.; McSweeney, T. Parallel computing in railway research. *Int. J. Rail Trans.* **2018**, 1–24. [CrossRef]
9. Santur, Y.; Karaköse, M.; Akin, E. An adaptive fault diagnosis approach using pipeline implementation for railway inspection. *Turk. J. Electr. Eng. Comput. Sci.* **2018**, *26*, 987–998. [CrossRef]
10. Nitisiri, K.; Gen, M.; Ohwada, H. A parallel multi-objective genetic algorithm with learning based mutation for railway scheduling. *Comput. Ind. Eng.* **2019**, *130*, 381–394. [CrossRef]
11. Zawidzki, M.; Szklarski, J. Effective multi-objective discrete optimization of Truss-Z layouts using a GPU. *Appl. Soft Comput.* **2018**, *70*, 501–512. [CrossRef]

12. Li, Y.; Liu, Z.; Xu, K.; Yu, H.; Ren, F. A GPU-Outperforming FPGA Accelerator Architecture for Binary Convolutional Neural Networks. Available online: https://dl.acm.org/doi/abs/10.1145/3154839 (accessed on 10 February 2020).
13. Choi, K.H.; Kim, S.W. Study of Cache Performance on GPGPU. *IEIE Trans. Smart Process. Comput.* **2015**, *4*, 78–82. [CrossRef]
14. Oh, C.; Yi, Y. CPU-GPU^2 Trigeneous Computing for Iterative Reconstruction in Computed Tomography. *IEIE Trans. Smart Process. Comput.* **2016**, *5*, 294–301. [CrossRef]
15. Burgess, J. RTX on—The NVIDIA Turing GPU. *IEEE Micro* **2020**, *40*, 36–44. [CrossRef]
16. Wittenbrink, C.M.; Kilgariff, E.; Prabhu, A. Fermi GF100 GPU Architecture. *IEEE Micro* **2011**, *31*, 50–59. [CrossRef]
17. Kundur, P.; Balu, N.J.; Lauby, M.G. *Power System Stability and Control*; McGraw-hill: New York, NY, USA, 1994; Volume 7.
18. Stott, B.; Alsac, O. Fast decoupled load flow. *IEEE Trans. Power Appar. Syst.* **1974**, *PAS-93*, 859–869. [CrossRef]
19. Glover, J.D.; Sarma, M.S.; Overbye, T. Power System Analysis & Design, SI Version. Available online: https://books.google.co.kr/books?hl=ko&lr=&id=XScJAAAAQBAJ&oi=fnd&pg=PR7&dq=Power+System+Analysis+%26+Design,+SI+Version&ots=QEYUnLwXnN&sig=fofumZoNynEPX0YuRaKiO2glnA8#v=onepage&q=Power%20System%20Analysis%20%26%20Design%2C%20SI%20Version&f=false (accessed on 16 February 2020).
20. Alvarado, F.L.; Tinney, W.F.; Enns, M.K. Sparsity in large-scale network computation. *Adv. Electr. Power Energy Convers. Syst. Dyn.* **1991**, *41*, 207–272.
21. Zhou, G.; Zhang, X.; Lang, Y.; Bo, R.; Jia, Y.; Lin, J.; Feng, Y. A novel GPU-accelerated strategy for contingency screening of static security analysis. *Int. J. Electr. Power Energy Syst.* **2016**, *83*, 33–39. [CrossRef]
22. Wu, J.Q.; Bose, A. Parallel solution of large sparse matrix equations and parallel power flow. *IEEE Trans. Power Syst.* **1995**, *10*, 1343–1349.
23. Lau, K.; Tylavsky, D.J.; Bose, A. Coarse grain scheduling in parallel triangular factorization and solution of power system matrices. *IEEE Trans. Power Syst.* **1991**, *6*, 708–714. [CrossRef]
24. Amano, M.; Zecevic, A.; Siljak, D. An improved block-parallel Newton method via epsilon decompositions for load-flow calculations. *IEEE Trans. Power Syst.* **1996**, *11*, 1519–1527. [CrossRef]
25. El-Keib, A.; Ding, H.; Maratukulam, D. A parallel load flow algorithm. *Electr. Power Syst. Res.* **1994**, *30*, 203–208. [CrossRef]
26. Fukuyama, Y.; Nakanishi, Y.; Chiang, H.D. Parallel power flow calculation in electric distribution networks. In Proceedings of the 1996 IEEE International Symposium on Circuits and Systems, Circuits and Systems Connecting the World (ISCAS 96), Atlanta, GA, USA, 15 May 1996; Volume 1, pp. 669–672.
27. Davis, T.A. Direct Methods for Sparse Linear Systems. Available online: https://books.google.co.kr/books?hl=ko&lr=&id=oovDyJrnr6UC&oi=fnd&pg=PR1&dq=Direct+Methods+for+Sparse+Linear+Systems&ots=rQcLPxz3GE&sig=jYD7MDkkW_KGkK6rU6AWOstk3j8#v=onepage&q=Direct%20Methods%20for%20Sparse%20Linear%20Systems&f=false (accessed on 20 February 2020).
28. Zhou, G.; Feng, Y.; Bo, R.; Chien, L.; Zhang, X.; Lang, Y.; Jia, Y.; Chen, Z. GPU-accelerated batch-ACPF solution for N-1 static security analysis. *IEEE Trans. Smart Grid* **2017**, *8*, 1406–1416. [CrossRef]
29. Golub, G.H.; Van Loan, C.F. Matrix Computations. Available online: https://books.google.co.kr/books?hl=ko&lr=&id=5U-l8U3P-VUC&oi=fnd&pg=PT10&dq=Matrix+Computations&ots=7_JDJm_Lfp&sig=FP_n3ws8CRIxseHwoh97znxmpmE#v=onepage&q=Matrix%20Computations&f=false (accessed on 1 March 2020).
30. Dag, H.; Semlyen, A. A new preconditioned conjugate gradient power flow. *IEEE Trans. Power Syst.* **2003**, *18*, 1248–1255. [CrossRef]
31. Garcia, N. Parallel power flow solutions using a biconjugate gradient algorithm and a Newton method: A GPU-based approach. In Proceedings of the Power and Energy Society General Meeting, Providence, RI, USA, 25–29 July 2010; pp. 1–4.
32. Li, X.; Li, F. GPU-based power flow analysis with Chebyshev preconditioner and conjugate gradient method. *Electr. Power Syst. Res.* **2014**, *116*, 87–93. [CrossRef]
33. Rafian, M.; Sterling, M.; Irving, M. Decomposed Load-Flow Algorithm Suitable for Parallel Processor implementation. In IEE Proceedings C (Generation, Transmission and Distribution). Available online: https://digital-library.theiet.org/content/journals/10.1049/ip-c.1985.0047 (accessed on 3 March 2020).

34. Wang, L.; Xiang, N.; Wang, S.; Huang, M. Parallel reduced gradient optimal power flow solution. *Electr. Power Syst. Res.* **1989**, *17*, 229–237. [CrossRef]
35. Enns, M.K.; Tinney, W.F.; Alvarado, F.L. Sparse matrix inverse factors (power systems). *IEEE Trans. Power Syst.* **1990**, *5*, 466–473. [CrossRef]
36. Huang, G.; Ongsakul, W. Managing the bottlenecks in parallel Gauss-Seidel type algorithms for power flow analysis. *IEEE Trans. Power Syst.* **1994**, *9*, 677–684. [CrossRef]
37. Ezhilarasi, G.A.; Swarup, K.S. Parallel contingency analysis in a high performance computing environment. In Proceedings of the 2009 International Conference on Power Systems, Kharagpur, India, 27–29 December 2009; pp. 1–6. doi:10.1109/ICPWS.2009.5442650. [CrossRef]
38. Huang, Z.; Chen, Y.; Nieplocha, J. Massive contingency analysis with high performance computing. In Proceedings of the 2009 IEEE Power & Energy Society General Meeting, Calgary, AB, Canada, 26–30 July 2009; pp. 1–8.
39. Smith, S.; Van Zandt, D.; Thomas, B.; Mahmood, S.; Woodward, C. *HPC4Energy Final Report: GE Energy*; Technical Report; Lawrence Livermore National Laboratory (LLNL): Livermore, CA, USA, 2014.
40. Konstantelos, I.; Jamgotchian, G.; Tindemans, S.H.; Duchesne, P.; Cole, S.; Merckx, C.; Strbac, G.; Panciatici, P. Implementation of a massively parallel dynamic security assessment platform for large-scale grids. *IEEE Trans. Smart Grid* **2017**, *8*, 1417–1426. [CrossRef]
41. Guo, C.; Jiang, B.; Yuan, H.; Yang, Z.; Wang, L.; Ren, S. Performance comparisons of parallel power flow solvers on GPU system. In Proceedings of the 2012 IEEE International Conference on Embedded and Real-Time Computing Systems and Applications, Seoul, Korea, 19–22 August 2012; pp. 232–239.
42. Demmel, J.W.; Eisenstat, S.C.; Gilbert, J.R.; Li, X.S.; Liu, J.W. A supernodal approach to sparse partial pivoting. *SIAM J. Matrix Anal. Appl.* **1999**, *20*, 720–755. [CrossRef]
43. Schenk, O.; Gärtner, K. Solving unsymmetric sparse systems of linear equations with PARDISO. *Future Gener. Comput. Syst.* **2004**, *20*, 475–487. [CrossRef]
44. Christen, M.; Schenk, O.; Burkhart, H. General-Purpose Sparse Matrix Building Blocks Using the NVIDIA CUDA Technology Platform. In First Workshop on General Purpose Processing on Graphics Processing Units. Available online: https://scholar.google.co.kr/scholar?hl=ko&as_sdt=0%2C5&q=General-purpose+sparse+matrix+building+blocks+using+the+NVIDIA+CUDA+technology+platform&btnG= (accessed on 10 March 2020).
45. Ren, L.; Chen, X.; Wang, Y.; Zhang, C.; Yang, H. Sparse LU factorization for parallel circuit simulation on GPU. In Proceedings of the 49th Annual Design Automation Conference, San Francisco, CA, USA, 3–7 June 2012; ACM: New York, NY, USA, 2012; pp. 1125–1130.
46. Jalili-Marandi, V.; Zhou, Z.; Dinavahi, V. Large-scale transient stability simulation of electrical power systems on parallel GPUs. In Proceedings of the 2012 IEEE Power and Energy Society General Meeting, San Diego, CA, USA, 22–26 July 2012; pp. 1–11.
47. Chen, D.; Li, Y.; Jiang, H.; Xu, D. A parallel power flow algorithm for large-scale grid based on stratified path trees and its implementation on GPU. *Autom. Electr. Power Syst.* **2014**, *38*, 63–69.
48. Gnanavignesh, R.; Shenoy, U.J. GPU-Accelerated Sparse LU Factorization for Power System Simulation. In Proceedings of the 2019 IEEE PES Innovative Smart Grid Technologies Europe (ISGT-Europe), Bucharest, Romania, 29 September–2 October 2019; pp. 1–5.
49. Roberge, V.; Tarbouchi, M.; Okou, F. Parallel Power Flow on Graphics Processing Units for Concurrent Evaluation of Many Networks. *IEEE Trans. Smart Grid* **2017**, *8*, 1639–1648. [CrossRef]
50. Li, X.; Li, F.; Yuan, H.; Cui, H.; Hu, Q. GPU-based fast decoupled power flow with preconditioned iterative solver and inexact newton method. *IEEE Trans. Power Syst.* **2017**, *32*, 2695–2703. [CrossRef]
51. Jiang, H.; Chen, D.; Li, Y.; Zheng, R. A Fine-Grained Parallel Power Flow Method for Large Scale Grid Based on Lightweight GPU Threads. In Proceedings of the 2016 IEEE 22nd International Conference on Parallel and Distributed Systems (ICPADS), Wuhan, China, 13–16 December 2016; pp. 785–790.
52. Liu, Z.; Song, Y.; Chen, Y.; Huang, S.; Wang, M. Batched Fast Decoupled Load Flow for Large-Scale Power System on GPU. In Proceedings of the 2018 International Conference on Power System Technology (POWERCON), Guangzhou, China, 6–8 November 2018; pp. 1775–1780.
53. Zhou, G.; Feng, Y.; Bo, R.; Zhang, T. GPU-accelerated sparse matrices parallel inversion algorithm for large-scale power systems. *Int. J. Electr. Power Energy Syst.* **2019**, *111*, 34–43. [CrossRef]

54. Huang, S.; Dinavahi, V. Performance analysis of GPU-accelerated fast decoupled power flow using direct linear solver. In Proceedings of the Electrical Power and Energy Conference (EPEC), Saskatoon, SK, Canada, 22–25 October 2017; pp. 1–6.
55. Zhou, G.; Bo, R.; Chien, L.; Zhang, X.; Shi, F.; Xu, C.; Feng, Y. GPU-based batch LU factorization solver for concurrent analysis of massive power flows. *IEEE Trans. Power Syst.* **2017**, *32*, 4975–4977. [CrossRef]
56. Eigen. Eigen 3 Documentation. Available online: http://eigen.tuxfamily.org (accessed on 12 April 2020).
57. Li, X.; Li, F.; Clark, J.M. Exploration of multifrontal method with GPU in power flow computation. In Proceedings of the Power and Energy Society General Meeting (PES), Vancouver, BC, Canada, 21–25 July 2013; pp. 1–5.
58. MATPOWER. MATPOWER User's Manual. Available online: http://matpower.org (accessed on 3 February 2020).
59. MathWorks Inc. MATLAB. Available online: http://www.mathworks.com (accessed on 15 March 2020).
60. Huang, S.; Dinavahi, V. Fast batched solution for real-time optimal power flow with penetration of renewable energy. *IEEE Access* **2018**, *6*, 13898–13910. [CrossRef]
61. Zhou, G.; Bo, R.; Chien, L.; Zhang, X.; Yang, S.; Su, D. GPU-accelerated algorithm for online probabilistic power flow. *IEEE Trans. Power Syst.* **2018**, *33*, 1132–1135. [CrossRef]
62. Su, X.; He, C.; Liu, T.; Wu, L. Full Parallel Power Flow Solution: A GPU-CPU Based Vectorization Parallelization and Sparse Techniques for Newton-Raphson Implementation. *IEEE Trans. Smart Grid* **2019**, *11*, 1833–1844. [CrossRef]
63. Araújo, I.; Tadaiesky, V.; Cardoso, D.; Fukuyama, Y.; Santana, Á Simultaneous parallel power flow calculations using hybrid CPU-GPU approach. *Int. J. Electr. Power Energy Syst.* **2019**, *105*, 229–236. [CrossRef]
64. Chen, X.; Ren, L.; Wang, Y.; Yang, H. GPU-accelerated sparse LU factorization for circuit simulation with performance modeling. *IEEE Trans. Parallel Distrib. Syst.* **2015**, *26*, 786–795. [CrossRef]
65. Davis, T. SUITESPARSE: A Suite OF Sparse Matrix Software. Available online: http://faculty.cse.tamu.edu/davis/suitesparse.html (accessed on 1 April 2020).
66. Schenk, O.; Gärtner, K. User Guide. Available online: https://www.pardiso-project.org/ (accessed on 15 April 2020).
67. Wang, M.; Xia, Y.; Chen, Y.; Huang, S. GPU-based power flow analysis with continuous Newton's method. In Proceedings of the 2017 IEEE Conference on Energy Internet and Energy System Integration (EI2), Beijing, China, 26–28 November 2017; pp. 1–5.
68. Yoon, D.H.; Kang, S.K.; Kim, M.; Han, Y. Exploiting Coarse-Grained Parallelism Using Cloud Computing in Massive Power Flow Computation. *Energies* **2018**, *11*, 2268. [CrossRef]

Article

Preventive Security-Constrained Optimal Power Flow with Probabilistic Guarantees

Hang Li *, Zhe Zhang, Xianggen Yin and Buhan Zhang

State Key Laboratory of Advanced Electromagnetic Engineering and Technology, Huazhong University of Science and Technology, 1037 Luoyu Road, Wuhan 430074, China; zz_mail2002@163.com (Z.Z.); xgyin@mail.hust.edu.cn (X.Y.); zhangbuhan@hust.edu.cn (B.Z.)

* Correspondence: leehang@outlook.com; Tel.: +86-1862-711-0403

Received: 17 March 2020; Accepted: 30 April 2020; Published: 8 May 2020

Abstract: The traditional security-constrained optimal power flow (SCOPF) model under the classical N-1 criterion is implemented in the power industry to ensure the secure operation of a power system. However, with increasing uncertainties from renewable energy sources (RES) and loads, the existing SCOPF model has difficulty meeting the practical requirements of the industry. This paper proposed a novel chance-constrained preventive SCOPF model that considers the uncertainty of power injections, including RES and load, and contingency probability. The chance constraint is used to constrain the overall line flow within the limits with high probabilistic guarantees and to significantly reduce the constraint scales. The cumulant and Johnson systems were combined to accurately approximate the cumulative distribution functions, which is important in solving chance-constrained optimization problems. The simulation results show that the model proposed in this paper can achieve better performance than traditional SCOPF.

Keywords: security-constrained optimal power flow; chance-constrained optimization; probability of contingency; renewable energy source

---

1. Introduction

The growth in renewable energy sources (RES) and charging loads in recent years, such as wind power, photovoltaics and electric vehicle, has brought considerable economic benefits; however, the uncertainty of power injections has increased, which leads to increased operational risks [1–3], especially for highly-loaded power systems. The increasing uncertainty of operation increases the need for new criteria, dispatch tools and control methods to better balance operational security and costs [4].

Optimal power flow (OPF) is the fundamental dispatch and planning tool that is used to minimize operational costs while ensuring the security of the normal state, and security-constrained optimal power flow (SCOPF) [4–7] is an extended form of OPF that considers the classical N-1 criterion. Unlike OPF, which only considers a single system topology (normal state), SCOPF typically ensures that the system state remains within the operational limits when unexpected component outages (contingency set) occur. However, with the emergence of uncertainties in the power system, several drawbacks of traditional SCOPF have become apparent and these need to be addressed. These include:

1. Traditional SCOPF does not consider the influence of the uncertainty of RES and loads, and it cannot provide a robust solution because increasing uncertainty makes the operational state more stochastic and may lead to frequent violations of the N-1 criterion.

2. Traditional SCOPF disregards the probability of a contingency occurring; in other words, it considers the occurrence probability to be 1 for every contingency in a contingency set [4]. Obviously, this does not match the actual situation because the probability of a contingency is usually very low.

3. The scale of the SCOPF problem is highly related to the scale of the power system and the number of contingencies. This means that for a large power system where a large number of

contingencies are considered, the calculation burden is high, and directly solving a SCOPF problem in a short time is quite challenging.

1.1. Literature Review

Numerous studies have attempted to address the drawbacks of the traditional SCOPF model.

To the best of the authors' knowledge, there are currently two strategies to reduce the calculation burden of traditional SCOPF and make it easier to solve. One strategy uses a contingency filtering (CF) [8,9] technique to reduce the number of contingencies. Usually, an index that ranks the severity of a contingency is used to filter contingencies; thus, only the contingency that exceeds the severity threshold is included in the contingency set. However, choosing the severity threshold itself is a challenge, for example, a very severe contingency may have a very low probability of occurring, and controlling it through SCOPF may result in excessive costs. The second strategy is to use Benders decomposition (BD) [10–12] to decompose the original SCOPF problem into a master problem and several subproblems. In this way, parallel computing technology can be used to improve the computing efficiency; however, BD requires convexity of the feasible region, which is not guaranteed in an SCOPF problem [12].

The concept of risk-based SCOPF [13–15] has been proposed as a method that comprehensively considers the probability and severity of contingencies. The risk of a contingency is defined as the product of the probability and severity of a contingency. Risk-based SCOPF uses risk as constraints to achieve a tradeoff between economic and security. This method relaxes the constraints of a single contingency [14] but controls the total risk of a contingency set to a certain level. Although the security and economy of power system operations are enhanced by risk-based SCOPF, the uncertainty of RES and load are not taken into consideration because measuring system risk under uncertainty is a challenging task. Moreover, the optimization formulation of risk-based SCOPF is complicated, and the calculation time is 4–7 times that of traditional SCOPF [14], which makes it difficult to apply in a real-time dispatch.

Chance-constrained optimization (CCO) [16–26] is a promising method to handle the uncertainty in power systems and it has been successfully applied to many problems. Instead of rigid constraint, CCO ensures a certain level of probability that the constraint is satisfied. The work of Bienstock [19] provides a solid foundation for incorporating CCO with OPF. This model was further extended in [22] to incorporate corrective SCOPF. Li et al. [23] provided a novel transmission expansion planning approach based on CCO and BD. Liu et al. [24], proposed solutions based on CCO for peak power shaving and frequency regulation in microgrids. Based on CCO, a day-ahead scheduling approach is proposed in [25] and a volt/var control approach is provided in [26]. Although CCO has been successfully applied to a variety of problems, the probability distribution of the uncertainty source is usually assumed to follow a Gaussian distribution. Studies [27–29] have indicated that the distribution of wind power forecast error and photovoltaic power is very different from a Gaussian distribution; therefore, the existing models should be improved so that they are able to handle arbitrary distributions. Moreover, there are few CCO models that consider contingency probability.

1.2. Contributions

This paper proposes a novel chance-constrained preventive SCOPF model (CC-PSCOPF) that is an improvement on the traditional preventive SCOPF (PSCOPF) model. The main contributions are as follows:

1. A novel CC-PSCOPF model is proposed to improve the overall operational reliability. The model considers contingency probability and the uncertainty of power injections (including RES and load).

2. Instead of using the large-scale line flow limits of traditional PSCOPF, the probability distribution of the overall line flow is obtained and constrained in the proposed optimization model, which significantly reduces the constraint scale.

3. The cumulant and Johnson systems are combined in this paper to accurately approximate the cumulative distribution function (CDF) of an arbitrary distribution random variable, which only requires the first four orders of moment information.

The remainder of this paper is organized as follows: Section 2 reviews the traditional PSCOPF model. Section 3 describes the formulation of the proposed CC-PSCOPF, and a cumulative distribution function (CDF) approximation method based on cumulants and the Johnson system is also introduced. A case study is presented in Section 4 to test the performance of the proposed model. Section 5 presents our discussion and conclusions.

2. Review of Traditional PSCOPF

There are two types of SCOPF: PSCOPF and corrective SCOPF (CSCOPF). Using PSCOPF, pre-contingency controls are the only measures allowed to ensure that the system always operates in a state where any single component outage does not lead to constraint violations. This indicates that the operational state determined by PSCOPF simultaneously satisfies the pre- and post-contingency constraints. Different from PSCOPF, CSCOPF determines an operational state that allows post-contingency constraint violations, and it ensures that there are adequate post-contingency control measures, e.g., generator re-dispatch, topology reconfiguration and load shedding to eliminate post-contingency constraint violations. PSCOPF is safer, while CSCOPF is more economical [4].

This paper focuses on improving the traditional PSCOPF, and the proposed optimization model attempts to improve the overall security performance of the system operation through pre-contingency controls.

The DC-based PSCOPF model is reviewed in this section, as it provides the foundation for the optimization model proposed in this paper. DC approximation is used in this paper because it provides a convex guarantee that the optimization problem is tractable [20].

The objective function of DC-based PSCOPF minimizes the system's operational cost in the normal state, and it is expressed as follows:

$$\min \sum_{Gi=1}^{N_G} P_{Gi}^T c_{2i} P_{Gi} + c_{1i}^T P_{Gi} + c_{0i} \tag{1}$$

where N_G is the number of generators; P_{Gi} is the ith generator output in the normal state, which is the control variable of the optimization model; and c_{2i}, c_{1i}, and c_{0i} are the quadratic, linear and constant cost coefficients, respectively.

The equality and inequality constraints of the PSCOPF model are as follows:

$$\sum_{Gi=1}^{N_G} P_{Gi} + \sum_{Ri=1}^{N_R} P_{Ri} = \sum_{Di=1}^{N_D} P_{Di} \tag{2}$$

$$\underline{P}_{Gi} \leq P_{Gi} \leq \overline{P}_{Gi} \qquad \forall i \tag{3}$$

$$\sum_{Gi=1}^{N_G} A_{Gi}^k P_{Gi} + \sum_{Ri=1}^{N_R} A_{Ri}^k P_{Ri} - \sum_{Di=1}^{N_D} A_{Di}^k P_{Di} \leq \overline{P}_l \quad \forall i, \forall k, \forall l \tag{4}$$

$$\underline{P}_l \leq \sum_{Gi=1}^{N_G} A_{Gi}^k P_{Gi} + \sum_{Ri=1}^{N_R} A_{Ri}^k P_{Ri} - \sum_{Di=1}^{N_D} A_{Di}^k P_{Di} \quad \forall i, \forall k, \forall l \tag{5}$$

These include the power balance of the system (2), the generator output limits (3) and the line flow limits (4) and (5). N_R and N_D are the number of RES and loads, respectively, in the system; P_{Ri} and P_{Di} are the forecast power injections of the ith RES and load; $\underline{P}_{Gi}$ and $\overline{P}_{Gi}$ are the ith generator's minimal output and maximum output, respectively; $\underline{P}_l$ and $\overline{P}_l$ are the lower and upper limits of the

ith line flows; superscript k is the index of system topology; $k = 0$ indicates the normal state system topology, while $k \geq 1$ indicates the contingency system topology; and A^k_{Gi}, A^k_{Ri} and A^k_{Di} are the power transmission distribution factors (PTDFs) of the ith generator, RES and load under system topology k, respectively. The PTDF can be obtained from the line susceptance matrix and bus susceptance matrix, and details can be found in [6].

The traditional DC-based PSCOPF optimization model is a typical quadratic programming problem that can be solved by common commercial solvers. As discussed in the introduction, the uncertainties of power injections and contingency probability are not considered in this model; therefore, the operational state obtained by this model is not robust to uncertainty and may be very costly. The constraint number of this model is $1 + N_G + 2 \times N_l \times N_k$, where N_l is the number of lines and N_k is the scale of the contingency set. Obviously, when the system scale is large with a large contingency set, the constraint number of this model is quite high, which significantly increases the calculation burden.

3. Formulation of the Proposed Optimization Model

3.1. Modeling of Uncertainties

The forecast error of the RES and load is the main source of uncertainty, which is the deviation of the forecast value from the actual value. The forecast error can be seen as a continuous random variable and described by a continuous probability distribution model. Therefore, the actual power injection of RES and loads can be modeled as a forecast value plus a continuous random variable that represents forecast error:

$$\begin{cases} \widetilde{P}_{Ri} = P_{Ri} + \delta_{Ri} \\ \widetilde{P}_{Di} = P_{Di} + \delta_{Di} \end{cases} \quad \forall i \tag{6}$$

where $\widetilde{P}_{Ri}$ and $\widetilde{P}_{Di}$ are the actual power injection of the ith RES and load, respectively, and δ_{Ri} and δ_{Di} are random variables that represent the forecast error of the power injection of the ith RES and load, respectively.

The proper distribution model to describe forecast error depends on the type of power injection, forecasting scale [27], etc. For instance, the forecast error of a load is usually assumed to follow a Gaussian distribution, while the beta distribution [28] is an appropriate choice to describe the short-term forecast error of wind power. Although there are many distribution models that can be used to describe a forecast error, the optimization model proposed in this paper is not sensitive to the distribution used. The first four order moments of a distribution model is the only information that is required and this can be obtained from historical data This moment information is used to approximate the CDF of a random variable, which is discussed in the following section.

The line flow under a single system topology is linearly dependent on power injections; when considering uncertainty, its expression is:

$$\begin{aligned} P^k_l &= \sum_{Gi=1}^{N_G} A^k_{Gi} P_{Gi} + \sum_{Ri=1}^{N_R} A^k_{Ri} \widetilde{P}_{Ri} - \sum_{Di=1}^{N_D} A^k_{Di} \widetilde{P}_{Di} \\ &= \Big(\sum_{Gi=1}^{N_G} A^k_{Gi} P_{Gi} + \sum_{Ri=1}^{N_R} A^k_{Ri} P_{Ri} - \sum_{Di=1}^{N_D} A^k_{Di} P_{Di}\Big) + \Big(\sum_{Ri=1}^{N_R} A^k_{Ri} \delta_{Ri} - \sum_{Di=1}^{N_D} A^k_{Di} \delta_{Di}\Big) \end{aligned} \quad \forall l, \forall k \tag{7}$$

where P^k_l is the lth line flow under the kth system topology. The term in the first bracket of Equation (7) is the line flow part formed by forecast power injections, which is consistent with the traditional PSCOPF. The term in the second bracket of Equation (7) is the uncertainty part of a line flow, which is the linear combination of the forecast error of power injections of RES and load.

The overall line flow probability distribution, which comprehensively considers the influence of the uncertainty of power injection and system topology is our concern. Therefore, the occurrence probability of a system topology, or the so-called contingency probability, should be obtained.

The Poisson distribution [30] is used in this paper to describe the occurrence probability of a system topology:

$$\mathrm{P_r}(s_k) = 1 - e^{-\lambda_t} \tag{8}$$

where s_k is the kth contingency's system topology, $\mathrm{P_r}(s_k)$ is the corresponding occurrence probability, e is the base of the natural logarithm and λ_t is the failure rate of the component. Note that the failure rate can be modified according to the external weather conditions. An approach to calculate the failure rate under different weather conditions (normal, adverse and major adverse) is provided in [31].

For each contingency, an occurrence probability can be obtained from Equation (8). We assume that contingencies outside the contingency set will have little impact on system operation, so the probability of the normal state is approximated and expressed as:

$$\mathrm{P_r}(s_0) = 1 - \sum_{k=1}^{N_k} \mathrm{P_r}(s_k) \tag{9}$$

Considering the uncertainty of power injections and the probability of contingency, the overall line flow P_l can be obtained through the law of total probability theory, and it is expressed as:

$$\begin{aligned} P_l &= \mathrm{P_r}(s_0)P_l^0 + \mathrm{P_r}(s_1)P_l^1 + \cdots \mathrm{P_r}(s_{N_k})P_l^{N_k} \quad \forall l \\ &= \sum_{k=0}^{N_k}\sum_{Gi=1}^{N_G} A_{Gi}^k P_{Gi} \mathrm{P_r}(s_k) + \sum_{k=0}^{N_k}\sum_{Ri=1}^{N_R} A_{Ri}^k P_{Ri} \mathrm{P_r}(s_k) - \\ &\quad \sum_{k=0}^{N_k}\sum_{Ri=1}^{N_R} A_{Di}^k P_{Di} \mathrm{P_r}(s_k) + \sum_{k=0}^{N_k}\sum_{Ri=1}^{N_R} A_{Ri}^k \delta_{Ri} \mathrm{P_r}(s_k) - \sum_{k=0}^{N_k}\sum_{Ri=1}^{N_R} A_{Di}^k \delta_{Di} \mathrm{P_r}(s_k) \end{aligned} \quad \forall l \tag{10}$$

Obviously, the probability distribution of the overall line flow P_l can be regarded as the weighted average of the line flow probability distribution under each system topology.

3.2. Chance-Constrained Optimization

In this section, we briefly introduce the CCO, which also underpins the model proposed in this paper.

CCO is an important tool proposed by Charnes and Cooper [16,17] for solving optimization problems under uncertainties. The general form of a CCO problem is expressed as follows:

$$\begin{aligned} \min \quad & f(x,\xi) \\ \text{s.t.} \quad & g(x,\xi) = 0 \\ & \Pr\{h(x,\xi) \geq 0\} \geq 1-\alpha \end{aligned} \tag{11}$$

where $f(\cdot)$ is the objective function, $g(\cdot)$ is the equality constraint, $h(\cdot)$ is the inequality constraint, x is the decision variable, ξ is the uncertainty variable and α is the reliability parameter representing the allowed constraint violation level.

Under the CCO, the inequality constraint is formed as the chance constraint and ensures that the constraint $h(\cdot)$ is satisfied with probability $1-\alpha$ at least. The original CCO problem is often transformed into an equivalent deterministic form to facilitate the solution [20,24].

3.3. Chance-Constrained PSCOPF Model

In this section, we present a novel CC-PSCOPF model that considers the uncertainties of power injections and the probability of contingency.

The goal function, power balance constraint and generator output limit of the proposed CC-PSCOPF are the same as those of the traditional PSCOPF model, as Equations (1)–(3) show. The key improvement is the modeling of the line flow inequality constraints.

Instead of line flow constraints under each system topology used in traditional PSCOPF, the overall line flow P_l is constrained in CC-PSCOPF. As analyzed in the previous section, P_l is a random variable; therefore, the chance constraint is used to place it in a certain range with a high probability, and it is expressed as follows:

$$\Pr\{P_l \geq \overline{P}_l\} \leq \alpha_l^+ \quad \forall l \tag{12}$$

$$\Pr\{P_l \leq \underline{P}_l\} \leq \alpha_l^- \quad \forall l \tag{13}$$

where α_l^+ and α_l^- are predefined violation levels. Considering the low occurrence probability of contingencies, α_l^+ and α_l^- should be carefully defined.

A comparison of Equations (12) and (13) to Equations (4) and (5) shows that the optimization model proposed in this paper has the following significant advantages:

1. The uncertainty of power injections and contingency occurrence probability are considered through P_l. Even with the influence of various uncertainties, the operational state obtained by CC-PSCOPF has a high probability of ensuring that constraints are not violated. Obviously, the operational state is more robust to uncertainties compared to traditional PSCOPF.

2. The violation level for different lines is adjustable; for critical lines, the violation level could be adjusted lower to ensure operational safety, while for noncritical lines, the violation level could be increased to save control costs.

3. The scale of the line flow constraint is significantly reduced, which is only related to the number of lines in the system, allowing the optimization problems to be solved more efficiently.

4. As the occurrence probabilities of the contingency and chance constraints are used in this model, some contingencies that have quite low probability but high control costs are ignored in the optimization model, which helps to reduce the control costs. The control measures of these low probability contingencies can be solved using a separate accident plan.

However, solving an optimization problem with chance constraints directly is a challenging task. In this paper, chance constraints are transformed into deterministic linear constraints based on the cumulant and Johnson systems in the following section, which ensures that the CC-PSCOPF is tractable and solved efficiently.

3.4. Deterministic Reformulation of CC-PSCOPF

The main challenge of solving the proposed model is how to handle the two chance constraints related to overall line flows. In this section, these two constraints are converted into deterministic linear constraints. Through conversion, the optimization model becomes a linear constrained convex optimization problem that is easy to solve.

For convenience, the factors that determine the overall line flow P_l are divided into two terms:

$$\begin{aligned} P_{l,control} &= \sum_{k=0}^{N_k} \sum_{Gi=1}^{N_G} A_{Gi}^k P_{Gi} \mathrm{P_r}(s_k) \\ P_{l,uncertainty} &= \sum_{k=0}^{N_k} \sum_{Ri=1}^{N_R} A_{Ri}^k P_{Ri} \mathrm{P_r}(s_k) - \sum_{k=0}^{N_k} \sum_{Ri=1}^{N_R} A_{Di}^k P_{Di} \mathrm{P_r}(s_k) \quad \forall l \\ &+ \sum_{k=0}^{N_k} \sum_{Ri=1}^{N_R} A_{Ri}^k \delta_{Ri} \mathrm{P_r}(s_k) - \sum_{k=0}^{N_k} \sum_{Ri=1}^{N_R} A_{Di}^k \delta_{Di} \mathrm{P_r}(s_k) \end{aligned} \tag{14}$$

where $P_{l,control}^k$ is the line flow part determined by the control variable, which varies with the output of the generators, and $P_{l,uncertainty}$ is the line flow part determined by power injections of the RES and load, which is a random variable.

Substituting Equation (14) into Equation (12) yields:

$$\Pr\{P_{l,uncertainty} \leq \overline{P}_l - P_{l,control}\} \geq 1 - \alpha_l^+ \quad \forall l \tag{15}$$

Note that $P_{l,uncertainty}$ is the uncertainty part of the overall line flows, and we can obtain:

$$\begin{array}{l} CDF_{l,uncertainty}(\overline{P}_l - P_{l,control}) \geq 1 - \alpha_l^+ \\ \Downarrow \\ \overline{P}_l - P_{l,control} \geq CDF_{l,uncertainty}^{-1}(1 - \alpha_l^+) \end{array} \quad \forall l \tag{16}$$

where $CDF_{l,uncertainty}$ and $CDF_{l,uncertainty}^{-1}$ are the CDF and inverse CDF of $P_{l,uncertainty}$, respectively. Similar to Equation (16), by substituting Equation (14) into (13), we can obtain:

$$\underline{P}_l - P_{l,control} \leq CDF_{l,stochastic}^{-1}(\alpha_l^-) \quad \forall l \tag{17}$$

To replace Equations (12) and (13) with deterministic linear constraints Equations (16) and (17), $CDF_{l,uncertainty}^{-1}$ should be obtained. Traditionally, $CDF_{l,uncertainty}$ is obtained from Monte Carlo simulation (MCS), which performs a numerical search for the $CDF_{l,uncertainty}^{-1}$ [20]; however, this is time consuming. Moreover, MCS is difficult to implement in the absence of samples. Here, this procedure is improved by using the cumulant and Johnson system to obtain $CDF_{l,uncertainty}^{-1}$.

3.4.1. The Cumulant

The cumulant [32,33] is an alternative moment of a continuous probability distribution, and the relationship between the cumulant κ and moment m is as follows:

$$\begin{array}{l} \kappa_1 = m_1 \\ \kappa_2 = m_2 - m_1^2 \\ \kappa_3 = m_3 - 3m_1m_2 + 2m_1^3 \\ \kappa_4 = m_4 - 2m_2^2 - 4m_1m_3 + 12m_1^2m_2 - 6m_1^4 \\ \vdots \end{array} \tag{18}$$

Cumulants have two important characteristics. One is homogeneity. For a random variable x, the nth cumulant is homogeneous of order r:

$$\kappa_r(ax) = a^r\kappa_r(x) \tag{19}$$

The other is additivity; for two independent random variables x and y, and $z = x + y$, then:

$$\kappa_r(z) = \kappa_r(x) + \kappa_r(y) \tag{20}$$

The uncertainty part of line flow under a single system topology is the linear combination of δ_{Ri} and δ_{Di}. As we know the first four order moments of δ_{Ri} and δ_{Di} in advance, using Equation (18) and the characteristics of the cumulant, the first four order moments of the uncertainty part of the line flow under a single system topology can be obtained.

3.4.2. The Johnson System

Previous works [34,35] have shown that the Johnson system is a reliable and accurate method for obtaining CDF compared to the commonly used Gram-Charlier series [32] or Cornish-Fisher series [33]; therefore, it is used here to obtain the CDF of the uncertainty part of the line flow under a single system topology.

The Johnson system is a 4-parameter transformation system that uses the following function to transform the standard Gaussian variable u into a variable x that follows an unknown arbitrary probability distribution:

$$x = c + d \times f^{-1}(\frac{u - a}{b}) \tag{21}$$

where a and b are the shape parameters, c is the position parameter, and d is the scale parameter. The function $f^{-1}(\cdot)$ takes 4 forms to distinguish different distribution families:

$$\begin{cases} S_L : f^{-1}(\frac{u-a}{b}) = e^{(u-a)/b} \\ S_U : f^{-1}(\frac{u-a}{b}) = (e^{(u-a)/b} - e^{-(u-a)/b})/2 \\ S_B : f^{-1}(\frac{u-a}{b}) = 1/(1 + e^{-(u-a)/b}) \\ S_N : f^{-1}(\frac{u-a}{b}) = \frac{u-a}{b} \end{cases} \quad (22)$$

where S_L is the family of lognormal distributions, S_U is the family of unbounded distributions, which means the range of variables is unlimited, S_B is the family of bounded distributions, and S_N is the family of Gaussian distributions.

If the first four order moments of random variable x (the uncertainty part of the line flow in this paper) are available, the moment-based algorithm [36] can be used to obtain the parameters of the Johnson system, and the CDF of x is a function of the CDF of the standard Gaussian variable, which can be obtained easily.

The CDF of the uncertainty part of the line flow under each system topology is obtained. Based on the law of total probability, $CDF_{l,uncertainty}$ is calculated through the weighted average of each CDF of the uncertainty part of the line flow, and its inverse function $CDF^{-1}_{l,uncertainty}$ can be efficiently calculated.

4. Case Study

In this section, we discuss the performances of traditional PSCOPF and CC-PSCOPF, which were tested on two test systems. The optimization problem was solved by Ipopt [37] on a PC with a 2.8-GHz CPU and 16 GB RAM.

4.1. Description of the Test System

A modified IEEE-30 test system [38] was used here to analyze the characteristics of these two optimization formulations and a modified IEEE-108 test system [38] was used to evaluate the efficiency of these two methods. The IEEE-30 test system was modified to add two wind power generators at bus 7 and 12, which are representative of a RES power injection. Similarly, the IEEE-108 test system was modified to add four power generators at bus 44, 50, 88 and 98. The rated power of the wind power generators is 80 MW, the forecast outputs of wind power generators are assumed to be 0.8 p.u., and forecast error is assumed to follow a beta distribution: Beta (0.83, 1.82). The forecast error of the load at each bus is assumed to follow a Gaussian distribution, with the means of the power injections equal to those of the base case data and standard deviations equal to 5% of the means.

All N-1 contingencies are included in the contingency set, the occurrence probability of each contingency is assumed to be 0.01, and the violation level α_l^+ is set at 1% for all lines.

4.2. CDF Approximation Performance of the Proposed Method

Obtaining an accurate CDF curve, especially in the tail area, is the basis for accurately solving the chance-constrained optimization. This section presents the results to show that the proposed cumulant + Johnson system can accurately and efficiently approximate a CDF curve.

The proposed method is compared with 10,000 MCS, the commonly used cumulant + Gram-Charlier series proposed in [25], and a CDF curve based on the assumption that wind power forecast error follows a Gaussian distribution.

The CDF curve of line 4-6 flow and line 16-17 flow under the normal state is chosen here as representative and to visually show the approximation ability of the evaluated method. The CDF curves are shown in Figure 1a,b.

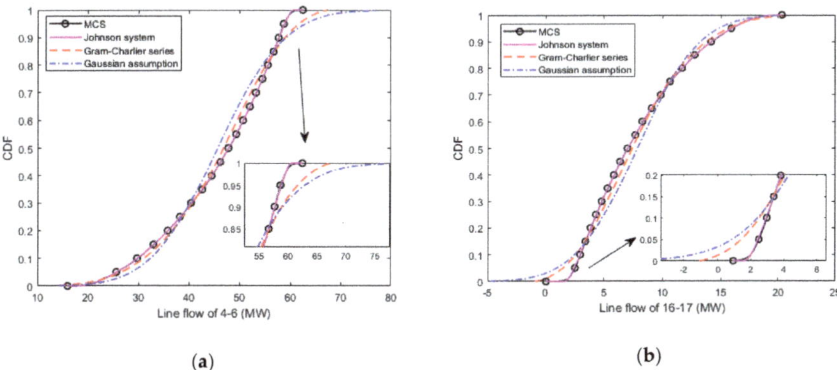

Figure 1. Cumulative distribution function (CDF) curve for line 4-6 flow (**a**) and line 16-17 flow (**b**).

The average root mean square (ARMS) [18] is also introduced here to quantitatively indicate the accuracy of the CDF approximation. The smaller the ARMS value is, the more accurately the method approximates the CDF curve. Table 1 shows the ARMS values of the evaluated methods.

Table 1. The average root mean square (ARMS) value of evaluated methods.

Lines	ARMS of Cumulant + Johnson System	ARMS of Cumulant + Gram-Charlier	ARMS of Gaussian Assumption
Line 4–6	0.0031	0.1294	0.1312
Line 16–17	0.0038	0.1288	0.1306

The results in Figure 1 and Table 1 show that the cumulant + Johnson system approximates the CDF curve best and only a small amount of error exists, while the curve approximated by cumulant + Johnson and Gaussian assumptions shows significant deviation, especially in the tail area.

Moreover, the method proposed in this paper has an advantage over MCS; that is, MCS cannot be implemented when only the moment information of power injections is available, while the method proposed in this paper can provide a reliable CDF curve.

4.3. Solutions of Different Optimization Formulations

The generation costs for the two optimization approaches are listed in Table 2. The cost of CC-PSCOPF is higher because it considers the uncertainty of power injections and gives a high probabilistic guarantee that line flows limit violations will not happen.

Table 2. Generation Cost.

	PSCOPF	CC-PSCOPF
Cost ($/hr)	7488.1	8196.5

Because the chance constraint is a soft constraint and the Gaussian distribution is an unbounded distribution, there are always extreme values that cause line flow violations. Based on the generator output schemes given by these two problem formulations, we implemented MCS with 10,000 samples to observe the actually probability of line flow violations.

The average violation probability and the maximum violation probability were introduced to illustrate the effectiveness of the proposed method. These two indices are listed in Table 3. It is obvious that the chance constraint works; the maximum violation probability under CC-PSCOPF equals 0.01, which is equal to the preset violation level, while under PSCOPF, the same index significantly exceeds the violation level. Both the average and maximum violation probability are smaller under CC-PSCOPF, which indicates that the proposed method provides a more robust operational state than PSCOPF.

Moreover, although the goal of the proposed method is to improve the operational reliability by controlling the violation probability of line flows in the overall situation rather than the violation probability under a specific contingency system topology, it is interesting to note that CC-PSCOPF effectively reduces violations under a single contingency system topology. We counted the number of contingencies N_V where the line flow violation probability exceeds the violation level. For contingencies with violations, the average number of line flows N_{alv} and maximum number of line flows N_{mlv} that exceeds the violation level were also counted. The statistical data are listed in Table 3. Obviously, CC-PSCOPF effectively reduces N_V, N_{alv} and N_{mlv}, and this indicates that more contingencies are effectively controlled.

Table 3. Constraint violation statistics.

	PSCOPF	**CC-PSCOPF**
Average violation probability	0.0045	0.0010
Maximum violation probability	0.0779	0.0100
N_V	22	6
N_{alv}	1.95	1.5
N_{mlv}	5	2

The line flow of 15–18 was chosen as representative to visually show the flow probability distributions under these two formulations. The histograms of line flow are shown in Figure 2a,b, respectively. It is clear that more PSCOPF samples fall outside of the line limits.

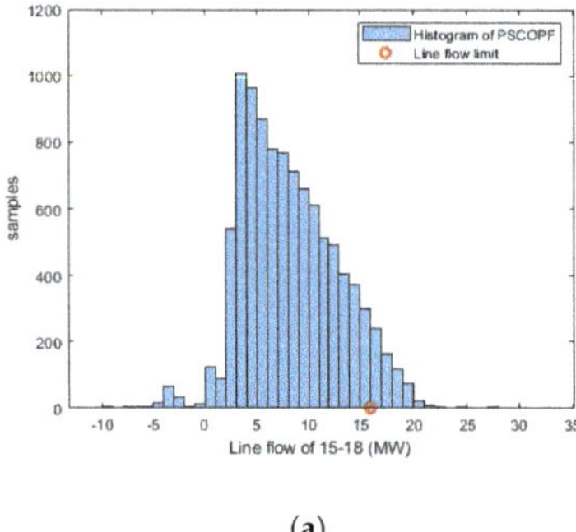

(**a**)

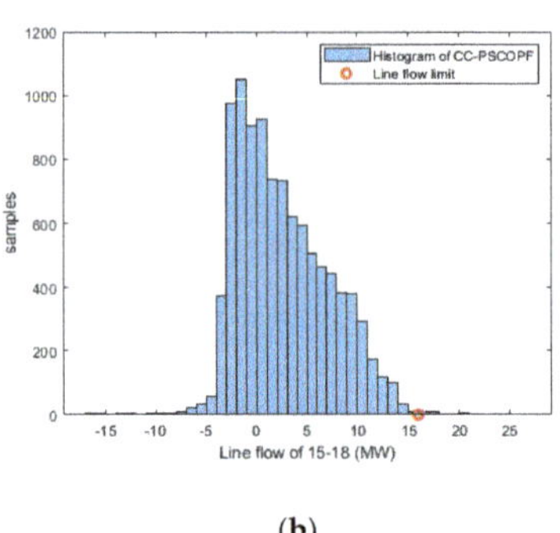

(**b**)

Figure 2. Histograms of line flow 15–18 under preventive security-constrained optimal power flow (PSCOPF) (**a**) and chance-constrained PSCOPF (CC-PSCOPF) (**b**).

4.4. Influence of the Value of Violation Level

Changing the violation level influences the solution of CC-PSCOPF. Figure 3 shows the change in the generation cost and violation probability of CC-PSCOPF with different violation levels.

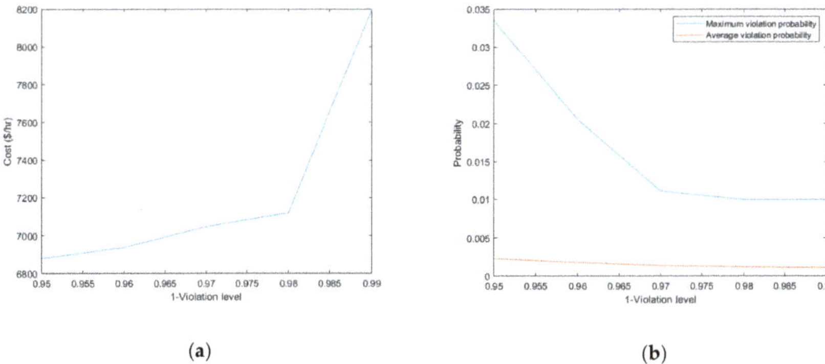

(a) (b)

Figure 3. Generation cost (**a**) and violation probability (**b**) of CC-PSCOPF with different violation levels.

As Figure 3 shows, with increasing violation level, the generation cost of CC-PSCOPF decreases, and inevitably, this causes a larger probability of constraint violations.

Obviously, the contradiction between power generation costs and operational security can be balanced by adjusting the violation level. As different transmission system operators have different requirements for line constraint violation probability, how to determine the optimal violation will be explored in subsequent studies. Note that, theoretically, the violation level can be set at 0 to completely eliminate the constraint violations, but this is not worth the gains. On the one hand, this will cause a surge in control costs, and on the other hand, it is easy to make the problem infeasible, especially when the uncertainty of power injection is large. Moreover, the violation level of different lines can be set to different values. For the critical lines, appropriately reducing the violation level can improve safety, and for non-critical lines, increasing the violation level appropriately can reduce generation costs. Obviously, CC-PSCOPF is more flexible compared with traditional PSCOPF, which can only balance the control costs and security by adjusting the contingency set.

4.5. Efficiency of the Proposed Method

The time consumption of PSCOPF and CC-PSCOPF was tested on the IEEE-30 test system and a larger-scale system, the IEEE-118 test system. The constraint numbers and time consumption of these two optimization formulations are listed in Table 4. It is evident that the constraint numbers significantly influence the efficiency of the optimization model. CC-PSCOPF is much faster than PSCOPF because the line limit constraints of all the system topologies are reduced to the same number of lines by the law of total probability.

Table 4. Time consumption.

Test System		PSCOPF	CC-PSCOPF
IEEE-30	Constraint Numbers	1600	42
	Time (s)	1.23	0.13
IEEE-118	Constraint Numbers	33109	187
	Time (s)	319.57	0.67

5. Conclusions

Currently, and in the future, increasing uncertainty brings challenges for power systems. The traditional SCOPF under strict N-1 criterion can barely meet the requirements of a secure operation. To obtain an operational state that is robust to uncertainties and to improve the overall operational reliability, a novel CC-PSCOPF was proposed in this paper. The probability distribution of the overall line flow is obtained and constrained within the limits with a high probability guarantee in the proposed optimization model. This type of constraint greatly reduces the number of constraints for the entire optimization problem, and additionally, the violation probability of each line can be flexibly adjusted as needed. In addition, the cumulant and Johnson systems are proposed to approximate the CDF curves, so the chance-constraint optimization model proposed in this paper is not limited to the Gaussian distribution assumption.

The proposed CC-PSCOPF can be used to improve the safety level of a system's operation, especially for a system with a high level of RES penetration. How to determine the optimal violation level and correlations between uncertainty sources will be investigated in subsequent studies.

Author Contributions: H.L. conceived and designed the study; this work was performed under the guidance of Z.Z., B.Z. and X.Y. All authors have read and agreed to the published version of the manuscript.

Funding: This research received no external funding.

Acknowledgments: The authors gratefully acknowledge the support of the National High Technology Research and Development Program of China (863 Program): No. 2015AA050201.

Conflicts of Interest: The authors declare no conflict of interest.

References

1. Li, X.; Zhang, X.; Wu, L.; Lu, P.; Zhang, S. Transmission Line Overload Risk Assessment for Power Systems with Wind and Load-Power Generation Correlation. *IEEE Trans. Smart Grid.* **2015**, *6*, 1233–1242. [CrossRef]
2. Duan, Y.; Zhang, B. Security risk assessment using fast probabilistic power flow considering static power-frequency characteristics of power systems. *Int. J. Electr. Power Energy Syst.* **2014**, *6*, 53–58. [CrossRef]
3. Da Silva, A.M.L.; De Castro, A.M. Risk Assessment in Probabilistic Load Flow via Monte Carlo Simulation and Cross-Entropy Method. *IEEE Trans. Power Syst.* **2019**, *34*, 1193–1202. [CrossRef]
4. Capitanescu, F. Critical review of recent advances and further developments needed in AC optimal power flow. *Electr. Power Syst. Res.* **2016**, *136*, 57–68. [CrossRef]
5. Monticelli, A.; Pereira, M.V.F.; Granville, S. Security-constrained optimal power flow with post-contingency corrective rescheduling. *IEEE Trans. Power Syst.* **1987**, *2*, 175–182. [CrossRef]
6. Hinojosa, V.H.; Gonzalez-Longatt, F. Preventive Security-Constrained DCOPF Formulation Using Power Transmission Distribution Factors and Line Outage Distribution Factors. *Energies* **2018**, *11*, 1497. [CrossRef]
7. Xu, Y.; Dong, Z.Y.; Zhang, R.; Wong, K.P.; Lai, M.Y. Solving Preventive-Corrective SCOPF by a Hybrid Computational Strategy. *IEEE Trans. Power Syst.* **2014**, *29*, 1345–1355. [CrossRef]
8. Alsac, O.; Bright, J.; Prais, M.; Stott, B. Further developments in LP-based optimal power flow. *IEEE Trans. Power Syst.* **1990**, *5*, 697–711. [CrossRef]
9. Capitanescu, F.; Glavic, M.; Ernst, D.; Wehenkel, L. Contingency filtering techniques for preventive security-constrained optimal power flow. *IEEE Trans. Power Syst.* **2007**, *22*, 1690–1697. [CrossRef]
10. Li, Y.; McCalley, J.D. Decomposed SCOPF for improving efficiency. *IEEE Trans. Power Syst.* **2009**, *24*, 494–495.
11. Fu, Y.; Shahidehpour, M.A.C. Contingency dispatch based on security-constrained unit commitment. *IEEE Trans. Power Syst.* **2006**, *21*, 897–908. [CrossRef]
12. Granville, S.; Lima, M.C.A. Application of decomposition techniques to VAr planning: Methodological and computational aspects. *IEEE Trans. Power Syst.* **1994**, *9*, 1780–1787. [CrossRef]
13. Wang, Q.; McCally, J.D. Risk and "N-1" Criteria Coordination for Real-Time Operations. *IEEE Trans. Power Syst.* **2013**, *28*, 3505–3506. [CrossRef]
14. Wang, Q.; McCally, J.D.; Zheng, T.X.; Litvinov, E. Solving corrective risk-based security-constrained optimal power flow with Lagrangian relaxation and Benders decomposition. *Int. J. Electr. Power Energy Syst.* **2016**, *75*, 255–264. [CrossRef]
15. Roald, L.; Vrakopoulou, M.; Oldewurtel, F.; Andersson, G. Risk-based optimal power flow with probabilistic guarantees. *Int. J. Electr. Power Energy Syst.* **2015**, *72*, 66–74. [CrossRef]
16. Charnes, A.; Cooper, W.; Symonds, G. Cost horizons and certainty equivalents: An approach to stochastic programming of heating oil. *Manag. Sci.* **1958**, *4*, 235–263. [CrossRef]
17. Charnes, A.; Cooper, W. Deterministic equivalents for optimizing and satisficing under chance constraints. *Oper. Res.* **1963**, *11*, 18–39. [CrossRef]

18. Zhang, Z.S.; Sun, Y.Z.; Gao, D.W.; Lin, J.; Cheng, L. A Versatile Probability Distribution Model for Wind Power Forecast Errors and Its Application in Economic Dispatch. *IEEE Trans. Power Syst.* **2013**, *28*, 3114–3125. [CrossRef]
19. Bienstock, D.; Chertkov, M.; Harnett, S. Chance-Constrained Optimal Power Flow: Risk-Aware Network Control under Uncertainty. *SIAM Rev.* **2014**, *56*, 461–495. [CrossRef]
20. Wang, Z.W.; Shen, C.; Liu, F.; Wu, X.Y.; Liu, C.C.; Gao, F. Chance-Constrained Economic Dispatch With Non-Gaussian Correlated Wind Power Uncertainty. *IEEE Trans. Power Syst.* **2017**, *32*, 4880–4893. [CrossRef]
21. Venzke, A.; Halilbasic, L.; Markovic, U.; Hub, G.; Chatzivasileiadis, S. Convex Relaxations of Chance Constrained AC Optimal Power Flow. *IEEE Trans. Power Syst.* **2018**, *33*, 2829–2841. [CrossRef]
22. Roald, L.; Misra, S.; Krause, T.; Andersson, G. Corrective Control to Handle Forecast Uncertainty: A Chance Constrained Optimal Power Flow. *IEEE Trans. Power Syst.* **2017**, *32*, 1626–1637.
23. Li, Y.; Wang, J.; Ding, T. Clustering-based chance-constrained transmission expansion planning using an improved benders decomposition algorithm. *IET Gener. Transm. Distrib.* **2018**, *12*, 935–946. [CrossRef]
24. Liu, J.; Chen, H.; Zhang, W.; Yurkovich, B.; Rizzoni, G. Energy Management Problems Under Uncertainties for Grid-Connected Microgrids: A Chance Constrained Programming Approach. *IEEE Trans. Smart Grid.* **2017**, *8*, 2585–2596. [CrossRef]
25. Wu, H.; Shahidehpour, M.; Li, Z.; Tian, W. Chance-constrained day-ahead scheduling in stochastic power system operation. *IEEE Trans. Power Syst.* **2014**, *29*, 1583–1591. [CrossRef]
26. Nazir, F.; Pal, B.; Jabr, R. A two-stage chance constrained volt/var control scheme for active distribution networks with nodal power uncertainties. *IEEE Trans. Power Syst.* **2019**, *34*, 314–325. [CrossRef]
27. Wu, J.; Zhang, B.; Li, H.; Li, Z.; Chen, Y.; Miao, X. Statistical distribution for wind power forecast error and its application to determine optimal size of energy storage system. *Int. J. Electr. Power Energy Syst.* **2014**, *55*, 100–107. [CrossRef]
28. Bludszuweit, H.; Dominguez-Navarro, J.A.; Llombart, A. Statistical analysis of wind power forecast error. *IEEE Trans. Power Syst.* **2008**, *23*, 983–991. [CrossRef]
29. Kaur, A.; Pedro, H.T.C.; Coimbra, C.F.M. Impact of onsite solar generation on system load demand forecast. *Energy Convers. Manag.* **2013**, *75*, 701–709. [CrossRef]
30. Li, Y.; McCally, J.D. Risk-Based Optimal Power Flow and System Operation State. In Proceedings of the Power & Energy Society General Meeting 2009 IEEE, Calgary, BC, Canada, 26–30 July 2009.
31. Billinton, R.; Singh, G. Application of adverse and extreme adverse weather: Modelling in transmission and distribution system reliability evaluation. *IEE Proc. Gener. Transm. Distrib.* **2006**, *153*, 115–120. [CrossRef]
32. Yuan, Y.; Zhou, J.; Ju, P.; Feuchtwang, J. Probabilistic load flow computation of a power system containing wind farms using the method of combined cumulants and Gram-Charlier expansion. *IET Gener. Transm. Distrib.* **2011**, *5*, 448–454. [CrossRef]
33. Usaola, J. Probabilistic load flow with correlated wind power injections. *Electr. Power Syst. Res.* **2010**, *80*, 528–536. [CrossRef]
34. Zhang, L.B.; Cheng, H.Z.; Zhang, S.X.; Zeng, P.L.; Yao, L.Z. Probabilistic power flow calculation using the Johnson system and Sobol's quasi-random numbers. *IET Gener. Transm. Distrib.* **2016**, *10*, 3050–3059. [CrossRef]
35. Soukissian, T. Use of multi-parameter distributions for offshore wind speed modeling: The Johnson S-B distribution. *Appl. Energy* **2013**, *111*, 982–1000. [CrossRef]
36. Hill, I.D.; Hill, R.; Holder, R.L. Algorithm AS 99: Fitting Johnson Curves by Moments. *Appl. Sci.* **1980**, *25*, 180–189. [CrossRef]
37. Wachter, A.; Biegler, L.T. On the implementation of an interior-point filter line-search algorithm for large-scale nonlinear programming. *Math. Program.* **2016**, *106*, 25–57. [CrossRef]
38. Power System Test Case Archive. Available online: https://labs.ece.uw.edu/pstca/ (accessed on 1 March 2020).

Article

Multi-Objective Tabu Search for the Location and Sizing of Multiple Types of FACTS and DG in Electrical Networks

Othón Aram Coronado de Koster * and José Antonio Domínguez-Navarro *

Department of Electrical Engineering, University of Zaragoza, María de Luna 3, 50018 Zaragoza, Spain
* Correspondence: 544299@unizar.es (O.A.C.d.K.); jadona@unizar.es (J.A.D.-N.); Tel.: +52-834-175-6334 (O.A.C.d.K.)

Received: 21 April 2020; Accepted: 22 May 2020; Published: 28 May 2020

Abstract: Flexible AC transmission systems and distributed generation units in power systems provide several benefits such as voltage stability, power loss minimization, thermal limits enhancement, or enables power system management close to the limit operation points; and by extension, economic benefits such as power fuel cost and power loss cost minimization. This work presents a multi-objective optimization algorithm to determine the location and size of hybrid solutions based on a combination of Flexible AC transmission systems devices and distributed generation. Further, the work expands the types of FACTS usually considered. The problem is solved by means of a Tabu search algorithm with good results when tested in a network of 300 nodes.

Keywords: flexible AC transmission systems; tabu search; multi-objective; power systems

---

1. Introduction

In deregulated markets, power transactions in transmission and distribution networks lead the power systems to operate close to their limits to maximize the benefits [1–4]. Furthermore, climate change and other environmental concerns force the installation of distributed generation (DG) units, renewable and conventional, close to load centres to feed the demand growth [5]. There is no doubt that the installation of DG has benefits for both the consumer, the supplier and the network [6], but the increase in the penetration of the DG can also cause several problems in the operation of the network [7] (voltage profile, stability, wave quality, harmonics, imbalances, ...). The creation of a microgrid architecture [8] together with a suitable energy management system [9] is the most advantageous mode of operation for both the consumer and the network.

Flexible AC transmission systems (FACTS) and DG installations in the transmission lines have the capacity to enhance power systems [10]. However, finding the optimal FACTS and DG devices location for power system enhancement is a non-linear complex problem which involves economical, environmental and electrical variables. In the technical literature, authors have used different FACTS to control the power network attributes [3,11,12] as summarized in Table 1. Several authors utilise intelligent systems to improve the voltage stability in real time, during the operation of saturated electrical networks. Devaraj et al. [13] use a radial basis function network model to estimate the voltage stability level of the power system based on the L-index, and this way, detect how far the nodes are from voltage collapse. Tomin et al. [14] present an automatic intelligent system for voltage security control based on a decision trees model, and use the L-index for the localisation of critical nodes. Satheesh et al. [15] use a neural network to identify the optimal location of FACTS controllers and a Bees algorithm to calculate the operation point of these devices in the power system.

One way to solve the problem in consideration is by optimizing an economic objective function, formulated with fixed kVAr costs [16,17] or with quadratic formulations [18–25]. Other authors include the annual investment cost [26], capital recovery factor [27,28], annual cost device [29], or a combination of them; as well as the active power generation fuel cost and the reactive power generation fuel cost, if the DG units are combustion machines [30–32]. The optimization problem is usually subject to common power flow constraints, such as bus voltage limits, thermal limits, feeders power transfer capability, real and reactive power generation limits, among others [33,34].

Table 1. FACTS control attributes.

FACTS Controller	Control Attributes								
	1	2	3	4	5	6	7	8	9
STATCOM	x	x	x	x					
SVC	x	x	x	x	x				
SSSC			x		x	x	x		
TCSC			x	x	x	x	x		
UPFC	x	x	x	x	x		x	x	x

(1) Voltage control, (2) VAr compensation, (3) damping oscillations, (4) voltage stability, (5) transient and dynamic stability, (6) current control, (7) fault current limiting, (8) active power control, (9) reactive power control.

Another way is to solve it as a multi-objective optimization problem and obtain a set of non-dominated solutions. Authors in [35] optimize the location of thyristor controlled series capacitors (TCSC) and/or static VAR compensators (SVC), considering the investment and power generation cost as objective functions, by means of genetic algorithms (GA), successive linear programming and Benders decomposition, maintaining the voltage profile within its limits. Another multi-objective formulation of FACTS costs has been developed by [36], averaging investment and generation costs, and solving with a GA technique to find the optimal location of unified power flow control (UPFC), TCSC, thyristor controlled phase shifting transformer and SVC devices in power systems, where the FACTS candidate nodes and lines are selected using a randomization method. Voltage profile enhancement and TCSC device number minimization are used as objectives to improve line congestion in [37], and solved through simulated annealing and sequential quadratic programming (SQP) algorithms. A model for finding the optimal location of TCSC and SVC devices using a hybrid GA-SQP algorithm with a fuzzy multi-objective function, that includes power loss, investment cost (quadratic costs functions), peak point power generation, voltage deviation, as well as security margin minimization, is presented by [38]; this work is addressed in [39], adding Pareto optimal solutions to obtain faster results. Other authors [40,41] determine the maximum loading factor possible, implementing FACTS devices in power systems, taking into account the voltage deviation and the real power loss minimization, finding the optimal parameters settings and locations of coordinated SVC and TCSC devices, and selecting the best compromise solution of the Pareto optimal solutions in non-dominated sorting particle swarm optimization. The optimal FACTS location problem is solved in [42] considering the power system total cost, where Akaike's information criterion is minimized and the expected security is maximized. A multi-objective non-dominated sorting improved harmony search is proposed by [43] for voltage stability improvement, considering the optimal placement of TCSC and/or SVC devices in power systems through loading factor maximization, and voltage deviation and real power loss minimization. The gravitational search algorithm is introduced and compared with particle swarm optimization for reactive power planning, considering FACTS implementation in power systems, by [44]; in said work, the goal is to minimize both real power loss and FACTS investment cost, while increasing the reactive load. The effectiveness of the harmony search algorithm is used in [45] to find optimal TCSC and static synchronous series compensator (SSSC) locations, considering power system loading factor maximization.

Presently, Tabu search has been used in the location and sizing of DG [46] or the FACTS [47–50] with mono-objective models, and with multi-objective models for the location and sizing of DG [51]. In this paper, a multi-objective Tabu search (MOTS) algorithm is carried out to find the optimal location and size of FACTS devices and DG units in a power system network. The problem has also been generalized by expanding the types of FACTS considered, including their hybrid use with other solutions, such as the installation of DG and high-voltage direct current (HVDC) systems. The FACTS devices considered in this work are: HVDC, STATCOM, SSSC, SVC, TCSC and UPFC. Section 2 depicts the multi-objective function, and the tabu search algorithm method, including the description of the permanency and the recency effect in the memory, and the selection of nodes and branches through analytical methods. Section 3 presents the test results obtained for a modified IEEE 300-bus system. Finally, in Section 4 the conclusions are presented.

2. Methodology

In the proposed model, an analytical method is applied to select the lines or buses to install DG units or FACTS devices to enhance the power system stability, taking into consideration the technical power flow constraints and the total power system cost analysis. Costs are presented in annualized values, with the objective to obtain benefits regarding the total investment cost (*TIC*) and the total generation cost (*TGC*), including the power losses cost (*PLC*). The algorithm is coded in MATLAB 2019 using the continuation power flow routine extracted from the power system analysis toolbox [52] to work embedded with tabu search [53]. The results are compared with the initial solution on a IEEE 300 bus test system.

2.1. Cost Functions

The economic analysis in this work employs the TIC, the TGC, and the PLC. To calculate the TIC, maintenance, operation and installation costs are considered. For the TGC, reference bus active power generation cost (*PGC*), and DG units *PGC* and reactive power generation cost (*QGC*) are taken into account.

2.1.1. Power Losses Cost

The PLC is determined by the continuation power flow using the mathematical formulation presented in [54].

$$PLC = \sum_{p \epsilon N_p} \sum_{ij \epsilon N_L} Pl_{ij,p} \cdot C_{L,p} \cdot \alpha_p^{PLC} \tag{1}$$

where α_p^{PLC} (Equation (2)) is the power loss discount rate for the PLC determination in period p; $Pl_{ij,p}$ are the power losses in the line ij in period p; $C_{L,p}$ is the cost of losses in period p in $/kWh, N_p is the period set and N_L is the lines set.

The discount factors are determined by means of:

$$\alpha_p^{PLC} = \sum_{y=1}^{ny} \frac{(1 + LG_p)^y}{(1 + r)^{(ny)(p-1)+y}} \tag{2}$$

where ny are the periods in the planning horizon in years; LG_p is the load annual growth rate in period p; and r is the annual interest rate.

2.1.2. FACTS Costs

In this research, HVDC, TCSC, SSSC, STATCOM, SVC and UPFC are the devices considered to enhance the power system network, and thus, it is necessary to determine the FACTS investment cost (*FIC*).

$$FIC = \sum_{i \epsilon N_F} \text{CRF} \cdot C_i \tag{3}$$

where N_F is the FACTS set, the FACTS cost (C_i) and the capital recovery factor (CRF) [27–29,55–63] are:

$$C_i = F_{1,i} \cdot S_i^2 - F_{2,i} \cdot S_i + F_{3,i} \tag{4}$$

$$\text{CRF} = \frac{r\,(1+r)^{ny}}{(1+r)^{ny} - 1} \tag{5}$$

where S_i is the nominal apparent power of FACTS i.

The values of the coefficients $F_{1,i}, F_{2,i}$ and $F_{3,i}$ are specified in Table A1. These coefficients are retrieved from [18–25,64]. For the STATCOM function, the data curve recognition provided by [18] is employed; the values for the HVDC are obtained by adjusting the curve using the cost information calculated in [64].

2.1.3. Distributed Generation Costs

The DG units are considered as PV nodes and the costs are simulated as diesel generators. In [30–32,36,65–67], the simplified equation for PGC is presented, taking into account the equations developed by [68].

$$PGC = 8760 \cdot \left\{ \left[\sum_{i \in N_G} \left(\alpha_{2,i} \cdot P_{G,i}^2 + \alpha_{1,i} \cdot P_{G,i} + \alpha_{0,i} \right) \right] + 61.38 \cdot P_{G,SW} \right\} \tag{6}$$

where $\alpha_{2,i}$, $\alpha_{1,i}$ and $\alpha_{0,i}$ are the generators coefficients [57,62,69–72] (Table A2); the coefficient 61.38 is obtained from [73]; $P_{G,i}$ is the active power supplied by generator i and N_G is the generators set.

The QGC equations are developed in [65,74], where $\beta_{1,i} = 0.1\alpha_{1,i}$ and $\beta_{0,i} = 0.1\alpha_{0,i}$ [65]; and $Q_{G,i}$ is the reactive power supplied by generator i.

$$QGC = 8760 \cdot \left[\sum_{i \in N_G} \left(\beta_1 \cdot Q_{G,i} + \beta_0 \right) \right] \tag{7}$$

The distributed generation investment cost (GIC) function considers the CRF, DG installation cost, and operation and maintenance costs in the C_i value [75].

$$GIC = \sum_{i \in N_G} \text{CRF} \cdot C_i \tag{8}$$

2.1.4. Multi-Objective Function

The multi-objective function presented aims at minimizing the TIC and the TGC, where the TIC is formulated by means of the Equations (3) and (8). The TGC is calculated using Equations (1), (6) and (7).

$$\min(TGC) = \min\left(PGC + QGC + PLC\right) \tag{9}$$

$$\min(TIC) = \min\left(FIC + GIC\right) \tag{10}$$

Subject to the following constraints:

- Power balance with FACTS in each bus i:

$$P_i(\vartheta, V) - P_{G,i} + P_{D,i} + P_{F,i} = 0 \tag{11}$$

$$Q_i(\vartheta, V) - Q_{G,i} + Q_{D,i} + Q_{F,i} = 0 \tag{12}$$

- Thermal limit in each line ij:

$$|S_{ij}| \leq S_{ij}^{\max} \tag{13}$$

- Bus voltage and angle limits in each bus i:

$$V_i^{\min} \leq V_i \leq V_i^{\max} \tag{14}$$

$$\vartheta_i^{\min} \leq \vartheta_i \leq \vartheta_i^{\max} \quad (15)$$

- Power limits in each generator i:

$$P_{G,i}^{\min} \leq P_{G,i} \leq P_{G,i}^{\max} \quad (16)$$

$$Q_{G,i}^{\min} \leq Q_{G,i} \leq Q_{G,i}^{\max} \quad (17)$$

- Power limits in each FACT i:

$$P_{F,i}^{\min} \leq P_{F,i} \leq P_{F,i}^{\max} \quad (18)$$

$$Q_{F,i}^{\min} \leq Q_{F,i} \leq Q_{F,i}^{\max} \quad (19)$$

- Power relations between bus i and bus j with FACTS:

$$f(P_{F,i}, P_{F,j}, Q_{F,i}, Q_{F,j}) = 0 \quad (20)$$

where $P_{D,i}, Q_{D,i}$ represent the active and reactive power demand in bus i. The Power System Analysis Toolbox (PSAT) [52,76,77] is used to evaluate the power flow.

2.2. *Multi-Objective Tabu Search Algorithm*

Tabu search is the heuristic method used in this work, proposed by Glover [53]. It is based on local search with different strategies to escape the local optima, such as by means of long- and short-term memory analysis, giving the model the ability to change the search area.

The local search consists of making a *movement* between two interchangeable elements selected in the actual search area (called *neighbourhood*) to find a solution that satisfies an objective function. As stated above, the heuristic employs two types of memory structures to store movements: the first one, short-term memory, provides the capability to avoid movements that do not result in favorable solutions in a fixed number of iterations. The short-term memory forces the algorithm to search in other directions inside the actual neighbourhood using one or more strategies such as: aspiration plus, elite candidate list, successive filter strategy, sequential fan or bounded change candidate list. In respect of the long-term memory, this structure sequentially stores every movement in a frequency list, and is used to modify the neighbourhood search areas by means of different strategies such as: modifying the choice rules, restarting, strategic oscillation patterns and decisions or path re-linking techniques.

The process employed to solve the multi-objective function to optimize FACTS and DG units location using tabu search is described in Algorithm 1. As a result, a list is obtained with all the solutions that are part of the Pareto front. The following subsections address the explanation of each algorithm line.

2.2.1. Initialization

Network data are loaded in the PSAT and MOTS structures. The location of the swing bus is determined by the rotation-buses technique [78] that uses the shortest path function to determine the route from the slack bus to the end buses and initial solution is obtained. Then, a radial search tree is built by means of a minimum spanning tree using the *Kruskal* method [79].

Algorithm 1: Multi-Objective Tabu Search.

```
input: electrical network;
output: Pareto front list;
variables:
PFL: Pareto front list;
TL: Tabu list; (Section 2.2.4)
FL: Frequency list; (Section 2.2.5)
S_k, Snew_k: a solution of the problem;
X_k: a bus or line of the electrical network;
begin
    Initialization (Section 2.2.1)
    Obtain initial solution: S_0.
    Store in Pareto front list: PFL ← S_0.
    Initialize tabu list and frequency list: TL = {}; FL = {};
    repeat
        for ∀ S_k ∈ PFL do
            Solution neighbourhood (Section 2.2.3)
            Select the buses and lines candidates and store in CL: CL ← X_k
            for ((∀ X_k ∈ CL) && (X_k ∉ TL)) do
                Apply the possible movements and obtain new solutions: X_k → Snew_k
                Evaluate solution Snew_k (Section 2.2.2)
                if Snew_k is non-dominated in PFL then
                    Update PFL ← Snew_k
                end
                Update TL ← X_k
                Update FL ← X_k
            end
        end
        Diversification and intensification process (Section 2.2.6)
        if diversification = TRUE then
            Diversification_process
        end
        if intensification = TRUE then
            Intensification_process
        end
        Stop criteria (Section 2.2.8)
    until Stop_criteria = TRUE;
end
```

2.2.2. Solution Evaluation

The power flow of the actual network is evaluated with PSAT. The TIC and TGC objectives are evaluated after the continuation power flow routine reaches a feasible solution, considering the technical power flow constraints. Besides, others indices ($C_{ev,1}$, $C_{ev,2}$ and $C_{ev,3}$) are obtained:

$$C_{ev,1} = \frac{\Delta PLC}{\Delta TIC} \tag{21}$$

$$C_{ev,2} = \begin{cases} 0, & \text{if } TGC_{act} < TGC_{bst} \\ 1, & \text{otherwise} \end{cases} \tag{22}$$

$$C_{ev,3} = \Delta PGC + \Delta QGC + \Delta PGC_{SW} + \Delta PLC \tag{23}$$

where Δ represents the difference between the best and the actual solution values.

These indexes allow accepting a solution even if the economic objective function is worse than the current one; this happens when $C_{ev,1} > 0$ or when $C_{ev,2} = 1$ and $C_{ev,3} > 0$. This strategy allows escape from local minima.

All non-dominated solutions are stored in the Pareto Front list, PFL, and the list is then updated by removing the dominated solutions.

2.2.3. Solution Neighbourhood

Every modification in the location or size of a DG unit or FACTS device in the power system is considered a *movement* in the search process; the possible movements are enlisted as follows:

- Add, change or remove a generator.
- Add, change or remove a FACTS device.
- Remove a generator to add a FACTS device.
- Remove a FACTS device to add a generator.

These movements are applied to the candidate buses and lines which are selected according to the following rules for each device:

- DG units: worst bus voltage in non-end bus or transformer node.
- SVC: worst bus voltage and worst line with the highest power losses.
- TCSC: lines voltage or current flow out of the limits.
- STATCOM: weakest bus voltage and reactive power received control needs.
- UPFC: weakest bus voltage or reactance control needs.
- SSSC: weakest bus voltage, reactance or apparent power control needs.
- HVDC: voltage current or apparent power control needs, taking longest lines, between 500 and 800 km or 40 and 80 km if cables are used and power ranges up to 4000 MW at $\pm 500\ kV_b$, 4800 MW at $\pm 600\ kV_b$ in accordance with [80].

2.2.4. Tabu List

When any movement does not yield an improvement, in order to avoid repetitive movements, that movement is penalized and prohibited during several iterations. The list where these movements are stored is called the *tabu list*, and the number of iterations that the movement is stored in the tabu list is the *tabu tenure* (Equation (24)), whose length TT is a function of the number of buses N_{buses}.

$$TT = \sqrt[3]{10 \cdot \log\left(N_{buses}\right) + \frac{N_{buses}}{10}} + \frac{0.2 \cdot N_{buses}}{4} \tag{24}$$

2.2.5. Frequency List

The frequency list adds the current movement to the permanent memory to store the devices and branches or buses employed in the search process. These values provide the tabu search with less explored areas to select, when the improvements have stalled.

2.2.6. Diversification and Intensification Process

The diversification process takes into account the non-improvement movements to evaluate the HVDC, STATCOM, SSSC and UPFC devices installation in the power system. The intensification process considers the less used buses to either add or remove devices, or change DG units. The diversification and intensification processes are activated three times during the tabu search routine.

2.2.7. Aspiration Plus

The aspiration plus strategy is applied in this work, it consists of establishing a threshold for the quality of a move, based on the search pattern history [53]. When a good movement is found within the threshold, additional movements around this good movement are inspected to expand the search in that area and then select the best movement. Afterwards, the procedure continues with the next iteration in another neighbourhood area.

2.2.8. Stop Criteria

The proposed algorithm finishes when the maximum number of iterations is reached or the non-improvement moves reach 50% of the maximum number of iterations, and the diversification and intensification processes have been applied at least once.

3. Results

The proposed MOTS algorithm has been tested with a modified IEEE 300-bus power system network [81] with 69 PV nodes whose data are in Table 2. Voltage limits between 0.9 and 1.1 of the nominal voltage have been added in the nodes, and the thermal limit has been placed at 1.2 of the nominal current in the lines and transformers. In addition, DC lines and associated converters have been removed.

For the initial solution, *TIC* is zero, *TGC* is 4,541,173.09 USD/year and there are 45 violations of the proposed constraints (18 bus minimum voltage, 26 maximum and one minimum reactive power). Results were obtained using an AMD AM3+ FX 6300 CPU with 24 GB RAM and MATLAB 2019a.

Table 2. PV nodes data from the initial solution.

N	Cap	N	Cap	N	Cap	N	Cap	N	Cap	N	Cap
8	0.05	10	0.05	19	0.1	55	0	63	0	69	3.75
76	1.55	77	2.9	80	0.68	88	1.17	98	19.3	103	2.4
104	0	117	1.925	120	2.81	122	6.96	125	0.84	126	2.17
128	1.03	131	3.72	132	2.16	135	0	149	2.05	150	0
155	2.28	156	0.84	164	2	165	12	166	12	169	4.75
170	19.73	177	4.24	192	2.72	199	1	200	4.5	201	2.5
206	3.03	209	3.45	212	3	215	6	217	2.5	218	5.5
220	5.7543	221	1.7	222	0.84	247	4.67	248	6.23	249	12.1
250	2.34	251	3.72	252	3.3	253	1.85	254	4.1	255	5
256	0.37	258	0.45	259	1.65	260	4	261	4	261	1.16
263	12.92	264	7	265	5.53	267	0.042	292	0.3581	294	0.2648
295	0.5	296	0.08								

N: bus location; Cap: capacity of generator in p.u.

Three strategies were tested with the MOTS algorithm to improve the original case: the first strategy consists of adding only DG units, the second is to add only FACTS, and the third is a combined strategy that allows adding both DG units and FACTS. The comparison between these three strategies is analysed in this study by means of the Pareto optimal frontier, as shown in Figures 1–3. The algorithm minimises two objectives: *TIC* and *TGC*. "Black dots" are the solutions that are part of the Pareto front and the "star dot" is the initial solution. The range of variation of the *TGC* is similar, but not equal, for the three strategies, however there are significant differences when the *TIC* is compared. The DG installation is the cheapest strategy and the FACTS installation is the most expensive, while the combined strategy has intermediate costs.

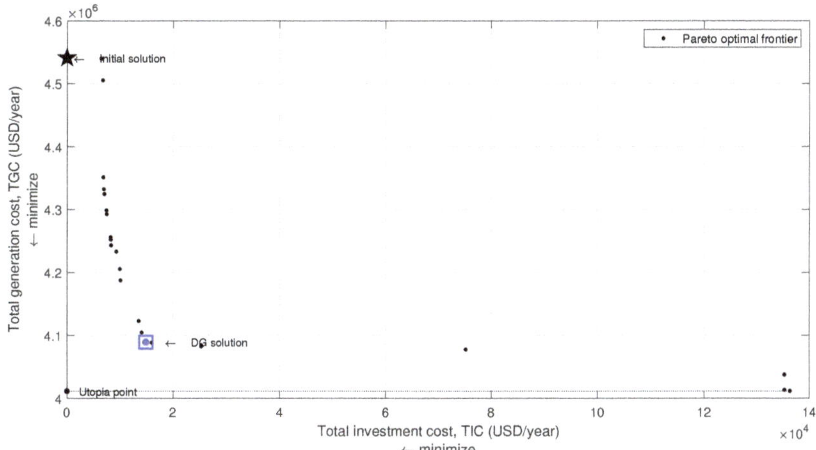

Figure 1. Pareto optimal frontier for the DG units case.

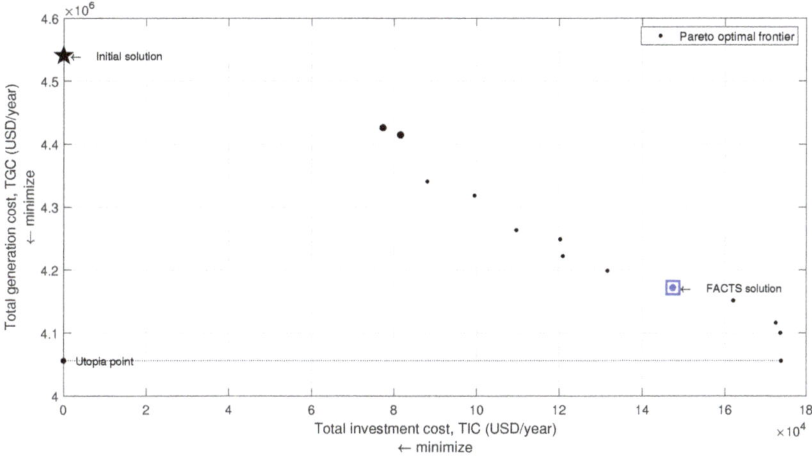

Figure 2. Pareto optimal frontier for the FACTS devices case.

First, considering the Pareto front for the DG units case in Figure 1, it can be seen that the generation costs are low with relatively small investments, in the range of 0 to 2×10^4 USD/year. However, from that point on, strong investments, about 14×10^4 USD/year, are needed to only slightly reduce the cost of generation. In other words, the installation of DG is cheap, and reduces losses in the network, but eventually the network is saturated. Furthermore, in the DG units case, it is possible to keep the voltages within the allowed limits, as can be seen in Figure 4b. In the FACTS devices case, as shown in Figure 4c, it is possible to obtain generation costs similar to those obtained in the DG units case, although higher investments are necessary. In this case, there is also the disadvantage that it is not possible to maintain the voltages at all buses within their limits, as shown in Figure 4c; i.e., the voltage profile is improved, but not enough at all buses. Finally, in the Pareto front for the Combined devices case in Figure 3, it can be observed that the cost of generation may be reduced to a point below the minimum obtained in the previously mentioned cases. The solution for this case requires greater investments than in the DG units case, but much smaller than in the FACTS devices case, in the range of 2×10^4 to 8×10^4 USD/year. That is, the joint installation of DG and FACTS allows to solve the problems that a high penetration of DG in the network produces, decreasing the losses in the network and keeping the voltage profile within the permissible limits, as presented in Figure 4d.

The algorithm monitors two technical indices during the search, shown in Figure 5, to determine if the technical characteristics of the solutions found with the different strategies are similar. These indices are: the L-index [82], which is a parameter that indicates the proximity of the node to voltage collapse, and the other is the difference between the reference voltage and the real voltage.

The L-index for add-only DG or add-only FACTS strategies approaches one, meaning that such solutions are near to the point of collapse. In contrast, for the combined strategy, a value further from the point of collapse is obtained. Thus, the inclusion of both DG and FACTS allows for the system to approach the limit operation point without collapse problems.

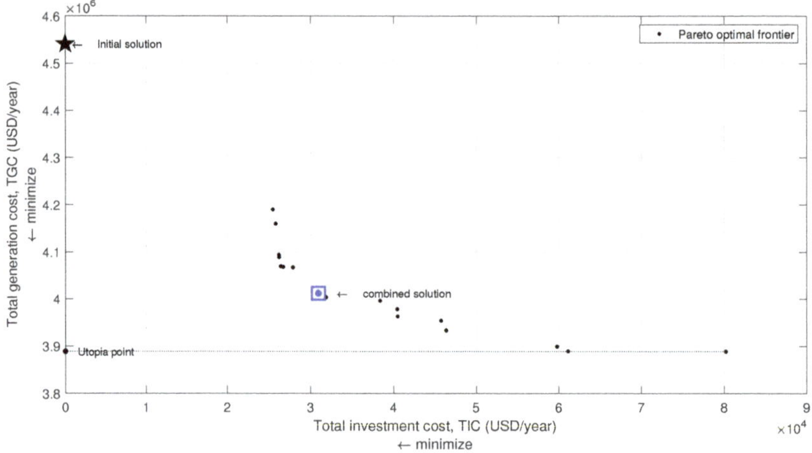

Figure 3. Pareto optimal frontier for the combined units case.

All the solutions of the Pareto curve are non-dominated, so to choose one, one must use an external criteria for the optimization to be carried out. The solution for each curve is the one with the best L-index and smallest voltage variation. These solutions are marked with a blue square in the Pareto curves and the description of location and characteristics of the installed devices are detailed in Table 3.

The TIC plus the TGC for the three strategies (DG, FACTS and combination) are presented in Table 4. The solution for only DG units has zero bus voltage violations and 38 lines are less saturated, obtaining a reduction of 437,064.48 USD/year compared to the initial solution. The solution with only FACTS devices obtains a 221,462.40 USD/year reduction in comparison with the initial solution, 47 lines are less saturated and the number of bus voltage violations is reduced to nine. The combination of DG and FACTS devices achieves a reduction of 497,787.92 USD/year in comparison to the initial solution, with zero bus voltage violations, and 53 lines less saturated.

Figure 4 depicts the bus voltage profiles comparison between the selected solutions for each strategy analysed in this work, where it can be seen that the first and third strategies provide better results. The second strategy (FACTS addition) is not able to reach an acceptable bus voltage profile improvement. The combination of DG and FACTS results in the best solution, finishing with more non-expenses and better bus voltage profile for all buses.

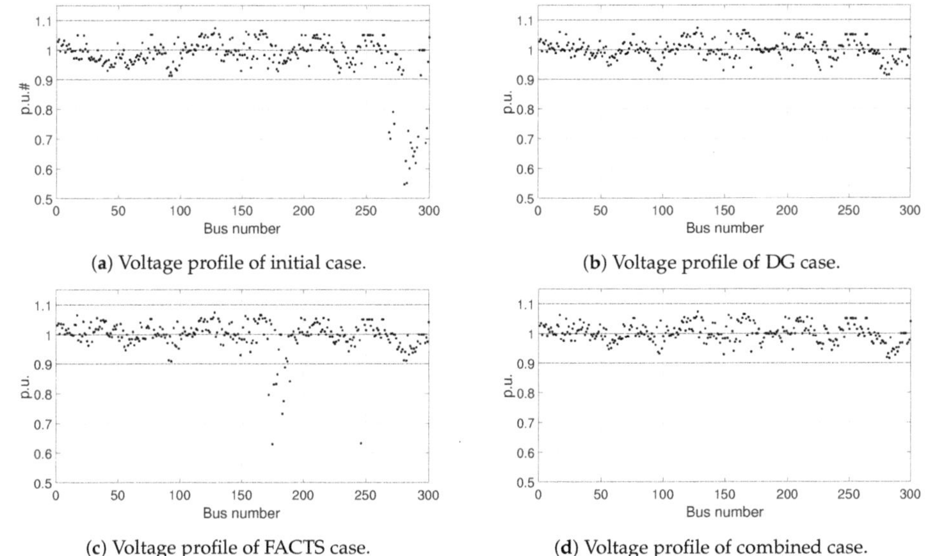

(**a**) Voltage profile of initial case.

(**b**) Voltage profile of DG case.

(**c**) Voltage profile of FACTS case.

(**d**) Voltage profile of combined case.

Figure 4. Bus voltage profile for the selected solutions on each strategy.

Table 3. Location and characteristics of installed devices.

Stragtegy	Dev.	N	Observ.	Dev.	N	Observ.	Dev.	N	Observ.
Only DG	G	268	Cap = 1	G	190	Cap = 2	G	179	Cap = 1
	G	183	Cap = 8	G	93	Cap = 2	G	180	Cap = 1.2
	G	168	Cap = 2.5	G	175	Cap = 8	G	204	Cap = 1
	G	246	Cap = 1						
Only FACTS	SVC	24	$\alpha = -0.6191$ $v_m = 0.99986$	SVC	81	$b = 1.5$	SVC	266	$\alpha = -0.61726$ $v_m = 1.0014$
	SVC	87	$\alpha = -0.4941$ $v_m = 0.99992$	SVC	54	$b = -0.30335$	SVC	272	$\alpha = 0.33423$ $v_m = 1.0001$
	SVC	38	$b = 1.5$	SVC	89	$\alpha = 0.62832$ $v_m = 1$	SVC	97	$b = 1.5$
	SVC	208	$b = -1$	SVC	7	$b = 1.5$	SVC	58	$b = 0.07184$
	SVC	12	$\alpha = -0.52826$ $v_m = 0.99999$	SVC	57	$\alpha = 0.95074$ $v_m = 0.99996$	SVC	188	$b = -0.35074$
	SVC	182	$\alpha = -0.19191$ $v_m = 0.99969$	SVC	168	$\alpha = 1.0107$ $v_m = 0.99997$	SVC	227	$\alpha = 0.81705$ $v_m = 0.99987$
	TCSC	267	$x = 0.03279$	SSSC	381	$v_{cs} = 2.5534$	UPFC	16	$v_p = 0$ $v_q = 0.00001$ $i_q = 0.80591$
DG-FACTS	G	172	Cap = 2	G	189	Cap = 2	G	116	Cap = 1
	G	176	Cap = 8	G	184	Cap = 2	G	187	Cap = 0.8
	G	183	Cap = 1	G	175	Cap = 8	G	91	Cap = 1
	TCSC	13	$x = 0.00323$	UPFC	179	$v_p = 0$ $v_q = 0.02027$ $i_q = -1$			

Dev.: device type; N: bus or line location; Observ.: characteristics of device; α: SVC firing angle; v_m: SVC measured voltage; b: SVC susceptance; x: TCSC series reactance; v_{cs}: SSSC voltage in quadrature with the line current; v_p: UPFC series voltage in phase with the line current; v_q: UPFC series voltage in quadrature with the line current; i_q: UPFC shunt current wich is in quadrature with the bus voltage in the line.

Table 4. Costs results for selected solutions in [USD/year].

Description	Initial Solution	Only DG	Only FACTS	Combined Devices
TIC	0.00	14,958.00	147,639.31	30,924.21
TGC	4,541,173.09	4,089,150.61	4,172,071.38	4,012,460.96
Total cost	4,541,173.09	4,104,108.61	4,319,710.69	4,043,385.17
Save	–	−437,064.48	−221,462.40	−497,787.92

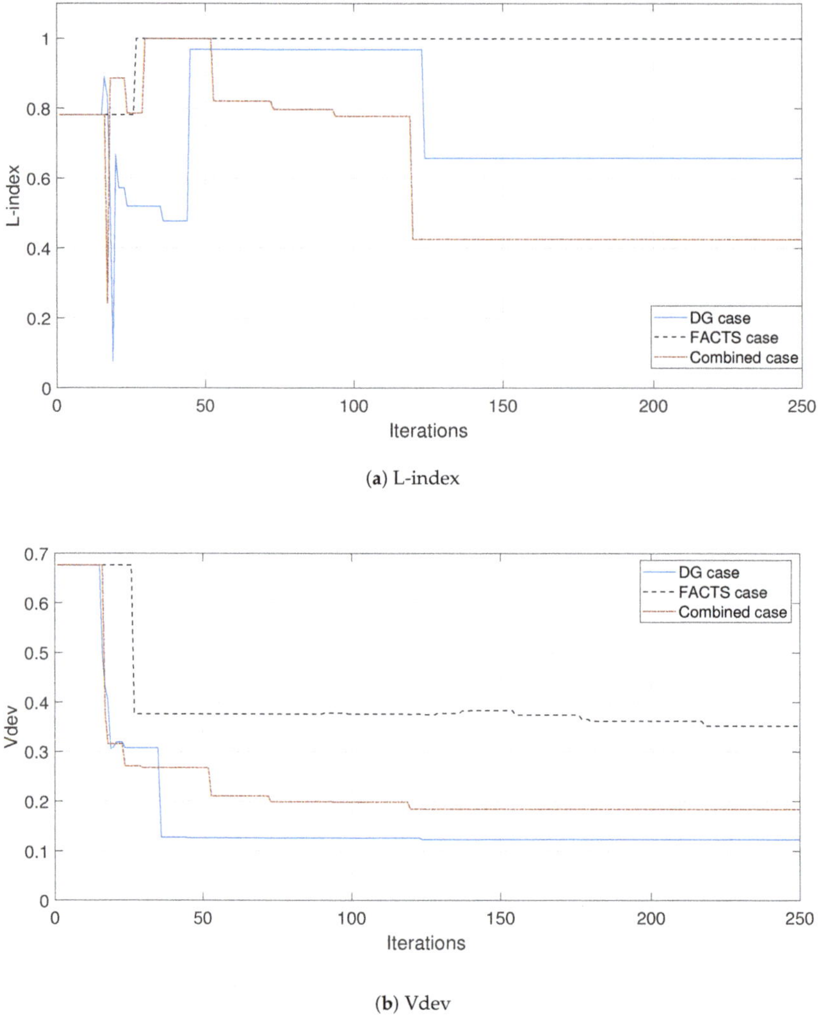

(**a**) L-index

(**b**) Vdev

Figure 5. Evolution of technical parameters in the MOTS search.

4. Conclusions

In this work, the capability of the proposed MOTS algorithm to find good solutions, in large-scale power systems, is demonstrated. By means of DG units implementation in the power system network, the bus voltage profile is improved, reaching zero violations. On the other hand, the add-only FACTS strategy is not able to reach zero bus voltage violations. Finally, the combination strategy with the installation of FACTS and DG units into the power system delivers better results than the other strategies analysed in this work. An excessive use of only FACTS devices or only DG units does not guarantee to enhance the power system stability nor to achieve power losses cost minimization; meanwhile, the combination of them reduces the total investment with improved expenses reduction. These conclusions are confirmed in the results presented in Table 4. The investment necessary to reach the solution of the DG units case is the lowest of the three cases, and an important amount of savings is obtained. However, it is not the best solution, because the solution of the Combined devices case, with only double the investment, allows to further decrease the generation cost and therefore amortize the investment, obtaining slightly higher savings than the DG units case. The FACTS devices case generates savings, but considerably less so than the other options. Maybe it could also be emphasized that with DG units alone, the saturation point is very close to the proposed solution, so that any generation cost improvement will require large investment costs, while the combined case allows for the reduction in generation cost to be more proportional to the investment cost.

These results are in line with the conclusions obtained in other papers. Singh's review of DG impact in power systems [83] indicates that installing DG improves the voltage profile and reduces losses, among other benefits, but also warns that high DG penetration causes network problems. The authors of [84] comment that DG is advantageous over FACTS to improve the voltage profile. And in the papers [10,85], the advantages of the joint installation of DG and FACTS are discussed.

Below the most important practical information that can be extracted from this paper is summarized:

- DG installation is the cheapest measure that can be used in unsaturated networks, both to improve the voltage profile and to reduce losses.
- The installation of FACTS improves the network, but at higher prices.
- When the network is saturated with DG, it is also necessary to use FACTS to improve the network in terms of losses, voltages or even to improve stability.
- Locating multiple DG and FACTS units in a network is a complex problem that needs the help of specialized algorithms.

Author Contributions: Conceptualization, O.A.C.d.K. and J.A.D.-N.; methodology, O.A.C.d.K.; software, O.A.C.d.K.; writing—original draft preparation, O.A.C.d.K.; writing—review and supervision, J.A.D.-N. All authors have read and agreed to the published version of the manuscript.

Funding: This research received no external funding.

Conflicts of Interest: The authors declare no conflict of interest.

Appendix A. FACTS and DG Coefficients

Table A1. Cost functions coefficients for FACTS devices.

Device	F_1	F_2	F_3
HVDC	0.1576	711.9	26,330
TCSC	0.0015	−0.7130	153.75
STATCOM	0.0001	−0.185	158
SSSC	0.00039	−0.3245	173.42
SVC	0.0003	−0.3051	127.38
UPFC	0.0003	−0.2691	188.22

Table A2. Coefficients for active power generation cost depending on their production capacity.

$P_{G,i}^{max}$	$P_{G,i}^{min}$	$Q_{G,i}^{max}$	$P_{G,i}^{min}$	α_2	α_1	α_0
MW	MW	MVAr	MVAr	$/MW^{2}h	$/MWh	$/h
250	45	150	−100	0.000820	11.00	692.32
250				0.11000	5.00	150
250				0.12250	1.00	33
200				0.08500	1.50	0
150	15	50	−40	0.000776	12.00	692.32
118.9				0.07000	30.00	0
55				0.00830	3.25	0
30				0.02500	3.00	0
8.0				0.00040	24.30	0
3.6	0.5	1.5	−0.4	0.0037	2	18

Table A2. *Cont.*

$P_{G,i}^{max}$	$P_{G,i}^{min}$	$Q_{G,i}^{max}$	$P_{G,i}^{min}$	α_2	α_1	α_0
MW	**MW**	**MVAr**	**MVAr**	**$/MW^{2}h**	**$/MWh**	**$/h**
3.0				0.00060	29.10	0
2.7	0.2	1.0	−0.8	0.01	10	100
2.5	0.2	0.8	−0.8	0.05	30	100
2.0	0.3	0.5	−0.5	0.02	15	100
2.0	0.2	0.8	−0.8	0.05	30	100
2.0	0.4	0.7	−0.7	0.03	20	100
2.0				0.00150	50.00	0
1.4	0.2	0.6	−0.4	0.0175	1	16
1.2				0.00067	6.90	0
1.0	0.15	0.4	−0.4	0.0625	1	14
1.0	0.10	0.4	−0.1	0.0083	3.25	12
1.0	0.10	0.24	−0.6	0.025	3	13
0.8				0.00026	6.90	0

References

1. Song, Y.H.; Johns, A.T. *Flexible AC Transmission Systems (FACTS)*; Institution of Engineering and Technology: London, UK, 1999.
2. Borbely, A.M.; Kreider, J.F. *Distributed Generation: The Power Paradigm for the New Millennium*; Mechanical and Aerospace Engineering Series; CRC Press: Boca Raton, FL, USA, 2001.
3. Acha, E.; Fuerte-Esquivel, C.R.; Ambriz-Pérez, H.; Angeles-Camacho, C. *FACTS: Modelling and Simulation in Power Networks*; John Wiley & Sons, Ltd.: Hoboken, NJ, USA, 2004. [CrossRef]
4. Vijayakumar, K.; Kumudinidevi, R.P. A new method for locating TCSC for congestion management in deregulated electricity markets. *Int. J. Power Energy Convers.* **2009**, *1*, 313–326. [CrossRef]
5. Pepermans, G.; Driesen, J.; Haeseldonckx, D.; Belmans, R.; D'haeseleer, W. Distributed generation: Definition, benefits and issues. *Energy Policy* **2005**, *33*, 787–798. [CrossRef]
6. Smallwood, C. Distributed generation in autonomous and nonautonomous micro grids. In Proceedings of the 2002 Rural Electric Power Conference. Papers Presented at the 46th Annual Conference (Cat. No. 02CH37360), Colorado Springs, CO, USA, 5–7 May 2002.
7. Hatziargyriou, N.; Meliopoulos, A. Distributed energy sources: Technical challenges. In Proceedings of the 2002 IEEE Power Engineering Society Winter Meeting. Conference Proceedings (Cat. No.02CH37309), New York, NY, USA, 27–31 January 2002.
8. Lasseter, R.; Piagi, P. Microgrid: A conceptual solution. In Proceedings of the 2004 IEEE 35th Annual Power Electronics Specialists Conference (IEEE Cat. No.04CH37551), Aachen, Germany, 20–25 June 2004.
9. Tian, W.; Zhang, Y.; Fu, R.; Zhao, Y.; Wang, G.; Winter, R. Modeling and Control Architecture of Source and Load Management in Islanded Power Systems. In Proceedings of the 2015 IEEE Energy Conversion Congress and Exposition (ECCE), Montreal, QC, Canada, 20–24 September 2015.
10. Singh, B.; Mukherjee, V.; Tiwari, P. A survey on impact assessment of DG and FACTS controllers in power systems. *Renew. Sustain. Energy Rev.* **2015**, *42*, 846–882. [CrossRef]
11. Zhang, X.P.; Rehtanz, C.; Pal, B. *Flexible AC Transmission Systems: Modelling and Control (Power Systems)*; Springer: Berlin, Germany, 2006.
12. Padiyar, K.R. *FACTS Controllers in Power Transmission and Distribution*; New Age International (P) Limited, Publishers: Delhi, India, 2007.
13. Devaraj, D.; Roselyn, P. On-line voltage stability assessment using radial basis function network model with reduced input features. *Int. J. Electr. Power Energy Syst.* **2011**, *33*, 1550–1555. [CrossRef]
14. Tomin, N.; Zhukov, A.; Kurbatsky, V.; Sidorov, D.; Negnevitsky, M. Development of automatic intelligent system for on-line voltage security control of power systems. In Proceedings of the 2017 IEEE Manchester PowerTech, Manchester, UK, 18–22 June 2017. [CrossRef]
15. Satheesh, A. Maintaining power system stability with FACTS controller using bees algorithm and NN. *J. Theor. Appl. Inf. Technol.* **2013**, *49*, 38–47.

16. Acharya, N.; Sode-Yome, A.; Nadarajah, M. Facts about flexible AC transmission systems (FACTS) controllers: Practical installations and benefits. In Proceedings of the Australian Universities Power Engineering Conference (AUPEC), Melbourne, Australia, 25–28 September 2005; Volume 2, p. 6.
17. Nagesh, H.B.; Puttaswamy, P.S. Enhancement of voltage stability margin using FACTS controllers. *Int. J. Comput. Electr. Eng.* **2013**, *5*, 261–265. [CrossRef]
18. Habur, K.; O'Leary, D. FACTS for Cost Effective and Reliable Transmission of Electrical Energy Siemens-World Bank document–Final Draft Report, Erlangen. 2004. Available online: https://sites.google.com/site/lawking/facts_siemens.pdf (accessed on 20 February 2020).
19. Rashed, G.I.; Shaheen, H.I.; Cheng, S.J. Optimal location and parameter settings of multiple TCSCs for increasing power system loadability based on GA and PSO techniques. In Proceedings of the Third International Conference on Natural Computation (ICNC 2007), Haikou, China, 24–27 August 2007; Volume 1, pp. 1–10. [CrossRef]
20. Saravanan, M.; Slochanal, S.M.R.; Venkatesh, P.; Abraham, J.P.S. Application of PSO technique for optimal location of FACTS devices considering cost of installation and system loadability. *Electr. Power Syst. Res.* **2007**, *77*, 276–283. [CrossRef]
21. El Metwally, M.M.; El Emary, A.A.; El Bendary, F.M.; Mosaad, M.I. Optimal allocation of FACTS devices in power system using genetic algorithms. In Proceedings of the 2008 12th International Middle-East Power System Conference, Aswan, Egypt, 12–15 March 2008; pp. 1–4. [CrossRef]
22. Idris, R.M.; Khairuddin, A.; Mustafa, M.W. A multi-objective Bees Algorithm for optimum allocation of FACTS devices for restructured power system. In Proceedings of the TENCON 2009 IEEE Region 10 Conference, Singapore, 23–26 January 2009; pp. 1–6. [CrossRef]
23. Peikherfeh, M.; Abapour, M.; Moghaddam, M.P.; Namdari, A. Optimal allocation of FACTS devices for provision of voltage control ancillary services. In Proceedings of the 2010 7th International Conference on the European Energy Market, Madrid, Spain, 23–25 June 2010; Volume 23, pp. 1–5. [CrossRef]
24. Ara, A.L.; Kazemi, A.; Niaki, S.A.N. Multiobjective Optimal Location of FACTS Shunt-Series Controllers for Power System Operation Planning. *IEEE Trans. Power Del.* **2012**, *27*, 481–490. [CrossRef]
25. Deb, T.; Siddiqui, A.S. Congestion management through optimal placement of SSSC using modified gravitational search algorithm. *Am. Int. J. Res. Sci. Technol. Eng. Math. (AIJRSTEM)* **2017**, *17*, 66–69.
26. Pisica, I.; Bulac, C.; Toma, L.; Eremia, M. Optimal SVC placement in electric power systems using a genetic algorithms based method. In Proceedings of the 2009 IEEE Bucharest PowerTech, Bucharest, Romania, 28 June–2 July 2009; pp. 1–6. [CrossRef]
27. Chang, Y.C. Multi-objective optimal SVC installation for power system loading margin improvement. *IEEE Trans. Power Syst.* **2012**, *27*, 984–992. [CrossRef]
28. Taher, S.A.; Afsari, S.A. Optimal location and sizing of DSTATCOM in distribution systems by immune algorithm. *Int. J. Electr. Power Energy Syst.* **2014**, *60*, 34–44. [CrossRef]
29. Baghaee, H.R.; Kaviani, A.; Mirsalim, M.; Gharehpetian, B.G. Short circuit level and loss reduction by allocating TCSC and UPFC using particle swarm optimization. In Proceedings of the 2011 19th Iranian Conference on Electrical Engineering, Tehran, Iran, 17–19 May 2011; pp. 1–6.
30. Basu, M. Optimal power flow with FACTS devices using differential evolution. *Int. J. Electr. Power Energy Syst.* **2008**, *30*, 150–156. [CrossRef]
31. Singh, R.P.; Mukherjee, V.; Ghoshal, S.P. Particle swarm optimization with an aging leader and challengers algorithm for optimal power flow problem with FACTS devices. *Int. J. Electr. Power Energy Syst.* **2015**, *64*, 1185–1196. [CrossRef]
32. Mukherjee, A.; Mukherjee, V. Solution of optimal power flow with FACTS devices using a novel oppositional krill herd algorithm. *Int. J. Electr. Power Energy Syst.* **2016**, *78*, 700–714. [CrossRef]
33. John Grainger, W.S.J. *Power System Analysis*; McGraw-Hill series in electrical and computer engineering: Power and energy; McGraw-Hill: New York, NY, USA, 1994.
34. Weedy, B.M.; Cory, B.J.; Jenkins, N.; Ekanayake, J.B.; Strbac, G. *Electric Power Systems*; John Wiley and Sons Ltd.: West Sussex, UK, 2012.
35. Yorino, N.; El-Araby, E.E.; Sasaki, H.; Harada, S. A new formulation for FACTS allocation for security enhancement against voltage collapse. *IEEE Trans. Power Syst.* **2003**, *18*, 3–10. [CrossRef]

36. Cai, L.J.; Erlich, I.; Stamtsis, G. Optimal choice and allocation of FACTS devices in deregulated electricity market using genetic algorithms. In Proceedings of the IEEE PES Power Systems Conference and Exposition, New York, NY, USA, 10–13 October 2004; pp. 201–207. [CrossRef]
37. Gitizadeh, M.; Kalantar, M. A new approach for congestion management via optimal location of FACTS devices in deregulated power systems. In Proceedings of the Third International Conference on Electric Utility Deregulation and Restructuring and Power Technologies, Nanjing, China, 6–9 April 2008; pp. 1592–1597. [CrossRef]
38. Gitizadeh, M.; Kalantar, M. Genetic algorithm based fuzzy multi-objective approach to FACTS devices allocation in FARS regional electric network. *Int. J. Sci. Technol.* **2008**, *15*, 534–546.
39. Gitizadeh, M. Allocation of multi-type FACTS devices using multi-objective genetic algorithm approach for power system reinforcement. *Electr. Eng.* **2010**, *92*, 227–237. [CrossRef]
40. Benabid, R.; Boudour, M.; Abido, M.A. Optimal location and setting of SVC and TCSC devices using non-dominated sorting particle swarm optimization. *Electr. Power Syst. Res.* **2009**, *79*, 1668–1677. [CrossRef]
41. Benabid, R.; Boudour, M.; Abido, M.A. Optimal placement of FACTS devices for multi-objective voltage stability problem. In Proceedings of the IEEE/PES Power Systems Conference and Exposition, Seattle, WA, USA, 15–18 March 2009. [CrossRef]
42. Wibowo, R.S.; Yorino, N.; Eghbal, M.; Zoka, Y.; Sasaki, Y. FACTS devices allocation with control coordination considering congestion relief and voltage stability. *IEEE Trans. Power Syst.* **2011**, *26*, 2302–2310. [CrossRef]
43. Laïfa, A.; Medoued, A. Optimal FACTS location to enhance voltage stability using multi-objective harmony search. In Proceedings of the 3rd International Conference on Electric Power and Energy Conversion Systems, Istanbul, Turkey, 2–4 October 2013; Volume 1, p. 6. [CrossRef]
44. Bhattacharyya, B.; Kumar, S. Reactive power planning with FACTS devices using gravitational search algorithm. *Ain Shams Eng. J.* **2015**, *6*, 865–871. [CrossRef]
45. Mohammadi, M.; Rezazadeh, A.; Sedighizadeh, M. Optimal placement and sizing of FACTS devices for loadability enhancement in deregulated power systems. *Recent Re. Artif. Intell. Database Manag.* **2019**, *12*, 148–156.
46. Sugimoto, J.; Yokoyama, R.; Niimura, T.; Fukuyama, Y. Tabu search based-optimal allocation of voltage control devices by connections ofdistributed generators in distribution systems. In Proceedings of the 39th International Universities Power Engineering Conference, UPEC 2004, Bristol, UK, 6–8 September 2004.
47. Bhasaputra, P.; Ongsakul, W. Optimal placement of multi-type FACTS devices by hybrid TS/SA approach. In Proceedings of the 2003 International Symposium on Circuits and Systems, 2003. ISCAS '03, Bangkok, Thailand, 25–28 May 2003.
48. Chansareewittaya, S.; Jirapong, P. Total transfer capability enhancement with optimal number of FACTS controllers using hybrid TSSA. In Proceedings of the IEEE Southeastcon, Orlando, FL, USA, 15–18 March 2012.
49. Mori, H.; Tani, H. Two-staged tabu search for determining optimal allocation of D-FACTS in radial distribution systems with distributed generation. In Proceedings of the IEEE/PES Transmission and Distribution Conference and Exhibition, Yokohama, Japan, 6–10 October 2002.
50. Ebrahimi, S.; Farsangi, M.; Nezamabadi-Pour, H.; Lee, K. Optimal allocation of static VAr compensators using modal analysis, simulatedannealing and tabu search. In Proceedings of the IFAC Proceedings Volumes, Kananaskis, AB, Canada, 25–28 June 2006.
51. Maciel, R.; Padilha-Feltrin, A. Distributed Generation Impact Evaluation Using a Multi-Objective Tabu Search. In Proceedings of the 2009 15th International Conference on Intelligent System Applications to Power Systems, Curitiba, Brazil, 8–12 November 2009.
52. Milano, F. An Open Source Power System Analysis Toolbox. *Trans. Power Syst.* **2005**, *20*, 1199–1206. [CrossRef]
53. Glover, F.W.; Laguna, M. *Tabu Search*; Number 1 in Tabu Search; Springer: Berlin, Germany, 1998. [CrossRef]
54. Pereira, B.R.; Cossi., A.M.; Contreras, J.; Mantovani, J.R.S. Multiobjective multistage distribution system planning using tabu search. *IET Gener. Transm. Distrib.* **2013**, *8*, 35–45. [CrossRef]
55. Singh, S.N.; David, A.K. Placement of FACTS devices in open power market. In Proceedings of the APSCOM 2000—5th International Conference on Advances in Power System Control, Operation and Management. Institution of Engineering and Technology, Hong Kong, China, 30 October–1 November 2000; pp. 173–177. [CrossRef]

56. Singh, S.N.; David, A.K. A new approach for placement of FACTS devices in open power markets. *IEEE Power Eng. Rev.* **2001**, *21*, 58–60. [CrossRef]
57. Ippolito, L.; Siano, P. Selection of optimal number and location of thyristor-controlled phase shifters using genetic based algorithms. *IEE Proc.-Gener. Transm. Distrib.* **2004**, *151*, 630–637. [CrossRef]
58. Ippolito, L. A novel strategy for selection of the optimal number and location of UPFC devices in deregulated electric power systems. In Proceedings of the 2005 IEEE Russia Power Tech, St. Petersburg, Russia, 27–30 June 2005; pp. 1–9. [CrossRef]
59. Ippolito, L.; Cortiglia, A.L.; Petrocelli, M. Optimal allocation of FACTS devices by using multi-objective optimal power flow and genetic algorithms. *Int. J. Emerg. Electr. Power Syst.* **2006**, *7*. [CrossRef]
60. Eghbal, M.; Yorino, N.; Zoka, Y. Comparative study on the application of modern heuristic techniques to SVC placement problem. *J. Comput.* **2009**, *4*. [CrossRef]
61. Joorabian, M.; Saniei, M.; Sepahvand, H. Locating and parameters setting of TCSC for congestion management in deregulated electricity market. In Proceedings of the 6th IEEE Conference on Industrial Electronics and Applications, Beijing, China, 21–23 June 2011; Volume 1, pp. 2185–2190. [CrossRef]
62. Taher, S.A.; Amooshahi, M.K. New approach for optimal UPFC placement using hybrid immune algorithm in electric power systems. *Int. J. Electr. Power Energy Syst.* **2012**, *43*, 899–909. [CrossRef]
63. Iqbal, F.; Khan, M.T.; Siddiqui, A.S. Optimal placement of DG and DSTATCOM for loss reduction and voltage profile improvement. *Alex. Eng. J.* **2017**, *57*, 755–765. [CrossRef]
64. Hartel, P.; Vrana, T.K.; Hennig, T.; von Bonin, M.; Wiggelinkhuizen, E.J.; Nieuwenhout, F.D.J. Review of investment model cost parameters for VSC HVDC transmission infrastructure. *Electr. Power Syst. Res.* **2017**, *151*, 419–431. [CrossRef]
65. Alabduljabbar, A.A.; Milanovic, J.V. Genetic algorithm based optimization for allocation of static VAr compensators. In Proceedings of the 8th IEE International Conference on AC and DC Power Transmission (ACDC 2006), London, UK, 28–31 March 2006; Volume 1, pp. 115–119. [CrossRef]
66. Chandrasekaran, K.; Jeyaraj, K.A.; Sahayasenthamil, L.; Saravanan, M. A new method to incorporate FACTS devices in optimal power flow using particle swarm optimization. *J. Theor. Appl. Inf. Technol.* **2009**, *5*, 67–74.
67. Nabavi, S.M.; Hajforoosh, S.; Hajforoosh, S.; Karimi, A.; Khafafi, K. Maximizing the overall satisfaction degree of all participants in the market using real code-based genetic algorithm by optimally locating and sizing the thyristor-controlled series capacitor. *J. Electr. Eng. Technol.* **2011**, *6*, 493–504. [CrossRef]
68. Zhong, J.; Bhattacharya, K. Toward a competitive market for reactive power. *IEEE Trans. Power Syst.* **2002**, *17*, 1206–1215. [CrossRef]
69. An, S.; Condren, J.; Gedra, T.W. An ideal transformer UPFC model, OPF first-order sensitivities, and application to screening for optimal UPFC locations. *IEEE Trans. Power Syst.* **2007**, *22*, 68–75. [CrossRef]
70. Blanco, G.; Olsina, F.; Garces, F.; Rehtanz, C. Real option valuation of FACTS investments based on the least square Monte Carlo method. *IEEE Trans. Power Syst.* **2011**, *26*, 1389–1398. [CrossRef]
71. Niknam, T.; Narimani, M.; Aghaei, J.; Azizipanah-Abarghooee, R. Improved particle swarm optimisation for multi-objective optimal power flow considering the cost, loss, emission and voltage stability index. *IET Gener. Transm. Distrib.* **2012**, *6*, 515. [CrossRef]
72. Tlijani, K.; Guesmi, T.; Abdallah, H.H.; Ouali, A. 0ptimal location and parameter setting of TCSC based on sensitivity analysis. In Proceedings of the 2012 First International Conference on Renewable Energies and Vehicular Technology, Hammamet, Tunisia, 26–28 March 2012; Volume 1, pp. 420–424. [CrossRef]
73. INEGI. Economic Information Bank. 2020. Available online: https://www.inegi.org.mx/sistemas/bie/ (accessed on 20 February 2020).
74. Chen, L.; Zhong, J.; Gan, D. Reactive power planning and its cost allocation for distribution systems with distributed generation. In Proceedings of the 2006 IEEE Power Engineering Society General Meeting, Montreal, QC, Canada, 18–22 June 2006; p. 6. [CrossRef]
75. CFE. Price per Requested Work. 2020. Available online: https://app.cfe.mx/Aplicaciones/OTROS/Aportaciones/MenuAportaciones.aspx (accessed on 20 February 2020).
76. Milano, F. *Power System Modelling and Scripting*; Power Systems; Springer: Berlin/Heidelberg, Germany, 2010. [CrossRef]
77. Milano, F. Power system analysis toolbox (PSAT). Available online: http://faraday1.ucd.ie/psat.html (accessed on 6 March 2019).
78. Matlab. *Matlab Documentation*; Mathworks: Natick, MA, USA, 2017.

79. Papachristoudis, G. Kruskal Algorithm. Mathworks File Exchange. 2014. Available online: https://la.mathworks.com/matlabcentral/fileexchange/41963-kruskal-s-algorithm (accessed on 20 February 2020).
80. Arcia-Garibaldi, G.; Cruz-Romero, P.; Gómez-Expósito, A. Future power transmission: Visions, technologies and challenges. *Renew. Sustain. Energy Rev.* **2018**, *94*, 285–301. [CrossRef]
81. University of Washington. Power Systems Tests Case Archive. 2020. Available online: https://labs.ece.uw.edu/pstca/pf300/pg_tca300bus.htm (accessed on 20 February 2020).
82. Kessel, P.; Glavitsch, H. Estimating the voltage stability of a power system. *IEEE Trans. Power Del.* **1986**, *1*, 346–354. [CrossRef]
83. Singh, B.; Sharma, J. A review on distributed generation planning. *Renew. Sustain. Energy Rev.* **2017**, *76*, 529–544. [CrossRef]
84. Chatterjee, S.; Nath, P.; Biswas, R.; Das, M. Advantage of DG for improving voltage profile over facts devices. *Int. J. Eng. Res. Appl.* **2013**, *3*, 2029–2032.
85. Bharathi Dasan, S.; Gayatri, S.; Lavanya, V.; Praveena, N.; Preethi, T. FACTS based voltage enhancement in hybrid DG distribution system with mixed load model. *J. Electr. Eng.* **2012**, *12*, 188–196.

Article

A Gini Coefficient-Based Impartial and Open Dispatching Model

Liang Sun [1,2], Na Zhang [3], Ning Li [2], Zhuo-ran Song [4] and Wei-dong Li [1,*]

1 School of Electrical Engineering, Dalian University of Technology, Dalian 116024, China; sunliang_3333@sina.com
2 State Grid Shenyang Electric Power Supply Company, Shenyang 110081, China; ln_dut@163.com
3 State Grid Liaoning Economic Research Institute, Shenyang 110015, China; zn_jyy@126.com
4 State Grid Liaoning Electric Power Company Limited, Shenyang 110006, China; zrsong_sg@126.com
* Correspondence: wdli@dlut.edu.cn; Tel.: +86-1394-269-8900

Received: 19 April 2020; Accepted: 5 June 2020; Published: 17 June 2020

Abstract: According to the existing widely applied impartial and open dispatching models, operation fairness was mainly emphasized, which severely restricted the optimization space of the economy of the overall system operation and affected the economic benefits. To solve the above problems, a scheduling model based on Gini coefficient under impartial and open dispatching principle is proposed in this paper, which can consider the balance between the fairness and economy of system operation. In the proposed model, the Gini coefficient is introduced to describe the fairness of electric energy completion rate among different generation units in the form of constraint conditions. Because the electricity production schedule can reflect the economic income of the electric power enterprise, and the Gini coefficient is used as an economic statistical indicator to evaluate the fairness in the overall distribution of income in social statistics, it is more appropriate to be used to measure the fairness of the power generation dispatching. The objective of the proposed model is to minimize the total operation costs. In the model, the balance between the system operation economy and fairness can be realized by adjusting the Gini coefficient value. The simulation results show that the proposed model is an extension of the traditional model. Compared with the traditional economic dispatching model and normal "impartial and open dispatching" model, the proposed model can better coordinate the relationship between fairness and economy. It could provide more choices for power generation dispatchers. It could also provide a reference for regulatory departments to formulate relevant policies by adjusting the threshold value of the Gini coefficient. Case studies show that the power dispatching decisions according to the proposed model can provide a scientific and fair reference basis for dispatching schemes, and could reduce the generation costs and also achieve optimal allocation of resources on the basis of ensuring fair dispatching.

Keywords: impartial and open dispatching; economy; Gini coefficient; generation scheduling; mixed integer quadratic programming

1. Introduction

In the past several decades, impartial and open dispatching has generally been implemented in some countries like China, although power dispatching must be guaranteed for it to be open and fair. The assessed object of the impartial and open dispatching is the completion progress of the annual power generation plan issued by the government department, which requires the annual electric energy production completion rate of all the power plants to be approximately equal [1,2]. According to the national regulation, deviation of no more than 2–3% between the annual schedule is required.

The impartial and open principle should be felicitously considered in the power system scheduling process. The unit's operation adjustment space would be limited strictly if the impartial and open dispatching principle is excessive emphasized, resulting in higher power generation costs of the power system. On the contrary, the excessive emphasis on operational economy will result in a serious imbalance among the units' power generation progresses, leading to conflicts and disputes among various power generation producers, which may disrupt the generation order, but also reduce market efficiency. Therefore, in the dispatching, the reliability, economy, and fairness of the system operation should be comprehensively considered to reasonably arrange the unit commitment and load distribution schedule of the power units [3–6].

Literature Review and Discussion

Many researchers all over the world have studied the electricity dispatching model. In past decades, an hourly scheduling model in a local day-ahead electricity market was presented in [7], to maximize the operational profits managed by an energy service company. A two-stage robust scheduling approach for a hydrothermal power system was studied in [8]. In [9], a day ahead optimal scheduling model of generators using dynamic programming method was proposed. In [10–12], an optimal generation scheduling strategy in micro-grids was studied. The studies above studied the scheduling problem under different requirements using different methods. Nevertheless, the generation fairness was not considered. The preparation process and allocation strategy for annual generation schedules based on the annual utilization hours under the impartial and open dispatching principle have been proposed in [13]. Scheduling models according to the impartial and open dispatching principle were presented in [14,15], respectively, aiming at minimizing the deviation between the actual output and the target output, and minimizing the deviation between the annual planned electricity energy and the contracted electric energy. Multi-objective optimization models were presented in [16–18], considering the economy of the system operation on the basis of the impartial and open dispatching principle, with the objective of minimizing the power generation costs and minimizing the contracted energy deviation. Fuzzy methods were used to solve the model. In the existing methods mentioned above, the fairness is considered using the standard deviation index of the actual electricity and the planned electricity.

The issue of fairness involves many fields such as philosophy, ethics, politics, society, economy, and so on. The criteria for measuring fairness are also difficult to determine. Although, in the existing studies, the indicator constructed by the standard deviation can be used to measure the difference among units' generation completion progresses [13–19], the threshold for judging fairness is difficult to determine, which leads to the following shortcomings:

(1) The fairness of the unit's generation schedule is over-emphasized, so that the system operation economic optimization space is reduced, which is not conducive to the resources optimizing configuration [20];
(2) There is a lack of coordination between fairness and economy, which limits the diversity of the optimization objectives of the dispatching. As a result, the different fairness requirement in the different processes of scheduling is difficult to meet.

Among the above analyses, it is not difficult to find that existing studies usually consider only fairness or safety. Especially for the fairness of scheduling, some fairness indicators are also mentioned in many literature, but the determination of the fairness index threshold still needs to be further considered, and there is no authoritative definition. So, whether or not fairness is justified remains to be discussed.

Currently, in the international arena, the Gini coefficient is used to comprehensively examine the differences in income distribution among residents as an important analytical indicator. Through theoretical research and practice for more than half a century, there has been a basic consensus on the corresponding relationship between the values of the Gini coefficient and the levels of fairness. The Gini coefficient has been widely applied in many fields including traffic systems [21,22], environmental resources [23], and power systems [24,25]. An indicator to measure the fairness of the annual generation schedule based on the Gini coefficient was proposed in [25]. The correlation and difference between the Gini coefficient and the standard deviation were analyzed in [26].

According to China's current electricity pricing system, the electric energy charge income is the main economic source of generation units. Therefore, the Gini coefficient is appropriate to be introduced into the impartial and open dispatching as a parameter to measure fairness, and the existing studies and practical results can be used as a basis for measuring the level of fairness when applying the Gini coefficient index. On the basis of the above understanding, an impartiality and openness of power dispatching model is proposed in this paper based on the Gini coefficient. According to the proposed model:

(1) The differences among units' generation energy progresses can be guaranteed in the specified range, which could give consideration to the interests of all the units;
(2) The balance between economy and fairness can be achieved by adjusting the value of the Gini coefficient, thereby the economic optimizing space in the dispatching could be effectively expanded, and the overall operational economy is improved.

The model proposed in this paper can provide more choices and solutions for the dispatching department, and also provide a reference for the regulatory authorities to formulate relevant policies. Avoid situations where dispatchers make power generation plans based on experience, and the power generation plan is more scientific and accurate. At the same time, the system operation economy, reliability, and fairness could be comprehensively considered using the method proposed in this paper, and the model reduces the total cost of power generation effectively, so that the fairness of the generation progresses of units could be ensured while the overall resources optimizing configuration is implemented.

2. Impartial and Open Dispatching and Gini Coefficient

2.1. Impartial and Open Dispatching

Impartial and open dispatching refers to electric power dispatching institutions following the national laws and regulations and treating all market entities equally in terms of dispatching management and information disclosure in accordance with the principle of fairness and transparency on the premise of satisfying the safety, stability, and economic operation of the power system. Impartial and open dispatching should follow the following principles:

(1) Abide by the relevant laws and regulations of the state, implement the national energy, environmental and industrial policies, and conscientiously implement the relevant national and industrial standards and regulations;
(2) Ensure the safety, quality, and economic operation of the power system and give full play to its capabilities to meet the needs of the community;
(3) Safeguard the legitimate rights and interests of power production enterprises, power grid operating enterprises, and power users;
(4) Give full play to the role of the market in regulating the allocation of electricity resources.

Impartial and open dispatching is the basic principle and working goal of power dispatching, which is of great significance to guarantee the economic interests of power generation enterprises. At present, according to the impartial and open dispatching, it requires that the power dispatching organization should make rolling adjustments to the completion progress of the annual electricity purchase contracts, and the completion progress of annual generation contracts in the same grid should be roughly equivalent [27].

2.2. Gini Coefficient

2.2.1. Concept of Gini Coefficient

The Gini coefficient is an economic statistics indicator that reflects the fairness in general, which is used to reflect whether the distribution of income in social statistics is fair [28,29]. The Lorenz curve is the basis of the Gini coefficient, as shown in Figure 1.

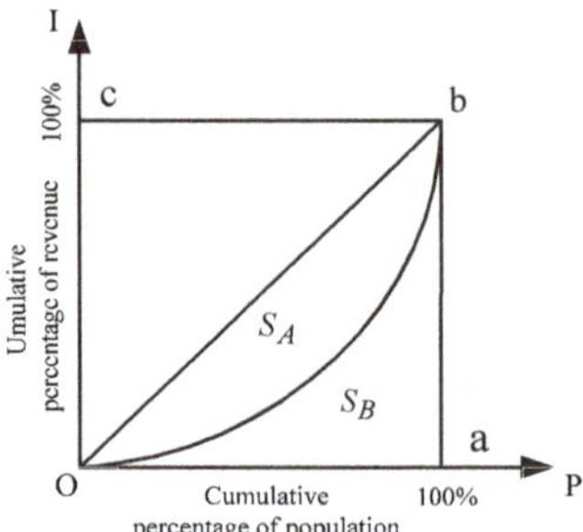

Figure 1. Lorenz curve.

On the basis of the Lorenz curve, the Gini coefficient is as follows:

$$G = \frac{S_A}{S_A + S_B} = \frac{2S_A}{2(S_A + S_B)} = \frac{2S_A}{1} = 1 - 2S_B \quad (1)$$

The value of the Gini coefficient is equal to twice the value of the area, and is converted to the value of area for the convenience of calculation. It can be found from Equation (1) that, when the area of is zero, the allocation is absolutely fair. The larger the area, the greater the unevenness of the distribution. It is generally believed that, if the Gini coefficient is less than 0.4, it means fairness is basically reached, while if the Gini coefficient is less than 0.3, it means fairness is better.

The calculation methods of Gini coefficient mainly include the direct calculation method, regression curve method, cardinal equal division method, and group decomposition method [19]. Considering the characteristics of the above calculation methods and practical applications, the direct calculation method is applied in this paper, which does not depend on the Lorenz curve. The direct calculation method can ensure that the calculated Gini coefficient value is completely true and accurate without any errors.

2.2.2. Characteristics of the Gini Coefficient

As an index to measure fairness, the Gini coefficient mainly has the following characteristics.

The fairness based on the Gini coefficient does not require all data samples to be identical, but meets certain fairness conditions on the whole, reflecting the overall fairness of samples.

(1) The Gini coefficient is of good observability, that is, fairness can be determined without comparing any data. When evaluating the fairness, the Gini coefficient ranges from 0 to 1, and the closer it is to 0, the fairer the distribution will be.
(2) The calculation process of Gini coefficient is the processing of all data collection, which is more universal and reasonable than some other fair indicators such as extreme poor.
(3) The Gini coefficient is an important international analysis index used to comprehensively investigate the income distribution differences among residents. It is an internationally recognized economic statistical index that can reasonably reflect the overall fairness.

According to the current electricity pricing system in China, the amount of power generation determined by the generation schedule directly determines the operating income of the units.

The completing rate of progress of each generator is determined by the power system dispatching. Therefore, it is appropriate to apply the Gini coefficient to the generation dispatching model in the power system to measure the equilibrium degree of economic income among the generating companies.

3. Mathematical Model

3.1. Mathematical Modeling

In order to comprehensively coordinate the two indicators, fairness and economy, the constraint method for mathematical modeling is used in this paper. This means the power generation costs are set as the objective function, and the fairness indicator is set as a constraint [30]. Using this method, the economic optimization space of the whole system can be increased to a certain extent, so as to improve the economic benefit of the whole system, on the basis of ensuring that the completion rate of each power generation unit is within the specified range, so as to safeguard the relevant interests of all parties.

3.1.1. Objective Function

According to the above modeling concept, the optimization objective of the impartiality and openness power dispatching model is to minimize the total power generation costs.

$$\min f = \sum_{i=1}^{N}\sum_{t=1}^{T}(a_i {p_{i,t}}^2 + b_i p_{i,t} + c_i u_{i,t} + s_{i,t}) \tag{2}$$

where, a_i, b_i, and c_i are the power generation cost coefficients of unit i; $p_{i,t}$ is the scheduled output of unit i at time interval t; $u_{i,t}$ is the commitment state of unit i at time interval t; $s_{i,t}$ is the startup and shutdown costs of unit i at time interval t; N is the total number of units; and T is the total number of time intervals.

3.1.2. Constraint Conditions

(1) The power balance constraint of the system

$$\sum_{i=1}^{N} u_{i,t}\, p_{i,t} - P_t^d = 0 \tag{3}$$

where P_t^d is the total system load demand at time interval t.

(2) The spinning reserve constraints of the system

$$\sum_{i=1}^{N} u_{i,t}\, P_{i,\max} \geq P_t^d + S_t^d \tag{4}$$

where $P_{i,\max}$ is the maximum output of the unit i at time t and S_t^d is the total spare capacity at time t.

(3) The maximum and minimum output constrains

$$u_{i,t}P_{i,\min} \leq p_{i,t} \leq u_{i,t}P_{i,\max} \tag{5}$$

where $P_{i,\max}$ and $P_{i,\min}$ are the maximum and minimum output of the unit i, respectively.

(4) The ramp rate constraints

$$p_{i,t+1} - p_{i,t} \leq u_{i,t} R_{U,i} + p_{i,\max}(1 - u_{i,t}) \tag{6}$$

$$p_{i,t} - p_{i,t+1} \leq u_{i,t+1} R_{D,i} + p_{i,\max}(1 - u_{i,t+1}) \tag{7}$$

where $R_{U,i}$ and $R_{D,i}$ are the maximum rate of upward ramping/downward ramping of unit i in each time interval, respectively.

(5) The minimum startup–shutdown time constraints

$$\begin{cases} (u_{i,t-1} - u_{i,t})\left(T_{i,t-1} - T_i^{\text{on}}\right) \geq 0 \\ (u_{i,t} - u_{i,t-1})\left(-T_{i,t-1} - T_i^{\text{off}}\right) \geq 0 \end{cases} \tag{8}$$

where T_i^{on} and T_i^{off} are the minimum operating time and minimum offtime of the unit i, respectively; and $T_{i,t}$ is the continuous offtime or continuous operating time of the unit i at time t.

(6) The startup–shutdown cost constraint

The start-up costs of cold start and hot start are different. By judging the length of continuous offtime, the specific constraint conditions for cold start and hot start can be determined as follows:

$$s_{i,t} = \begin{cases} C_i^{\text{hot}}, & 1 \leq t_{i,t}^{\text{off}} \leq T_i^{\text{off}} + T_i^{\text{cold}} \\ C_i^{\text{cold}}, & t_{i,t}^{\text{off}} > T_i^{\text{off}} + T_i^{\text{cold}} \end{cases} \tag{9}$$

where $t_{i,t}^{\text{off}}$ is the continuous offtime of unit i until time $t-1$; T_i^{cold} is the cold start time of unit i; and C_i^{hot} and C_i^{cold} are the costs of the unit i in the case of cold start and hot start, respectively.

(7) Generation fairness constraint

The Gini coefficient is introduced to model the fairness constraint of the complementation rate of electricity generation.

$$G = \frac{1}{2N(N-1)u} \sum_{i=1}^{N} \sum_{j=1}^{N} \left| X_i - X_j \right| \leq G_0 \tag{10}$$

where X_i and X_j are the scheduled power completion rates of unit i and unit j, respectively; u is the average power completion rates of all units; and G_0 is the threshold value of the Gini coefficient, which can be set according to the actual fairness requirement, and the value range is 0~1.

$$X_i = \frac{Q_i}{Q_i^0} = \sum_{t=1}^{T} P_{i,t} / \left(P_i^{\max} \times t_{\text{sh}}\right) \tag{11}$$

where Q_i^0 and Q_i are the completed electric energy according to the dispatching results and the daily planned energy of unit i, respectively.

$$u = \sum_{i=1}^{N} X_i / N \tag{12}$$

Substituting (12) into (10), the final fairness constraint formulation (13) can be obtained after shifting and simplifying.

$$\sum_{i=1}^{N} \sum_{j=1}^{N} \left| X_i - X_j \right| \leq 2G_0(N-1) \sum_{i=1}^{N} X_i \tag{13}$$

3.2. Solving Method

The model presented in this paper corresponds to the mixed integer quadratic programming problem, and the solving of such a problem usually adopts the branch and bound method [31,32]. In solving a large-scale mixed integer problem, as the quantity of discrete integral variable and complex constraints, the branch and bound method to solve the sub-problem of break out when the number of increased exponentially seriously affected the calculation efficiency; so, in solving combinatorial optimization such as a unit of this kind of large-scale mixed integer programming problem, the efficiency and the results of using a single method are usually unsatisfactory [33,34].

In order to solve the above problems, the business optimization package IBM ILOG CPLEX (IBM, Armonk, New York, NY, USA) is used to solve the model. Figure 2 depicts the schematic arrangement technique used to build wide projects such as the scheduling of hydrothermal power plants using IBM Ilog Cplex Optimization Studio. Two files are created, one for model statement and another for data. External data from Microsoft excel are connected to the optimization software by the use of the appropriate codes. Microsoft Excel is a good tool that can be easily used to write equation and formula; this leads to less effort in executing programming. At the same time, in order to solve the mixed integer programming problem, the CPLEX solver adds the cutting plane method on the branch and bound method to form an improved branch cutting plane method [35,36]. Its solving principle is based on the branch and bound method, and the branch optimization to solve the subproblem cutting plane method is used in the process of feasible solution searching. The binary tree optimization and the choice of branch were reduced, which can effectively reduce calculation time.

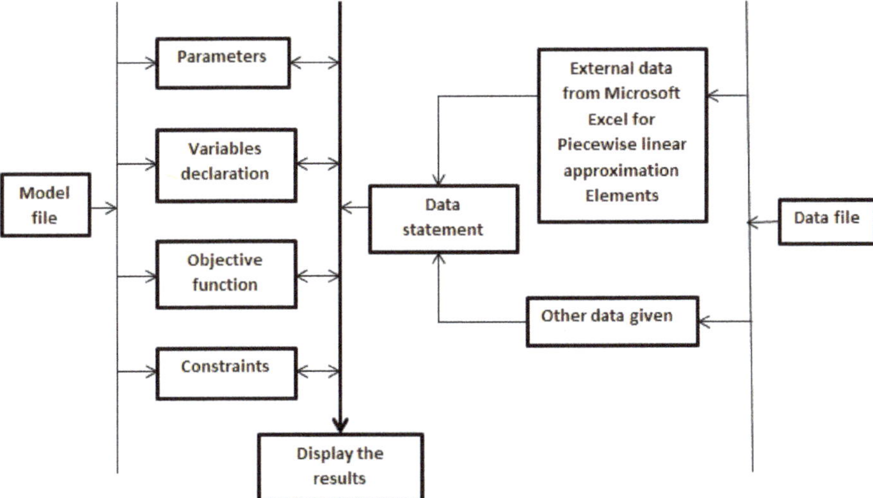

Figure 2. Schematic arrangement for building a model in IBM Ilog Cplex Optimization studio using the optimization programming language (OPL).

The solution flow of mixed integer programming by the CPLEX optimizer is shown in Figure 3. In order to solve the above problems, the paper uses the particular algorithm of the commercial optimization software ILOG CPLEX.

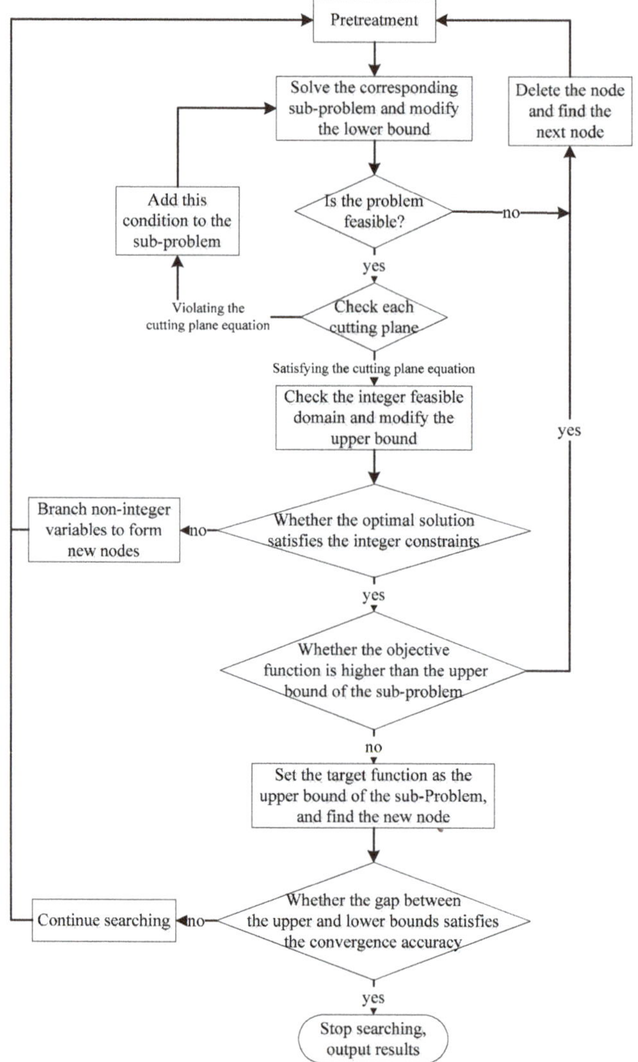

Figure 3. The optimization process flow of CPLEX branch cut plane method.

4. Simulation and Analysis

4.1. Simulation Condition

There are 10 units in the test system. The rationality and feasibility of the proposed model are verified by analyzing the simulation results, and the superiority is verified by comparing with the simulation results of other models. The parameters of units in the system are shown in Table 1, respectively. The simulation period is set to 30 days (one month), and the simulation step size (unit period) is set to 1 h. The calculation results of one typical day are shown below owing to page limit, the total system load of each period on this typical day is shown in Figure 4. The value of spinning reserve is set to 10% of the load demand, and the threshold value of Gini coefficient G0 is set to 0.4.

Table 1. Units' parameters. MW, h, $.

Unit No.	P_{max}	P_{min}	T_i^{on}	T_i^{off}	T_{inis}	T_{cold}	a	b	c	α_i	β_i	R_{up}
1	455	150	8	8	8	5	0.00048	16.19	1000	4500	9000	80
2	455	150	8	8	8	5	0.00031	17.26	970	5000	10,000	80
3	130	20	5	5	−5	4	0.00200	16.60	700	550	1100	20
4	130	20	5	5	−5	4	0.00211	16.50	680	560	1120	20
5	162	25	6	6	−6	4	0.00398	19.70	450	900	1800	30
6	80	20	3	3	−3	2	0.00712	22.26	370	170	340	10
7	85	25	3	3	−3	2	0.00079	27.74	480	260	520	10
8	55	10	1	1	−1	0	0.00413	25.92	660	30	60	5
9	55	10	1	1	−1	0	0.00222	27.27	665	30	60	5
10	55	10	1	1	−1	0	0.00173	27.79	670	30	60	5

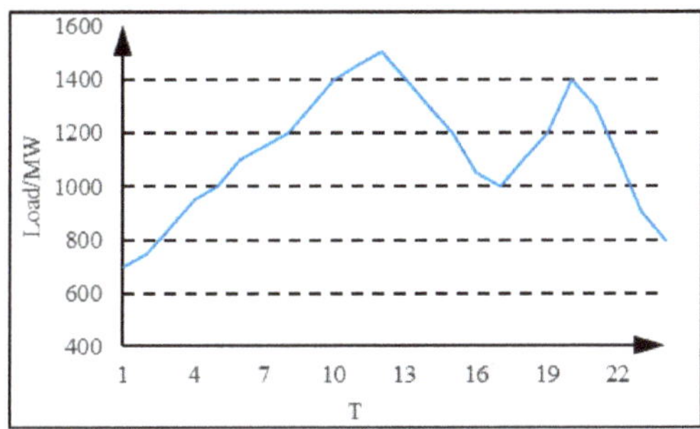

Figure 4. Total load demand curve.

The simulation is programmed, run, and processed on a PC with a CPU of Intel(R) Pentium(R) CPU G630 @ 2.70 GHz, and the RAM is 4.0 G.

4.2. Simulation Results

4.2.1. Unit Commitment and Output Schedule

First, it is simulated with the Gini coefficient set to 0.3. The total power generation cost was $17,468,276, and the program running time was 4.3 s. For the typical day, the unit commitment schedule and load dispatching schedule are shown in Figures 5 and 6, respectively. From the 24 h generation schedule, it can be seen that, for large-scale units (units 1 and 2), there is no shutdown and the generation output did not change significantly during load following. The stability of output power can give full play to the operating efficiency of large-scale units. The small unit can meet the constraint demand of system load change through progressive output adjustment. Because of its lower start-up and shutdown cost, it is more suitable for following load fluctuation. Thus, the economy of the overall system operation can be ensured.

	1	2	3	4	5	6	7	8	9	10	11	12	13	14	15	16	17	18	19	20	21	22	23	24
Unit1	1	1	1	1	1	1	1	1	1	1	1	1	1	1	1	1	1	1	1	1	1	1	1	1
Unit2	1	1	1	1	1	1	1	1	1	1	1	1	1	1	1	1	1	1	1	1	1	1	1	1
Unit3	0	0	0	0	0	1	1	1	1	1	1	1	1	1	0	0	0	0	0	1	1	1	1	1
Unit4	0	0	0	0	0	0	0	0	1	1	1	1	1	1	1	1	1	1	1	1	1	1	0	0
Unit5	0	0	0	1	1	1	1	1	1	1	1	1	1	1	1	1	1	1	1	1	1	1	0	0
Unit6	0	0	0	0	0	0	0	0	0	1	1	1	1	1	1	1	0	0	0	0	1	1	1	0
Unit7	0	0	0	0	0	0	1	1	1	1	1	1	1	1	1	0	0	0	0	0	1	1	1	0
Unit8	0	0	0	0	0	0	0	0	0	0	0	0	1	0	1	0	0	0	0	0	1	0	1	0
Unit9	0	0	0	0	0	0	0	0	0	0	0	1	1	1	1	0	0	0	1	0	0	0	0	0
Unit10	0	0	0	0	0	0	0	0	0	0	1	1	1	0	0	1	0	0	0	0	0	0	0	0

Figure 5. The unit commitment schedule.

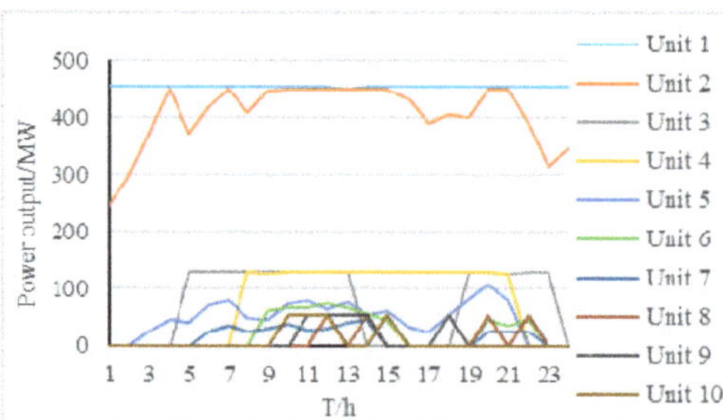

Figure 6. The unit generation output schedule.

4.2.2. Simulation Results with Different Threshold Values of Gini Coefficient

Generally, the Gini coefficient value 0.4 is taken as the "warning line" to measure the fairness of economic income. Therefore, in the case studies, the Gini coefficient value is set around 0.4 to study the changes in fairness and operation economy. Specifically, within the range of 0.3~0.5, the threshold value of Gini coefficient is selected at an interval of 0.02, so as to study the changes in the objective function value (total cost). The changes of the total costs of power generation with different threshold value of Gini coefficient can be found in Figure 7. In order to more intuitively reflect and display the relationship between the two, it is drawn as a line graph.

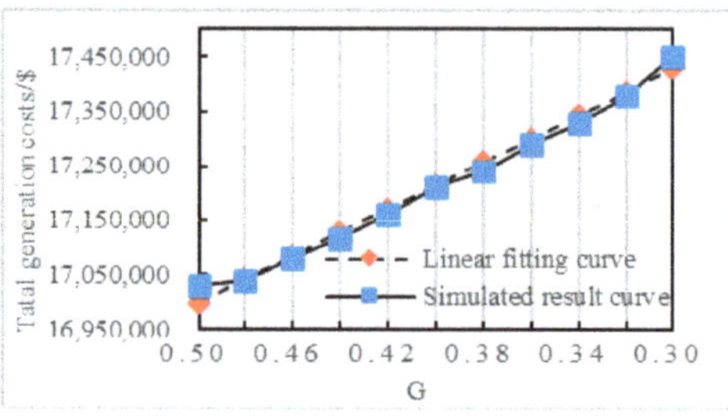

Figure 7. The relationship between the total cost and the Gini coefficient.

From Figure 7, it can be seen that the relationship between the total cost and the Gini coefficient is negatively correlated. That is, with the increase in Gini coefficient value, the optimization space of the model becomes larger, so that there could be a more economical way to revise the model. However, owing to the increase in Gini coefficient, the fairness of power system operation is weaker.

The numerical simulation results show that the relationship between the Gini coefficient and the total power generation cost is approximately linear. Specifically, with the Gini coefficient being reduced by 0.05, correspondingly, the total cost increases by about $135,130, accounting for 1% to 2% of the total power generation cost.

4.2.3. Simulation Results with Different Dispatching Modes

In order to compare the effect of three different dispatching methods, which are the traditional economic dispatching model, the existing impartial and open scheduling model and the Gini coefficient-based impartial and open dispatching model are proposed in this paper, the power system operation is simulated according to the three methods respectively, and the results and effect of fairness and economy are compared and analyzed.

In the traditional economic dispatching, minimizing total power generation cost is taken as the objective function, and the fairness is not considered. The objective function of existing impartial and open dispatching model is to minimize the standard deviation of the completion rate of electric energy of all the units. The results of relevant fairness and economy can be obtained in Figure 7.

It can be seen from Figure 8 that, when using the traditional economic dispatching model, the lowest total cost of power generation can be obtained, while the generation overall fairness among the generators is not ideal. The results when using the traditional "impartial and open" dispatching model are more fair because it pays more attention to treat each unit equally, but the total costs are the highest. Using the Gini coefficient-based "impartial and open" dispatching model proposed in this paper, the operation economic could be improved on the basis of ensuring fairness to some extent because of the coordination of the two indicators.

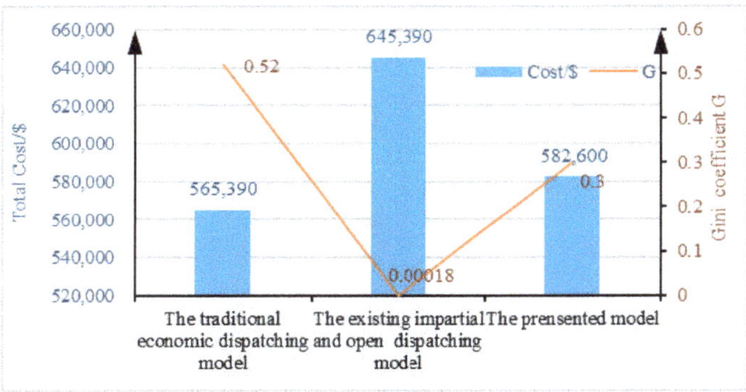

Figure 8. Simulation results with three different models.

It can be found combined with Figure 7 that, when the threshold value of Gini coefficient with the method presented in this paper is set as equal to the Gini coefficient values of the other two models, the total costs are the same, which shows that the model proposed in this paper is an extended model of the existing dispatching models. Different levels of balance between economy and equity can be achieved by adjusting threshold values of Gini coefficient according to different levels of fairness demand in actual system operation, so as to provide more choices for scheduling department, to realize the goal of fully and reasonably allocating scheduling resources under the "impartial and open scheduling principle".

5. Conclusions

In order to achieve the balance between the fairness and economy of power system operation, an impartial and open dispatching model based on the Gini coefficient is proposed in this paper. The theoretical analysis and simulation results show the following:

1. The proposed impartial and open dispatching model based on the Gini coefficient is an extended mode of the conventional dispatching model. The optimization space of system operation economy could be effectively improved on the premise of specifying fairness requirement by introducing the Gini coefficient in economics as an index to measure the fairness of electric energy completed progress in the form of constraint conditions.
2. In power system operation, different levels of balance between economy and equity can be achieved by adjusting the threshold value of the Gini coefficient according to different levels of fairness demand, so as to provide more choices for scheduling department. Better balance between fairness and economy could be realized to improve the overall system operation efficiency. The recognition of the generation companies to the impartial and open scheduling scheme could be improved because of the use of the internationally recognized Gini coefficient indicator, which has the recommended fairness value range.

3. The research in this paper is limited to the theoretical study of models and methods, and the proposed method cannot meet the computational requirements of optimal scheduling of large-scale power systems in terms of computational speed and computational convergence. However, the model presented in this paper has been put into trial operation in the monthly power generation scheduling in a provincial power grid in China. The next study content of the authors is to simplify the model and algorithm so that they could be applied to the actual dispatching work.

Author Contributions: Methodology, L.S.; Supervision, W.-d.L.; Writing—original draft, N.L.; Writing—review & editing, N.Z. and Z.-r.S. All authors have read and agreed to the published version of the manuscript.

Funding: This research was funded by The National Basic Research Program (China): grant number 51607021.

Conflicts of Interest: The authors declare no conflict of interest.

References

1. State Electricity Regulatory Commission. *Interim Measures for Promoting Open, Fair and Equitable Power Dispatching*; State Electricity Regulatory Commission: Beijing, China, 2004.
2. Long, H.; Huang, J.J. Development Direction Analysis of Coal-fired Power Units' Design Technology During the 13th Five-Year Plan. *Power Gener. Technol.* **2018**, *39*, 13–17.
3. Zhao, W.; Xu, J.; Wang, R. Security, economy and fairness of power systems under power market conditions. *China Electr. Power Educ.* **2008**, *S1*, 145–146.
4. Sun, J. Economic Analysis of Technology Supervision at Electric Power Generating Enterprises. *Power Gener. Air Cond.* **2017**, *38*, 72–77.
5. He, L.; Lv, H.F. Research on Optimal Dispatching of Multi-Microgrid Considering Economy. *Power Gener. Technol.* **2018**, *39*, 397–404.
6. Li, Y.J.; He, Z.C.; Zhang, Q. Test and Evaluation of Power Quality in Thermal Power Plant. *Power Gener. Technol.* **2018**, *39*, 135–139.
7. De la Nieta, A.A.S.; Gibescu, M. Day-ahead Scheduling in a Local Electricity Market. In Proceedings of the 2019 International Conference on Smart Energy Systems and Technologies (SEST), Porto, Portugal, 9–11 September 2019; pp. 1–6. [CrossRef]
8. Dashti, H.; Conejo, A.J.; Wang, R.J. Weekly Two-Stage Robust Generation Scheduling for Hydrothermal Power Systems. *IEEE Trans. Power Syst.* **2016**, *31*, 4554–4564. [CrossRef]
9. Jain, S.; Kanwar, N. Day ahead optimal scheduling of generators using Dynamic Programming method. In Proceedings of the 2019 8th International Conference on Power Systems (ICPS), Jaipur, India, 20–22 December 2019; pp. 1–6. [CrossRef]
10. Toma, L.; Tristiu, I.; Bulac, C. Optimal generation scheduling strategy in a microgrid. In Proceedings of the 2016 IEEE Transportation Electrification Conference and Expo, Asia-Pacific, Busan, Korea, 1–4 June 2016; pp. 491–496.
11. Nikmehr, N.; Ravadanegh, S.N. Optimal power dispatch of multi-microgrids at future smart distribution grids. *IEEE Trans. Smart Grid* **2015**, *6*, 1648–1657. [CrossRef]
12. Gregoratti, D.; Matamoros, J. Distributed energy trading: The multiple-microgrid case. *IEEE Trans. Ind. Electron.* **2015**, *62*, 2551–2559. [CrossRef]
13. Zhang, J.L.; Li, W.; Pan, Y. A Strategy for Balance of Medium-Term and Long-Term Electric Power and Energy in Fujian Province on the Dispatching Mode Conforming to Principles of Openness, Equity and Impartiality. In Proceedings of the Asia-Pacific Power and Energy Engineering Conference, Shanghai, China, 27–29 March 2012; pp. 1–7.
14. Li, L.l.; Geng, J.; Yao, J.G. Model and Algorithm for Security Constrained Economical Dispatch in Equilibrium Power Generation Dispatching Mode. *Autom. Electr. Power Syst.* **2010**, *34*, 23–27.
15. Yang, Z.H.L.; Tang, G.Q. A Generation Scheduling Optimization Model Suitable to Complete Period and Variable Intervals and Conforming to Principles of Openness, Equity and Justness. *Power Syst. Technol.* **2011**, *35*, 132–136.

16. Zhu, Z.L.; Zhou, J.Y.; Pan, Y. Security Constrained Economic Dispatch Considering Balance of Electric Power and Energy. *Proc. CSEE* **2013**, *33*, 168–176.
17. Pan, K. Coordination Optimization Dispatch and Its Evaluation in Wind Integrated Power Systems. Master's Thesis, Huazhong University of Science and Technology, Wuhan, China, 2014.
18. Jian, H.Y.; Kang, C.H.Q.; Du, B.Q. A Novel Approach to Assess Imparity and Openness of Power Dispatching in Electricity Market. *Power Syst. Technol.* **2005**, *32*, 26–32.
19. Zhang, C.H.G.; Wang, X.L. Impartiality Indexes of Power Dispatching Based on Modified Weighed Coefficient of Variation. *Electr. Power Technol. Econ.* **2009**, *21*, 5–9.
20. Li, W.D. On Frequency Response Control of Future Grid. *Power Gener. Technol.* **2018**, *39*, 84–89.
21. Sun, F.; Zuo, X.F.; Qin, Y. Road Network Equilibrium Analysis Based on Section Importance and Gini Coefficient. *Green Smart Connect. Transp. Syst.* **2020**, *617*, 1119–1134.
22. Dai, H.N.; Yao, E.J.; Liu, S.S. Flow inequality of freeway network based on Gini-coefficient. *J. Transp. Syst. Eng. Inf. Technol.* **2017**, *17*, 205–211.
23. Li, W.D. The Application of Gini Coefficient in Regional Environmental Pollutants Distribution Plan. *Gener. Technol.* **2018**, *39*, 84–89.
24. Wu, D.F.; Zeng, G.P.; Meng, L.G. Gini Coefficient-based Task Allocation for Multi-robot Systems with Limited Energy Resources. *Environ. Sci. Inf. Appl. Technol. ESIAT* **2010**, *1*, 426–429. [CrossRef]
25. Dai, J.L.; Wang, P.; Wang, X. Discussion on Impartiality Index of Power Dispatching Based on Gini Coefficient. *Autom. Electr. Power Syst.* **2008**, *32*, 1–4.
26. Zeng, F.; Wang, P. Fairness Coefficient Analysis of Power Dispatching in North China Area. *Mod. Electr. Power* **2010**, *27*, 78–81.
27. Wei, X.H.; Hu, Z.H.Y.; Yang, L. An Analysis and Suggestions for Existing Evaluation Indices of Open and Impartial Power Dispatching. *Autom. Electr. Power Syst.* **2012**, *20*, 109–112.
28. Li, S.H. Further explanation of the estimation method and decomposition of the Gini coefficient. *Econ. Res. J.* **2002**, *5*, 45–53.
29. Xiong, J. A Comparative Analysis of Appraisal Method of Gini Coefficient. *Res. Financ. Econ. Issues* **2003**, *1*, 79–82.
30. Ye, B.; Zhang, P.X.; Zhao, B. Multiobjective Hybrid Evolutionary Algorithm for Economic Load Dispatch. *Proc. CSU EPSA* **2007**, *2*, 66–72.
31. Geng, J.; Xu, F.; Yao, J. Performance Analysis of Mixed-integer Programming Based Algorithm for Security Constrained Unit Commitment. *Autom. Electr. Power Syst.* **2009**, *33*, 24–27.
32. Su, J.G.; Shu, X.; Xie, G.H. Linearization Method of Large Scale Unit Commitment Problem with Network Constraints. *Power Syst. Prot. Control* **2010**, *38*, 135–139.
33. Cheng, C.H.P.; Liu, C.H.W.; Liu, C.H. Unit commitment by Lagrangian relaxation and genetic algorithms. *IEEE Trans. Power Syst.* **2000**, *15*, 707–714.
34. Jiang, Z.H.M.; Guan, Q.M. The Process of Solving Optimization Problems Based on CPLEX and C++ Language. *Comput. Knowl. Technol.* **2015**, *11*, 49–50.
35. Young, G.O. Synthetic structure of industrial plastics. In *Plastics*, 2nd ed.; Peters, J., Ed.; McGraw-Hill: New York, NY, USA, 1964; Volume 3, pp. 15–64.
36. Chen, W.K. *Linear Networks and Systems. Belmont*; Wadsworth: Boston, MA, USA, 1993; pp. 123–135.

Article

A Novel Market Clearing and Safety Checking Method for Multi-Type Units That Considers Flexible Loads

Dong Hua [1,*], **Wutao Chen** [1] **and Cong Zhang** [2]

[1] School of Electric Power, South China University of Technology, Guangzhou 510641, China; 201821013949@mail.scut.edu.cn

[2] School of Electrical and Information Engineering, Hunan University, Changsha 410082, China; zcong@hnu.edu.cn

* Correspondence: dhua@scut.edu.cn; Tel.: +86-135-6001-7727

Received: 12 June 2020; Accepted: 20 July 2020; Published: 22 July 2020

Abstract: Flexible loads have flexibility and variability in time and space, and they have been widely studied by scholars. However, the research on the participation of flexible loads in market clearing and safety checking is still insufficient. We propose a market clearing and safety checking method for multi-type units that considers flexible loads. First, the flexible load is divided into reducible loads, shiftable loads, and convertible loads, and its mathematical model is established. Then, the convertible loads are considered in the market clearing model, and the power management agency executes the market clearing procedure to obtain the clearing result. When the line power exceeds the limit as a result of clearing, the power flow of the branches and sections is eliminated by adjusting the unit output and reducing the flexible load at the same time, and a safety checking model considering load reduction is established. The marginal electricity price of the nodes is obtained by the interior point method, and we solve the model by calling the CPLEX (v12.7.1) solver in GAMS (General Algebraic Modeling System v24.9.1). We use a regional power grid of 220 kV and above as an example for analysis; the results show that the proposed method can reduce the marginal electricity price of the nodes, reduce the cost of safety checking, and improve the safety of the market clearing.

Keywords: flexible loads; market clearing; safety checking; interior point method

1. Introduction

In recent years, flexible loads have become a research hotspot for scholars due to their corresponding speed, low carbon footprint, and low cost [1–3]. The continuous development of energy management technology has resulted in an increasing trend in the flexible load participation in demand response (DR) projects [4–6]. Many countries are allowing qualified, large consumers, or flexible load aggregators to participate directly in the electricity market [7,8]. In the study [9], authors propose a model for flexible aggregators flexibility provision in distribution networks, the model takes advantage of load flexibility resources allowing the re-schedule of shifting/real-time home-appliances to provision a request from a distribution system operator or a balance responsible party. In the mature stage of the spot market, the transition from the unilateral quotation to bilateral quotation mode will allow the flexible load to directly participate in the power market. In addition, due to the flexible variability in the time and space of the flexible load, the flexible load can also participate in the auxiliary service market [10,11]. Under the bilateral quotation model, how to participate in the market clearing of flexible loads and how to perform a safety checking of the power dispatching agency when the line overrun occurs as a result of the market clearing are issues that need to be resolved for the flexible load participation in the electricity market.

In order to maximize the benefits of integrated energy companies in an increasingly complex multi-participant energy market, some researchers [12] classified loads into three categories based on the elastic characteristics of the loads: reducible loads, convertible loads, and shiftable loads. The three types of loads participate in the demand response project, so that integrated energy companies get more benefits. In the work of [13], the authors proposed an innovative economic and engineering coupled framework to encourage typical flexible loads or load aggregators, such as parking lots with a high penetration of electric vehicles, to participate in the real-time retail electricity market based on an integrated e-voucher program directly. In order to integrate the flexible load in the distribution network, a new pricing mechanism was proposed in the literature [14]. The price can be calculated through the two-tier local distribution network market. Flexible load aggregators act as the price receiver, and the system operator is the manager; the proposed method can save the cost of the flexible load aggregator. Researchers have proposed a market clearing mechanism to minimize the costs (both the day-ahead and real-time adjustments) and maximize the social benefits in terms of the market clearing; this mechanism took into consideration the uncertainty of wind power generation and load forecasting [15]. Another market clearing mechanism was proposed in a different study [16], where the pricing uncertainty and generation reserve were determined by the uncertainty marginal price (UMP). At the same time, energy is priced by locational marginal price (LMP), and the uncertainty of the load is borne by the unit in the day-ahead market. UMP and LMP can be obtained through the robust optimization method. In the work of [17], the researchers adopted a two-stage market clearing method to weaken collusion among the power producers, established a causal relationship and a quantitative relationship between the bidding process of the power producers and the market clearing, and evaluated the effectiveness of the method using system dynamics.

The above-mentioned studies examined the flexible load and market clearing, and the related research is sufficient. However, there is very little research on the flexible load participation in market clearing. In the study [18], researchers assumed that users can participate in market competition (such as on the power generation) directly; they established a market equilibrium model that considered the DR and studied the impact of the DR implementation on the power market. However, in this literature, the classification of flexible loads is not specific, and the impact of flexible loads on the model is not analyzed. In order to ensure the safety of the market clearing results, the dispatching agency needs to perform the safety checking process. During the safety checking process, when the transmission line power exceeds the limit, the output of the line is generally adjusted by adjusting the output of the unit. In the safety checking of the power generation plan, the combined optimization technology is used to solve the power grid safety checking problem and the power generation plan in each time period, and it is widely used in the day-ahead market [19,20].

In this paper, we propose a market clearing and safety checking method for multi-type units that considers the flexible load. First, using the characteristics of the power load, the flexible load was divided into reducible loads, convertible loads and shiftable loads, and we created three mathematical models of the flexible load. In the day-ahead market, generators and demanders submit quotation curves separately, and so we established the market clearing model that considers the convertible loads. The power management agency executed the clearing process to obtain the clearing results; when the line transmission power breaks out as a result of clearing, the flow of the branches and sections can be eliminated by adjusting the output of the generator set and reducing the flexible loads at the same time.

The remainder of this paper is organized as follows: Section 2 discusses the three mathematical models of the flexible load and its mechanism. Section 3 includes the market clearing model that considers the convertible loads. In addition to conventional units, the model also considers the pumped storage and nuclear power units. In this section also discusses a safety checking model that considers the reducible loads. Section 4 introduces the solution method and steps of the model. Section 5 evaluates the effectiveness of the method proposed in this paper through actual cases.

2. Flexible Load Modeling and Its Mechanism

2.1. Flexible Load Modeling

In this paper, the flexible load is divided into reducible loads, convertible loads, and shiftable loads. The power consumption law of the flexible load can be accurately sensed by a real-time monitoring device, and the load control technology can be used to realize the interruption and translation of the load in a specific time period. From the perspective of the demand response, the flexible load can participate in the demand side management (DSM) and DR projects of the grid by adjusting its own power demand, and it can be dynamically adjusted with the balance of the supply and demand [21–24]. When the flexible loads participate in the control and responds to the demand, the load model can be expressed as:

$$P_{LD} = (t, P_{re}, P_{convert}, P_{shift}) = P_s(t) + P_{fl}(t, P_{re}, P_{convert}, P_{shift}) \tag{1}$$

where P_{LD} is the total load demand, $P_s(t)$ is the normal load demand, P_{fl} is the flexible load, P_{re} is the load demand reduction, $P_{convert}$ is the convertible load demand, and P_{shift} is the translatable load demand.

2.1.1. Reducible Loads

The meaning of reducible loads is that the power dispatch center can interrupt the load without affecting user comfort. Typical load reductions include air conditioning loads and lighting. Its mathematical model can be expressed as follows:

$$\Delta P_{re}(t) = f_1(P_0(t), \Delta p_{re}, k_{re}(t), v_{re}(t)) \tag{2}$$

where $\Delta P_{re}(t)$ is the response of the load during the t period; Δp_{re} is the change of the electricity price in the t period; $k_{re}(t)$ is the self-elastic coefficient of the load in the t period; and $v_{re}(t)$ is the rate of the load reduction.

2.1.2. Convertible Loads

Convertible loads mean that within a dispatch period, power demand can be transferred to any time, but the total amount of electricity used within one day remains unchanged. Typical convertible loads include energy storage and electric vehicle charging stations. Its mathematical model can be expressed as follows:

$$\Delta P_{convert}(t) = f_2(P_0(t), \Delta p_{convert}, k_{convert}(t), v_{convert}(t)) \tag{3}$$

where $\Delta P_{convert}(t)$ is the response of the transferable load during the t period; $P_0(t)$ is the base load of the t period; $\Delta p_{convert}(t)$ is the electric-difference vector between this period and other time periods; and $\Delta k_{convert}(t)$ is the mutual elastic vector of the t period relative to other periods. The load transfer rate T is the duty cycle of the load.

2.1.3. Shiftable Loads

The shiftable loads refer to the loads which power supply time can be changed according to the dispatching demand, and the entire period of time is shifted within a one-day scheduling period, such as when users are utilizing a washing machine or a water heater, or some large industrial users that are producing. The shiftable loads can be divided into multiple categories, according to the amount of electricity consumption and the duration of electricity consumption. Figure 1 shows the schematic diagram before and after the transfer of three different shiftable loads. During the transfer process, the power supply duration and power supply of the same shiftable load remain unchanged.

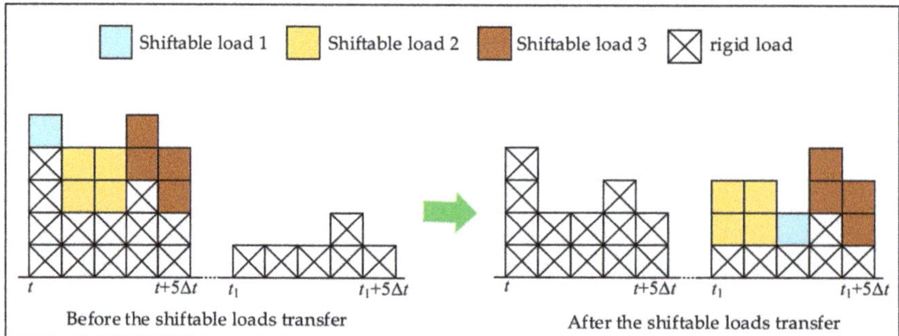

Figure 1. The schematic diagram of the transfer of three different shiftable loads.

Compared with the convertible loads, the power supply duration of the shiftable load is continuous, and the power supply within this period remain unchanged. Its mathematical model can be expressed as follows:

$$\Delta P_{\text{shift}}(t) = f_3(t + \Delta t(\Delta p)) - f_3(t) \tag{4}$$

where $\Delta P_{\text{shift}}(t)$ is the response of the shiftable loads during this period; $f_3(t)$ is its initial power consumption curve; and $\Delta t(\Delta p)$ is the duration time.

2.2. Mechanism Design

2.2.1. General Framework

The flexible load aggregator is an organic combination of multiple types of flexible loads. Through the energy management center, it participates in power grid operations and market transactions as a stakeholder. Flexible load aggregators determine the optimal bidding curve by maximizing their own interests. Figure 2 shows a schematic diagram of flexible load aggregators participating in the market composed of electric vehicles, temperature-controlled loads and distributed energy storage as typical flexible loads. The classification of flexible loads is conducive to the precise control of the power management department and to the flexible role of flexible loads. The framework of the market clearing and safety checking method in this paper is shown in Figure 3.

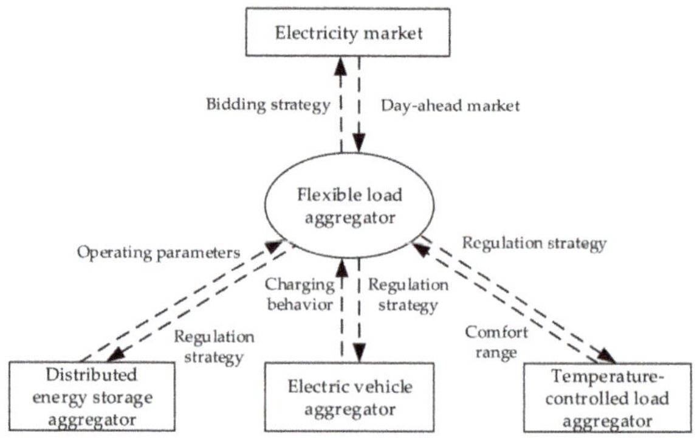

Figure 2. Diagram of flexible load aggregators participating in the market.

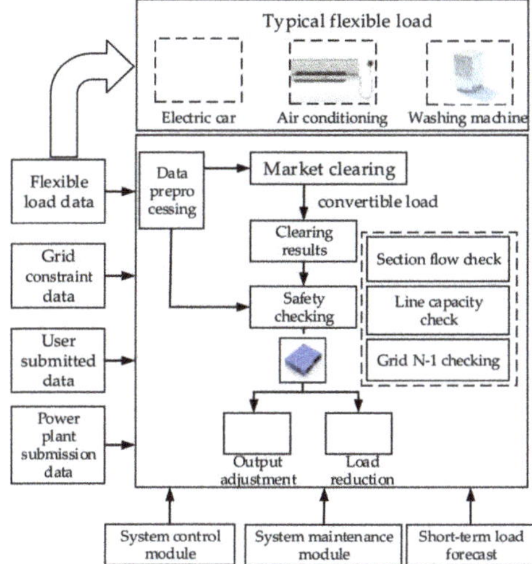

Figure 3. Market clearing and safety checking framework that considers the flexible load.

2.2.2. Business Process

Flexible loads participate in grid interaction through demand response, which can generally be divided into two basic modes: price-based demand response and incentive-based demand response. Price-based demand response refers to the behavior of users voluntarily avoiding high electricity prices to low electricity prices under the stimulation of electricity price signals; incentive-based demand response refers to users signing agreements with the grid side in advance in order to obtain economic compensation by accepting the direct control or interference control of electricity consumption from the grid side. The above operation requires the user to negotiate with the grid side in advance in the interactive contract to obtain appropriate data. The process is shown in Figure 4.

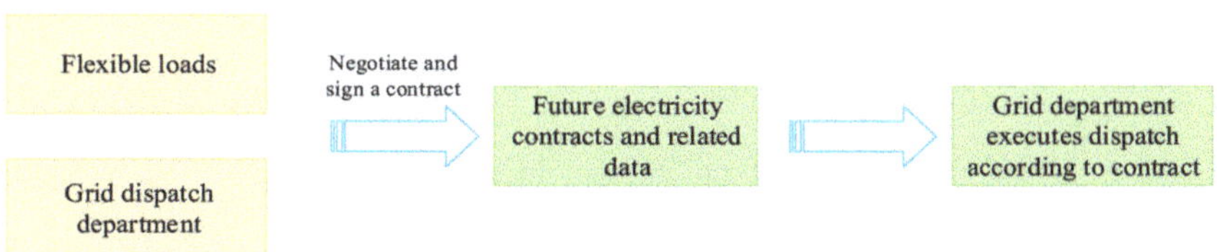

Figure 4. Real-time dispatching process of power grid considering flexible load.

The implementation of flexible load response projects should follow the principles of "safety, reliability, fairness, equality, openness, and transparency". Safety and reliability are the primary principles to be followed in the construction and implementation of demand response capabilities. It is necessary to ensure the stable and reliable operation of the power grid and the safe operation of enterprises; the principle of fairness and equality guarantees the effective development of demand response work, in strict accordance with relevant laws during the implementation process. The policies and agreed rules are implemented fairly and fair to all participating users; the principle of openness and transparency guarantees the continuous advancement of demand response work, the rules of participation are simple and clear, open to the society, and the majority of users are encouraged to participate voluntarily. During the implementation of demand response organization, choose a reasonable response range and capacity to ensure a basic balance between the peak electricity price increase fund and the power demand response subsidy expenditure. The process of flexible load participation in demand response is shown in Figure 5.

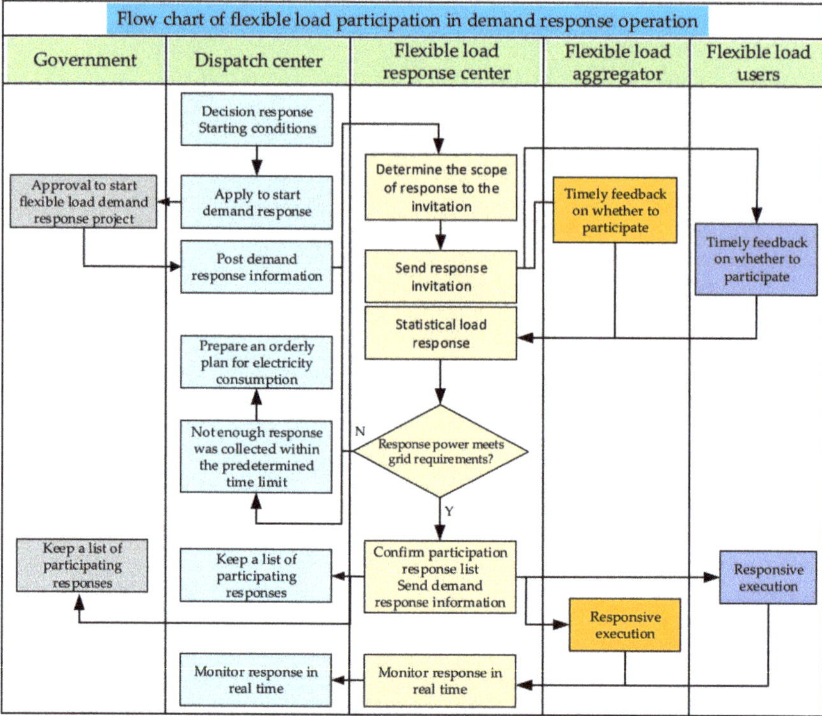

Figure 5. Flow chart of flexible load participation in demand response operation.

3. Market Clearing and Safety Checking Model

3.1. Market Clearing Model that Considers the Convertible Load

In the traditional power-dispatching model, the power producers need to fully generate power according to the dispatching instructions [25]. However, in the power market environment, the dispatching instructions must meet the needs of the power producers and the demand of the power load [26]. The electric load can be divided into the conventional load and the flexible load, and the conventional load can participate in the market bidding. The flexible load can increase the flexibility of the load by changing the demand for electric power. However, the flexible load cannot fully participate in the power adjustment, and the most basic power demands must be meet. The proportion of the flexible load participating in the market can be calculated by the following formula:

$$x = \frac{P_{fl}}{P_{fl} + P_s} \tag{5}$$

where P_{fl} and P_s represent the flexible load and the normal load, respectively. Due to the limited share of flexible load in the market, the scale factor x ranges from $[0, \theta]$ and the value of θ does not exceed 0.3.

3.1.1. Generator Bidding

In a period T, the cost of the generator $i \in S_G$ is assumed to be a quadratic curve [27]:

$$C(P_{Gi,t}) = \frac{1}{2}a_i P_{Gi,t}^2 + b_i P_{Gi,t} \tag{6}$$

where $P_{Gi,t}$ is the actual output of the generator, a_i and b_i are the quadratic coefficient and the primary coefficient of the cost curve of the generator, respectively. The power producer quotes according to the marginal cost, and uses the quotation strategy coefficient $k_{Gi,t}$ to affine the marginal cost function, and the quotation curve can be expressed as:

$$p_{Gi,t} = k_{Gi,t}(a_i P_{Gi,t} + b_i) \tag{7}$$

3.1.2. Demand Side Bidding

In a period T, the benefit function of the demand side $j \in S_L$ can be expressed as:

$$B(P_{Lj}) = -\frac{1}{2}\alpha_j P_{Lj,t}^2 + \beta_j P_{Lj.t} \tag{8}$$

where α_j and β_j are the quadratic coefficient and the primary coefficient of the benefit function, $P_{Lj,t}$ is the power demand of the load node. In the same way as the power producer, the demand side participates in the market bidding by adjusting the quotation strategy coefficient $k_{Li,t}$, and the submitted quotation curve is:

$$p_{Lj,t} = k_{Lj,t}(-\alpha_j P_{Lj,t} + \beta_j) \tag{9}$$

3.1.3. Market Clearing Model

In this paper, the demand side is divided into the normal load and flexible load. The corresponding benefit functions are $B(P_L)$ and $B(P_{FL})$, respectively. The difference between the demand side benefit and the generator cost $C(P_G)$ is the social welfare. From the perspective of the power market clearing, the purpose of electricity trading center is to maximize the social welfare. Therefore, the objective function is the maximization of the social welfare, and it can be expressed as:

$$\begin{aligned} \max_{P_G, P_L, P_{FL}} F_{ISO} &= B(P_L) + B(P_{FL}) - C(P_G) \\ &= \sum_{t=1}^{T}\sum_{j \in S_L} k_{Lj,t}(-\tfrac{1}{2}\alpha_j P_{Lj,t}^2 + \beta_j P_{Lj,t}) + \\ &\sum_{t=1}^{T}\sum_{j \in S_L}(-\tfrac{1}{2}\alpha_j^{FL} P_{FLj,t}^2 + \beta_j^{FL} P_{FLj,t}) - \sum_{t=1}^{T}\sum_{i \in S_G} k_{Gi,t}(\tfrac{1}{2}a_i P_{Gi,t}^2 + b_i P_{Gi,t}) \end{aligned} \tag{10}$$

where P_G, P_L, and P_{FL} are respectively the power generation output power, the conventional load demand, and the flexible load demand. Moreover, α_j^{FL} and β_j^{FL} are the quadratic coefficients and the primary coefficients of the flexible load benefit function, respectively. The clearing model that considers the convertible load needs to consider the transferable load constraint in addition to the conventional power balance constraints, network security constraints, and related constraints of the generator set. The transferable load does not change the total amount of electricity used in 1 power cycle, but the power consumption in each period can be flexibly adjusted.

(1) System power balance constraint

$$-\sum_{i \in S_G, j \in S_L} B_{ij}\theta_{ij} = P_{Gi,t} + P_{Ci,t} + P_{Hi,t} - P_{Lj,t} - P_{FLj,t} \tag{11}$$

where $P_{Gi,t}$ is the output of the conventional generator set, $P_{Ci,t}$ is the output of the pumped storage unit, $P_{Hi,t}$ is the output of the nuclear power unit, $P_{Lj,t}$ is the load of the conventional user, and $P_{FLj,t}$ is the load of the flexible load. When the power system is running, the power system power balance must be guaranteed. The pumped-storage generator set generates electricity when the electricity price is high during peak hours and transmits power to the grid; it draws water during the low-voltage period and consumes electricity. In one day, the sum of the power generation and electricity consumption is zero. The output of the nuclear power unit is stable, and its output in each period is considered constant, P_{H0}.

(2) Upper and lower limits of the unit

$$P_i^{\min} \leq P_{i,t} \leq P_i^{\max} \tag{12}$$

where, $P_{i,t}$ is the output of the unit in this period, which satisfies the upper and lower limits of the unit output. $P_i^{\min}$ and $P_i^{\max}$ are the minimum and maximum output of the unit, respectively.

(3) Climbing constraint of the unit

$$-\Delta P_{Gi,t,\mathrm{dw}} \leq P_{Gi,t} - P_{Gi,t-1} \leq \Delta P_{Gi,t,\mathrm{up}} \tag{13}$$

where $\Delta P_{Gi,t,\mathrm{up}}$ and $\Delta P_{Gi,t,dw}$ are the maximum up/down climbing power of unit. In two adjacent time periods, the climb rate of the unit must meet the minimum and maximum climb constraints.

(4) Convertible load constraint

In a period T, the total electricity consumption of the convertible load remains the same, satisfying the convertible load total constraint and the transferable load transfer interval constraint:

$$\begin{cases} \sum_{t=1}^{\mathrm{T}} P_{\mathrm{con},j,t} = \mathrm{T}P_{\mathrm{con},j} \\ -\tau\% P_{\mathrm{con},j} \leq \Delta P_{\mathrm{con},j,t} \leq -\tau\% P_{\mathrm{con},j} \\ t \in \mathrm{T}_{\mathrm{con},j} \end{cases} \tag{14}$$

where $\Delta P_{\mathrm{con},j,t}$ is the transfer amount of the convertible load at the time T of the node j. If it is positive, the load is transferred to time T, and if it is negative, the load is transferred from time T. The first equation is the transferable load total constraint, the second inequality is the upper and lower bounds of the transfer amount, and $\mathrm{T}_{\mathrm{con},j}$ is the convertible interval.

3.2. Safety Checking Model That Considers the Reducible Loads

The market clearing model aims to maximize the social welfare, but in the pursuit of maximizing profits, it will inevitably lead to a reduction in the system's security. To ensure that the power flow of the line does not exceed the limit, it is usually directly in the clearing model that considers the line safety constraint, to ensure the security of the line transmission. However, this method makes the clearing model very complicated and it is difficult to ensure the efficiency of solving the model [15,28,29]. In this paper, the market clearing process of this article actually includes two stages, the first stage does not consider the line safety constraints, and the second stage is the safety checking process considering safety constraints. In both stages, flexible loads are considered. Moreover, in the safety checking process, by simultaneously adjusting the output of the unit and reducing the load, the problem of power overrun is eliminated. This method divides the complex market clearing into two simple steps and uses the optimization algorithm to adjust the power generation output and the load reduction to eliminate the trend limit of the branch and section.

When using the mathematical optimized method to conduct the safety checking on the clearing result, it is necessary to establish an optimization model, and the unit adjustment amount and the load reduction amount can be used as the decision variables. The goal is to minimize the sum of the unit adjustment cost and the flexible load reduction cost. The objective function is:

$$\min f = C(\Delta P_{\mathrm{G}}) + C(P_{\mathrm{F}}) \tag{15}$$

where $C(\Delta P_{\mathrm{G}})$ and $C(P_{\mathrm{F}})$ are the generator set adjustment cost and the flexible load reduction cost, respectively. The unit adjustment cost can be expressed as:

$$C(\Delta P_{\mathrm{G}}) = \sum_{i \in S_{\mathrm{G}}} \delta_i |\Delta P_{\mathrm{G}i}| \tag{16}$$

The flexible load reduction costs can be expressed as:

$$C(P_{\mathrm{F}}) = \sum_{i \in S_{\mathrm{L}}} \gamma_i \left(P_{\mathrm{Fi}}^0 - P_{\mathrm{Fi}}\right) \tag{17}$$

where δ_i and γ_i express the cost factor of the unit adjusting and load reducing, respectively; here $\delta_i = 10\gamma_i$.

The constraint conditions consider the adjustment amount and the reduction amount of the balance constraint, the unit adjustment upper and lower limit constraints, the load reduction constraint, the line upper and lower limit constraints and the power distribution factor constraint.

(1) Balance between the adjustment and reduction

$$\sum \Delta P_{\mathrm{G}i} - P_{\mathrm{F}i} = 0 \tag{18}$$

where $\Delta P_{\mathrm{G}i}$ is the adjustment amount of the unit output, and $P_{\mathrm{F}j}$ is the flexible load reduction amount; in an ideal case, the flexible load reduction amount is equal to the unit output adjustment amount.

(2) Upper and lower limit constraints before and after the unit adjustment.

$$P_{\mathrm{G}i}^{\min} \le \Delta P_{\mathrm{G}i} + P_{\mathrm{G}i}^0 \le P_{\mathrm{G}i}^{\max} \tag{19}$$

where $P_{\mathrm{G}i}^0$ is the original output of the unit, and the sum of the output before the adjustment and after the adjustment must meet the upper and lower limits of the unit.

(3) Upper and lower limit constraint of the line transmission power..

$$-P_l^{\max} - P_{l0} \le \Delta P_l \le P_l^{\max} - P_{l0} \tag{20}$$

where $P_l^{\max}$ is the maximum value of the line transmission power, P_{l0} is the transmission power of the line before the safety checking, and ΔP_l is the adjustment amount of the line transmission power after the safety checking.

(4) Reduce load constraints.

$$P_{\mathrm{F}i,\min} \le P_{\mathrm{F}i} \le P_{\mathrm{F}i}^0 \tag{21}$$

where $P_{\mathrm{F}i,\min}$ is the minimum power demand for the flexible load and $P_{\mathrm{F}i}^0$ is the original fixed power demand for the flexible load user.

(5) Power distribution factor constraints

$$\Delta P_l = p_{l,i} \Delta P_{\mathrm{G},i} + q_{l,i}(P_{\mathrm{F}i0} - P_{\mathrm{F}i}) \tag{22}$$

where $p_{l,i}$ is the power distribution factor of line l with respect to $\Delta P_{\mathrm{G},i}$, $q_{l,i}$ is the power distribution factor of line l with respect to the flexible load reduction amount $P_{\mathrm{F}i0} - P_{\mathrm{F}i}$, and ΔP_l is the adjustment amount of the line transmission power after the safety checking.

In the process of safety checking, the power dispatch center implements the load reduction plan and needs to provide compensation fees F to the load reduction users. The compensation fee $F = C(P_{\mathrm{F}})$ in this paper is shown in (17).

4. Solution Method

4.1. Linearization of the Cost Function

In GAMS, nonlinear functions can be linearized by defining the SOS2 variables. The SOS2 variables consist of $\lambda_i(i = 1, 2, \cdots, n+1)$ elements, with up to two non-zero elements in λ_i, and satisfies $\lambda_1 + \lambda_2 + \cdots + \lambda_{n+1} = 1$. Suppose the fractional linear function $f(x)$ can be equally divided into linearized n segments on [a,b], and $a = x_1 < x_2 < \cdots < x_{n+1} = b$ are $n+1$ segmentation points [30,31]. The starting point of the paragraph i can be expressed as:

$$x_i = a + \frac{b-a}{n}(i-1) \tag{23}$$

Assuming $(\widetilde{x}, \widetilde{y})$ is a point in the segment i, and $\widetilde{x} = \lambda_i x_i + \lambda_{i+1} x_{i+1}$, λ_i, λ_{i+1} satisfies $\lambda_i + \lambda_{i+1} = 1$, then the linear approximation of $\widetilde{y}$ can be written as:

$$\widetilde{y} = \lambda_1 f(x_1) + \lambda_2 f(x_2) + \cdots + \lambda_{n+1} f(x_{n+1}) \tag{24}$$

Using the above method, the quadratic cost function in the objective function is linearized.

4.2. Market Clearing and Safety Checking Process

The method proposed in this paper can be divided into two stages. The first stage is the market clearing, which considers the convertible load during the market clearing; when the line transmission power is overrun in the clearing result, the second stage of the process is performed. By adjusting the output of the unit and reducing the flexible load at the same time, the purpose of eliminating the out-of-limit line flow is achieved. The safety checking can further ensure the safety of the market clearance. The specific process is shown in Figure 6.

The solution steps are as follows:

Step 1: Input unit data, line parameters, and other data. The generator and the demand side submit the quotation curve, and dispatch department execute the market clearing process. Then, initial power generation plan is formulated.

Step 2: Start the power flow calculation program to determine whether the power flow exceeds the limit. If it does not exceed the limit, end the calculation and output the power generator output; otherwise, proceed to the next step.

Step 3: Output the over-limit line number and over-limit power.

Step 4: Start the safety checking module, execute the safety checking procedure, calculate the generator adjustment of the generator, and reduce the load to reduce the power.

Step 5: Output the power of the generator and reduce the power by reducing the load.

The market clearing model discussed in this paper is a nonlinear programming model. After linearizing it, it is solved by calling the CPLEX solver in GAMS.

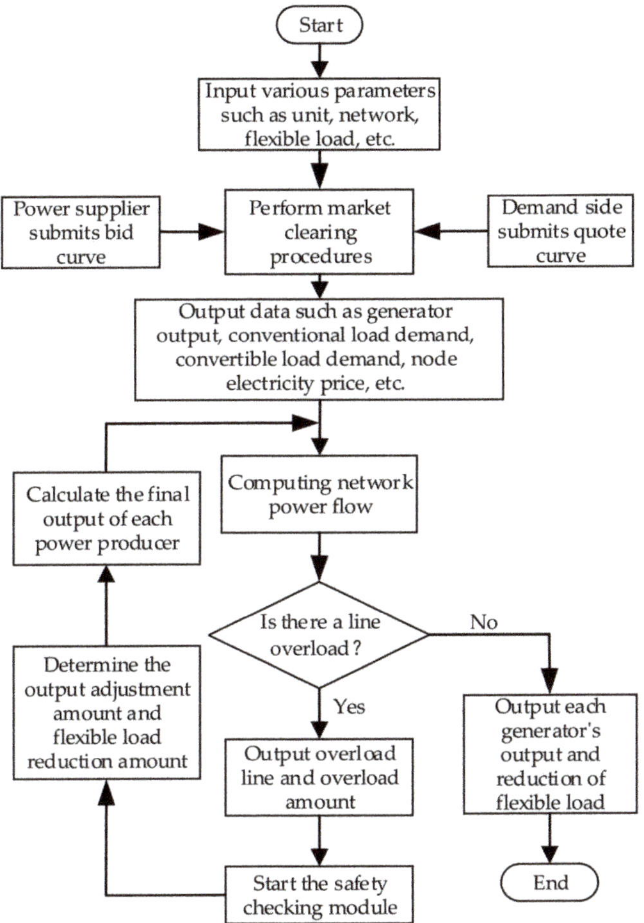

Figure 6. Market clearing and safety checking process that considers the flexible load.

5. Case Analysis

5.1. Basic Date

Here, we use a 220 kV and above line of a regional power grid as an example. There are 73 nodes and 93 lines in the regional power grid, including 5 thermal power plants, 1 pumped storage power plant, and 1 nuclear power plant. In addition, the external electricity is injected into the 1st node and 68th node, and these two nodes are regarded as units with adjustable output. The relevant parameters of the power plant are shown in Table 1. The network diagram is shown in Figure 7. Taking the typical daily load curve of summer in 2018, as an example, 85% of them are fixed loads to ensure basic electricity demand, 11.03% of the electricity demand is determined by participating in the market, and another 4.97% of the transferable load, the load curve, and the proportion of flexible load are shown in Figure 8.

Table 1. Unit parameters of the power plant.

Power Plant	a	b	Lower Limit(MW)	Upper Limit(MW)	Climbing Uphill (MW/h)	Downhill (MW/h)
1	0.00021	0.32	315.2	3152	2101	2101
24	0.00016	0.32	117	1170	780	780
26	6.6×10^{-5}	0.325	117	1170	780	780
36	0.00016	0.29	44	440	293	293
52	6.2×10^{-5}	0.35	196	1960	1307	1307
63	0.00005	0.01	−1200	1200	1200	1200
68	0.00021	0.225	500	5000	3333	3333
69	0.00016	0.29	268	2680	1787	1787
73	0.00005	0.09	415.2	4152	2768	2768

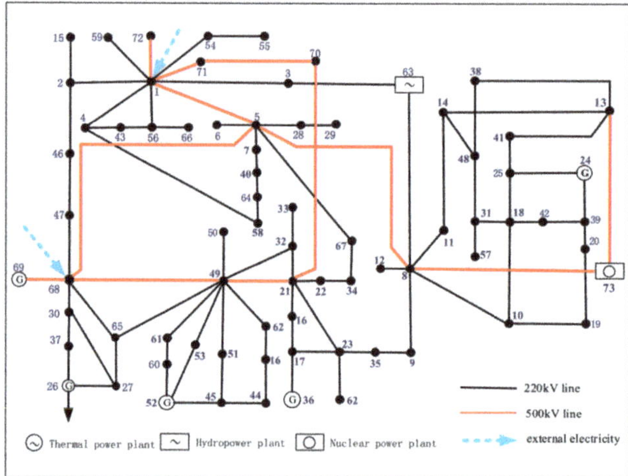

Figure 7. Grid connection diagram for a regional power grid.

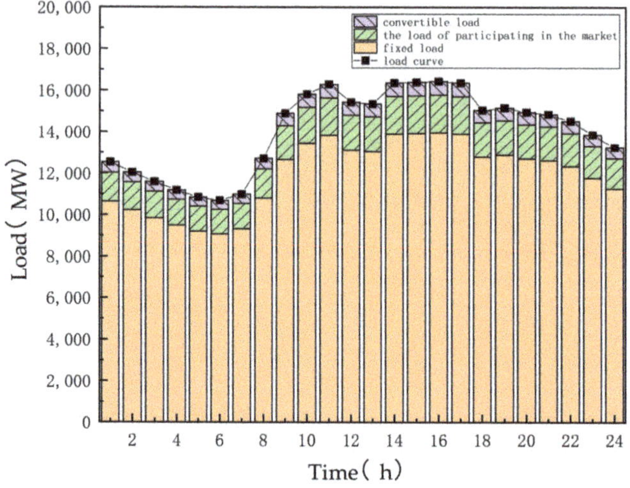

Figure 8. Load curve and proportion of the flexible load.

The spot market is based on the trading days, and each trading day can be divided into 24 or 48 time periods. Taking 24 time periods as an example, each trading time period is one hour. The market organizer is responsible for organizing the bidding and trading of the spot market, collecting the quotation information of both parties, and determining the transaction result. This example uses a bilateral quotation as an example. The bidding parties submit their quotation curves separately, as shown in Tables 2 and 3. The quote parameters in Tables 2 and 3 are randomly selected.

Table 2. Power plant bidding parameters.

Power Plant	Quote Factor	Power Plant	Quote Factor
1	1.03	63	1.01
24	1.02	68	1.02
26	0.98	69	0.98
36	0.96	73	1.00
52	0.99	-	-

Table 3. User's quote parameters.

Node	Quote Factor	Node	Quote Factor	Node	Quote Factor
2	1.03	25	1.01	48	1.03
3	1.02	26	0.99	49	1.02
4	0.98	27	1.01	50	0.98
5	0.96	28	1.03	51	0.96
6	0.99	29	1.02	52	0.99
7	1.01	30	0.98	53	1.01
8	1.03	31	0.96	54	1.03
9	1.02	32	0.99	55	1.02
10	0.98	33	1.01	56	0.98
11	0.96	34	1.03	57	0.96
12	0.99	35	1.02	58	0.99
13	1.01	36	0.98	59	1.01
14	1.03	37	0.96	60	1.03
15	1.02	38	0.99	61	1.02

5.2. Market Clearing

Based on the above data, we used two schemes to verify the impact of the flexible load on market clearing.

Case One: The market clearing model does not consider the flexible load, and the power consumption behavior of the flexible load does not change.

Case Two: 4.97% of the transferable load participates in the market clearing, and the transferable load can change the power usage behavior according to the cost of the electricity.

The marginal electricity price of a node is defined as the increase of the 1 MW load at the node. Under the premise of ensuring safety, the minimum production cost of the system can be obtained through the optimization model of market bidding [32]. The node's marginal price can be obtained by solving the Lagrange function, and the definition of the node's marginal electricity price can be applied in the market clearing model [33,34].

By comparing the two schemes, the impact of the flexible load on the market clearing model can be analyzed.

From Figures 9 and 10, we can know the marginal node electricity price of the market clearing. The difference is that Figure 9 does not consider flexible loads, and Figure 10 considers flexible loads. The results show that the node electricity price can be obtained successfully, and it is very close to the actual electricity price, which has great engineering application value.

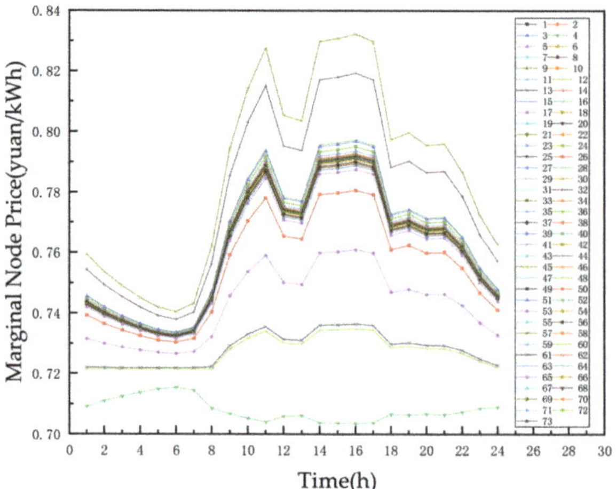

Figure 9. Marginal node price without considering the flexible load.

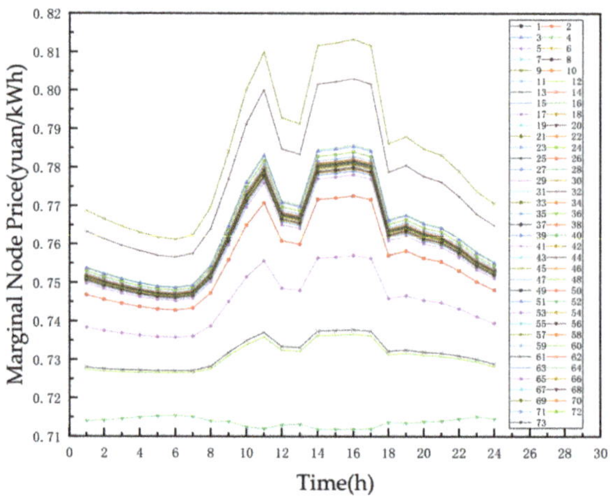

Figure 10. Marginal node price considering the flexible load.

When the convertible load is not considered, the highest price of the node is 0.832 yuan/kWh, and the lowest price is 0.704 yuan/kWh. Considering the participation of the convertible loads, the highest electricity price in all aspects is 0.813 yuan/kWh, and the lowest electricity price is 0.712 yuan/kWh. Therefore, the participation of the convertible loads in the market can clearly reduce the electricity price and reduce the peak-to-valley electricity price difference in the market. This result is in line with expectations, because the flexible load itself is economical.

During the market clearing process, the users of convertible loads are shifted and allocated to the time when the cost of electricity is lower. Because of this, the load curve will change, and the convertible load during the peak period of power consumption will shift to the lower period.

The load curve after the market clearing is shown in Figure 11. The peak-to-valley difference of the initial load is 33.71%. Compared with the initial load, the peak-to-valley difference of the load curve is reduced by 11.40%. This article is aimed at a specific urban power grid, which has a high proportion of load density. In other high-density load cities, the load characteristics have similar properties to the urban power grid. The research in this paper can be extended to similar urban power grids, which can effectively alleviate the challenges to the supply of electricity and improve the load curve.

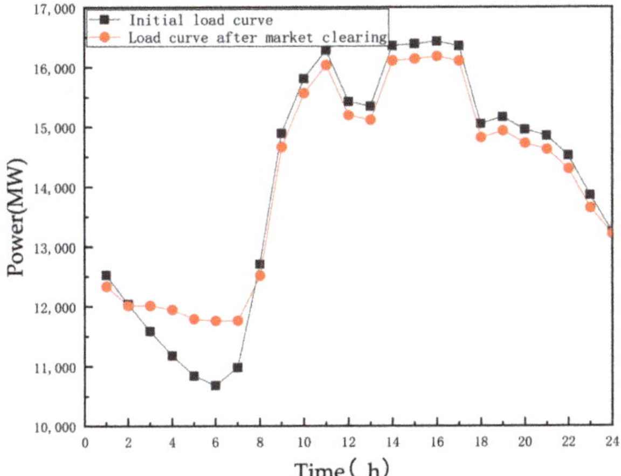

Figure 11. Load curve after the market clearing.

5.3. Safety Checking

In the safety checking of the power grid, the participation of the load reduction should be considered, and the maximum reduction is 20% [14]. As shown in Figure 12, the 76th line exceeds the limit, and the limit is 196 MW. In order to ensure the safety of the line flow, two cases were adopted. The cases are described below.

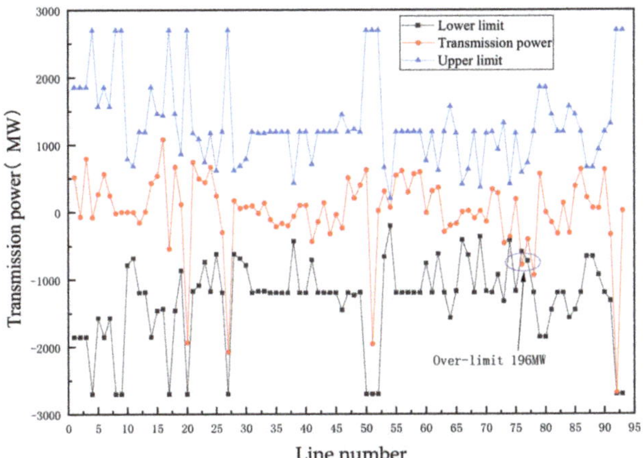

Figure 12. Transmission power of the 76th line.

The first case does not consider the participation of flexible loads and directly performs safety checking and correction through unit adjustment.

The second case considers that the load can be reduced to participate in the safety checking and correction. The load that can be reduced accounts for 10.09% of the total load.

The safety checking operation of the 76th line was carried out in the above two cases. We now analyze the differences between the two cases.

Based on the above data, the safety checking line power flow results obtained in Cases 1 and 2 are shown in Figure 13. The diagram shows that both cases can ensure that the line flow does not exceed the limit and can ensure the safe operation of the network; the transmission power of the line does not exceed its upper and lower limits.

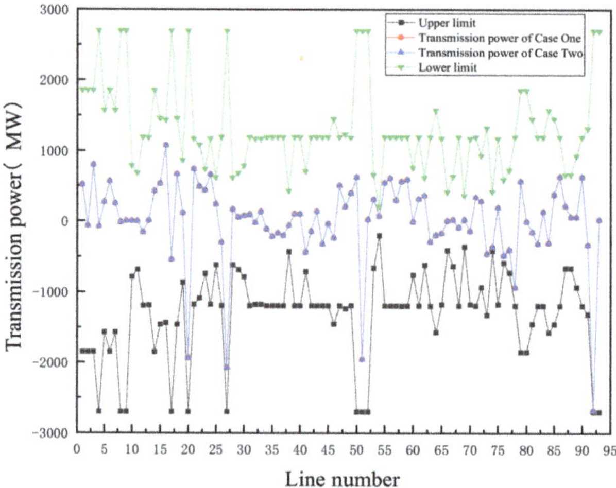

Figure 13. The safety checking line power flow results obtained in Cases One and Case Two.

The output adjustment results of the two cases are shown in Table 4. Table 4 shows that power plant 52 in Case Two emits 5.61 MW more power than Case One, and power plant 63 in Case One emits 137.89 MW more power than Case Two. The reason is that the reducible loads can participate in the safety checking in Case Two, and balance the power plant output by reducing the flexible load.

Table 4. The output adjustment results of the two cases.

Plant	Case One(MW)		Case Two(MW)		Lower Limit(MW)	Upper Limit(MW)
	Adjustment amount	Output after Adjustment	Adjustment Amount	Output after Adjustment		
1	0	2544. 52	0	2544.52	415.2	4152
24	0	1170.00	0	1170.00	117	1170
26	0	1170.00	0	1170.00	117	1170
36	0	440.00	0	440.00	44	440
52	−423.16	1536. 84	−417.55	1542.45	196	1960
63	423.16	−776.84	285.27	−914.73	−1200	1200
68	0	3039.67	0	3039.67	600	6000
69	0	2160.00	0	2160.00	268	2680
73	0	1660. 80	0	1660.80	415.2	4152

The adjustable amount of the reducible load results in Case Two are shown in Figure 14. The total amount of the reducible load after the reduction can be seen, where the total reduction in the flexible loads is 132.28 MW, which makes up for the power shortage of Unit 2 in Case Two.

From the above analysis, we know that the reducible load has a rapid response. By reducing the load and participating in the safety checking, the adjustment amount of the generator set can be reduced. In this way, the role of load reduction can be brought into full play, and the number of start–stops of the unit can also be reduced.

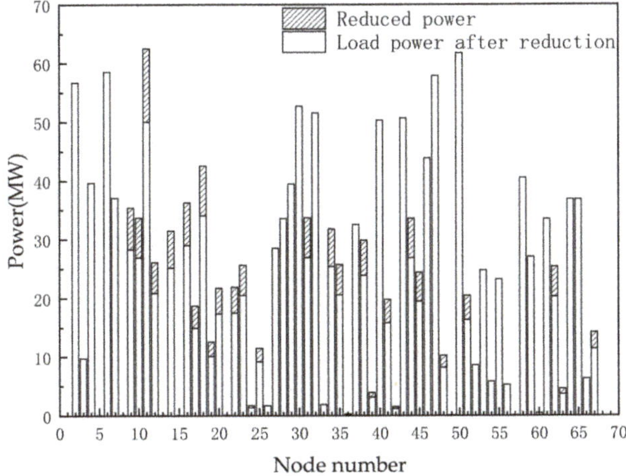

Figure 14. The flexible load reduction results in Case Two.

5.4. Economic Analysis

In order to reflect the impact of the flexible load on the safety checking, the safety checking costs of the two cases are compared in Table 5. The unit adjustment cost of the second case is less than that of the first case, mainly because it can reduce the load and make up the adjustment task of the unit, reducing the unit adjustment cost, and the load loss of the second option is lower than the unit output adjustment cost.

Table 5. The safety checking costs of the two cases.

Cost	Unit Adjustment Cost (yuan)	Out-Of-Load Cost (yuan)	Total Cost (yuan)
Case One	846,300	0	846,300
Case Two	702,800	132,300	835,100
Difference	143,500	−132,300	11,200

Table 5 shows that the total cost of the safety checking for the second case is smaller than the first case, and the cost savings are about 11,200 yuan. Table 6 reflects the impact of the load reduction on the safety checking. It can be seen from the table that the reducible load participates in the increased capacity that can increase the safety checking power adjustment.

Table 6. The impact of load reduction on safety checking.

Case	Increase Capacity (MW)	Reduce Capacity (MW)
Case One	6779.01	10,833.27
Case Two	7102.69	10,833.27
Difference	323.68	0

6. Conclusions

The paper proposed a market clearance and safety checking method for multi-type units that considers flexible loads. A market clearance model that considers the convertible loads and a safety checking model that considers the reducible loads were established. The safety checking can further eliminate the out-of-limit flow of the branches and sections and ensure the safety of the market clearing. A case study of a 220 kV and above power grid was used to analyze the results. The results show that the convertible loads participating in the market clearing can reduce the node's marginal electricity price and reduce the peak and valley electricity price. Reducible loads participate in the safety checking, compared with the single adjustment of the power of the generator set, the power adjustment amount of the generator set is reduced; at the same time, the cost of safety checking is reduced. However, in engineering applications, the establishment of a reasonable incentive/price mechanism is the prerequisite for guiding flexible loads to participate in market clearing and safety checking. In the next study, in-depth research will be conducted on this issue.

Author Contributions: Conceptualization, D.H.; Data curation, C.Z.; Formal analysis, C.Z.; Funding acquisition, D.H.; Investigation, W.C.; Methodology, C.Z.; Project administration, D.H.; Resources, D.H.; Software, C.Z.; Supervision, W.C.; Validation, D.H.; Visualization, W.C.; Writing—original draft, W.C.; Writing—review & editing, W.C. All authors have read and agreed to the published version of the manuscript.

Funding: This research was funded by the National Key Research and Development Program of China [2016YFB0900100].

Conflicts of Interest: The authors declare no conflict of interest.

References

1. Shareef, H.; Ahmed, M.S.; Mohamed, A.; Al Hassan, E. Review on Home Energy Management System Considering Demand Responses, Smart Technologies, and Intelligent Controllers. *IEEE Access* **2018**, *6*, 24498–24509. [CrossRef]
2. Bahrami, S.; Sheikhi, A. From Demand Response in Smart Grid toward Integrated Demand Response in Smart Energy Hub. *IEEE Trans. Smart Grid* **2015**, *7*, 650–658. [CrossRef]
3. Wang, K.; Hu, X.; Li, H.; Li, P.; Zeng, D.; Guo, S. A Survey on Energy Internet Communications for Sustainability. *IEEE Trans. Sustain. Comput.* **2017**, *2*, 231–254. [CrossRef]
4. Wang, D.; Liu, L.; Jia, H.; Wang, W.; Zhi, Y.; Meng, Z.; Zhou, B. Tianjin University Review of key problems related to integrated energy distribution systems. *CSEE J. Power Energy Syst.* **2018**, *4*, 130–145. [CrossRef]
5. Shi, Q.; Li, F.; Liu, G.; Shi, D.; Yi, Z.; Wang, Z.; Liu, G. Thermostatic Load Control for System Frequency Regulation Considering Daily Demand Profile and Progressive Recovery. *IEEE Trans. Smart Grid* **2019**, *10*, 6259–6270. [CrossRef]
6. Tu, J.; Zhou, M.; Cui, H.; Li, F. An Equivalent Aggregated Model of Large-Scale Flexible Loads for Load Scheduling. *IEEE Access* **2019**, *7*, 143431–143444. [CrossRef]
7. Liu, F.; Bie, Z.; Liu, S.; Li, G. Framework Design Transaction Mechanism and Key Issues of Energy Internet Market. *Autom. Electr. Power Syst.* **2018**, *42*, 114–123.
8. Lezama, F.; Soares, J.; Hernandez-Leal, P.; Kaisers, M.; Pinto, T.; Vale, Z. Local Energy Markets: Paving the Path Toward Fully Transactive Energy Systems. *IEEE Trans. Power Syst.* **2019**, *34*, 4081–4088. [CrossRef]
9. Lezama, F.; Soares, J.; Canizes, B.; Vale, Z. Flexibility management model of home appliances to support DSO requests in smart grids. *Sustain. Cities Soc.* **2020**, *55*, 102048. [CrossRef]
10. Borenstein, S.; Bushnell, J.; Stoft, S. The Competitive Effects of Transmission Capacity in a Deregulated Electricity Industry. *Natl. Bur. Econ. Res.* **2000**, *31*, 294–325. [CrossRef]
11. Di Somma, M.; Graditi, G.; Siano, P. Optimal Bidding Strategy for a DER Aggregator in the Day-Ahead Market in the Presence of Demand Flexibility. *IEEE Trans. Ind. Electron.* **2018**, *66*, 1509–1519. [CrossRef]
12. Liu, P.; Ding, T.; He, Y.; Chen, T. Integrated Demand Response in Multi-Energy Market Based on Flexible Loads Classification. In Proceedings of the 2019 IEEE Innovative Smart Grid Technologies—Asia (ISGT Asia), Chengdu, China, 21–24 May 2019; pp. 4346–4350.
13. Chen, T.; Pourbabak, H.; Liang, Z.; Su, W. An Integrated eVoucher Mechanism for Flexible Loads in Real-Time Retail Electricity Market. *IEEE Access* **2017**, *5*, 2101–2110. [CrossRef]
14. Hanif, S.; Creutzburg, P.; Gooi, H.B.; Hamacher, T. Pricing Mechanism for Flexible Loads Using Distribution Grid Hedging Rights. *IEEE Trans. Power Syst.* **2019**, *34*, 4048–4059. [CrossRef]
15. Reddy, S.S.; Bijwe, P.R.; Abhyankar, A.R. Optimal Posturing in Day-Ahead Market Clearing for Uncertainties Considering Anticipated Real-Time Adjustment Costs. *IEEE Syst. J.* **2013**, *9*, 177–190. [CrossRef]
16. Ye, H.; Ge, Y.; Shahidehpour, M.; Li, Z. Uncertainty Marginal Price, Transmission Reserve, and Day-Ahead Market Clearing With Robust Unit Commitment. *IEEE Trans. Power Syst.* **2017**, *32*, 1782–1795. [CrossRef]

17. Huang, C.; Yan, Z.; Chen, S.; Yang, L.; Li, X. Two-stage market clearing approach to mitigate generator collusion in Eastern China electricity market via system dynamics method. *IET Gener. Transm. Distrib.* **2019**, *13*, 3346–3353. [CrossRef]
18. Tan, F.Z.; Chen, G.J.; Zhao, J.B. Joint optimization model of peak-valley time-of-use electricity price based on energy-saving dispatch oriented electricity generation and electricity sales. *Proc. CSEE* **2009**, *29*, 55–62.
19. Streiffert, D.; Philbrick, R.; Ott, A. A mixed integer programming solution for market clearing and reliability analysis. In Proceedings of the IEEE Power Engineering Society General Meeting, San Francisco, CA, USA, 12–16 June 2005; Volume 3, pp. 2724–2731.
20. Xavier, Á.S.; Qiu, F.; Wang, F.; Thimmapuram, P.R. Transmission Constraint Filtering in Large-Scale Security-Constrained Unit Commitment. *IEEE Trans. Power Syst.* **2019**, *34*, 2457–2460. [CrossRef]
21. Chen, S.; Cheng, R.S. Operating Reserves Provision from Residential Users through Load Aggregators in Smart Grid: A Game Theoretic Approach. *IEEE Trans. Smart Grid* **2017**, *10*, 1588–1598. [CrossRef]
22. Bahrami, S.; Amini, M.H.; Shafie-Khah, M.; Catalao, J.P.S. A Decentralized Electricity Market Scheme Enabling Demand Response Deployment. *IEEE Trans. Power Syst.* **2018**, *33*, 4218–4227. [CrossRef]
23. Wang, S.; Bi, S.; Zhang, Y.J. Demand Response Management for Profit Maximizing Energy Loads in Real-Time Electricity Market. *IEEE Trans. Power Syst.* **2018**, *33*, 6387–6396. [CrossRef]
24. Kong, P.-Y.; Karagiannidis, G.K. Charging Schemes for Plug-In Hybrid Electric Vehicles in Smart Grid: A Survey. *IEEE Access* **2016**, *4*, 6846–6875. [CrossRef]
25. Baldick, R. Electricity Market Equilibrium Models: The Effect of Parameterization. *IEEE Power Eng. Rev.* **2002**, *22*, 53. [CrossRef]
26. De Paola, A.; Angeli, D.; Strbac, G. Price-Based Schemes for Distributed Coordination of Flexible Demand in the Electricity Market. *IEEE Trans. Smart Grid* **2017**, *8*, 3104–3116. [CrossRef]
27. Prabavathi, M.; Gnanadass, R. Electric power bidding model for practical utility system. *Alex. Eng. J.* **2018**, *57*, 277–286. [CrossRef]
28. Verbič, G.; Canizares, C.A. Probabilistic Optimal Power Flow in Electricity Markets Based on a Two-Point Estimate Method. *IEEE Trans. Power Syst.* **2006**, *21*, 1883–1893. [CrossRef]
29. Morales, J.M.; Ruiz, J.P. Point Estimate Schemes to Solve the Probabilistic Power Flow. *IEEE Trans. Power Syst.* **2007**, *22*, 1594–1601. [CrossRef]
30. Luenberger, D.G. *Linear and Nonlinear Programming*, 2nd ed.; Addison-Wesley: Reading, MA, USA, 1989.
31. Floudas, C.A. *Nonlinear and Mixed-Integer Optimization: Fundamentals and Applications*; Oxford Univ. Press: New York, NY, USA, 1995.
32. Chazarra, M.; Perez-Diaz, J.I.; Garcia-Gonzalez, J. Value of perfect information of spot prices in the joint energy and reserve hourly scheduling of pumped storage plants. In Proceedings of the 2016 13th International Conference on the European Energy Market (EEM), Porto, Portugal, 6–9 June 2016; pp. 1–6.
33. Sotkiewicz, P.M.; Vignolo, J.M. Nodal Pricing for Distribution Networks: Efficient Pricing for Efficiency Enhancing DG. *IEEE Trans. Power Syst.* **2006**, *21*, 1013–1014. [CrossRef]
34. Fernandez-Blanco, R.; Arroyo, J.M.; Alguacil, N. On the Solution of Revenue- and Network-Constrained Day-Ahead Market Clearing Under Marginal Pricing—Part II: Case Studies. *IEEE Trans. Power Syst.* **2017**, *32*, 220–227. [CrossRef]

Article

Optimal Selection of Time-Current Characteristic of Overcurrent Protection for Induction Motors in Drives of Mining Machines with Prolonged Starting Time

Jarosław Joostberens [1,*], Adam Heyduk [1], Sergiusz Boron [1] and Andrzej Bauerek [2]

1 Department of Electrical Engineering and Industrial Automation, Faculty of Mining, Safety Engineering and Industrial Automation, Silesian University of Technology, Akademicka 2 St., 44-100 Gliwice, Poland; Adam.Heyduk@polsl.pl (A.H.); Sergiusz.Boron@polsl.pl (S.B.)

2 Ośrodek Pomiarów i Automatyki S.A., Hagera 14 Street., 41-800 Zabrze, Poland; andrzejbauerek@opa.pl

* Correspondence: Jaroslaw.Joostberens@polsl.pl

Received: 20 July 2020; Accepted: 28 August 2020; Published: 29 August 2020

Abstract: The aim of this study is to analyze the effects of mine power network impedance on the starting time of induction motors, as well as on the operation of overcurrent protection relay. Proper selection of the time-current characteristic of overcurrent protection is crucial for the operation of the drive. A specific feature of mining power grids is their high impedance, which results from long cable lines with relatively small cross-sections. This causes relatively large voltage drops and significantly reduces the starting torque of the motor. Reduced starting torque increases the starting time and intensifies the motor overheating. This study analyzes a series of standardized time-current characteristics used in Invertim company protection devices. A simulation study of startup current and starting time was conducted for an exemplary medium-power motor with a large inertia fan at different values of power supply voltage below the rated value. Parameters of the motor equivalent circuit were calculated on the basis of manufacturer data. A new shape of the time-current characteristic has been proposed that would allow for prolonged starting at significant voltage drop in the mine network, ensuring protection from failed starting. This solution can be implemented in digital protection relays in addition to the standard characteristics.

Keywords: electric safety; induction motor; fan; overcurrent protection

---

1. Introduction

Overcurrent protection relays play an important role in providing the right conditions for safe and reliable electric drive system operation of mining machines. In accordance with Reference [1], electric power equipment and installations should be protected against the effects of line-to-earth short-circuits, line-to-line insulation faults and overloads in order to ensure the safety of the mine and its employees. The issue of safety becomes paramount in rooms which are at risk of methane and/or coal dust explosion. To minimize these risks, adequate protections and their settings in electric power installations must be selected in accordance with Reference [1] and with the principles of mining technology regarding safeguard and protective measures in mining power engineering. Modern microprocessor technology offers great opportunities for wide-range shaping of the time-current characteristics of those protection devices. The selection of appropriate curves of this characteristic must depend on the operating conditions of a given motor. This task is especially difficult in the case of motors which power high-moment inertia devices (e.g., fans). In such cases, the selection of settings must take into account the prolonged starting time and standard requirements. These difficulties are linked with the reliability of conducting a direct start of an electrical motor (with a grid voltage), while protecting the winding of the stator and rotor against the thermal effects of prolonged exposure to the starting current. The thermal effect of this current, which consists of temperature increase of the stator and rotor winding, may cause damage to the motor in case of the malfunctioning of protections or the cooling system. Another case that results in thermal damage to the winding of the motor is inadequate overcurrent protection. Proper selection of time-current characteristics is a complex issue that includes economic factors (linked to the costs of the motor itself, its replacement or repairs, but mostly costs linked to the interruption in operation), technological factors, as well as electric shock, fire or explosive safety issues (e.g., linked to a lack of ventilation). Proper thermal protection of the induction motor is particularly important for motors (and their cable lines) operating in locations at risk of explosion [2].

Challenges of coordinating characteristics of overload protection are presented in Reference [3], however this paper puts emphasis on comparing models defined in Reference [4] with machines and drives supplied from real power networks. This paper evaluates the selection method concerning time-current characteristics in relation to the auxiliary ventube fan motor and the cable network by which it is powered, as an example of a device with prolonged starting time. The following factors were included in the evaluation of the characteristics:

- Starting current course during start-up and
- The starting time.

These two factors have a great impact on both the reliability of the motor windings itself and on the reliable operation of overload protection relays. The detailed analysis is given below.

In case of double-cage motors, a protection against too-long or unsuccessful startups should be applied, because not only can a stator insulation temperature rise be an issue, but also the starting (outer) cage can exceed its temperature limit, even if the temperature of stator winding is below the permissible value [5,6]. In case of large medium voltage (over 1 kV) motors, this is usually accomplished by a separate protection relay (nowadays usually a part of a multifunction digital relay), but for smaller low voltage motors, it is desirable to use simple overcurrent relays, both for overloads and prolonged startups protection. It can be achieved by a careful selection of their time-current characteristics, but there is no suitable curve in international standards.

Our paper analyzes the high-inertia drive startup process in a comprehensive way—taking into account voltage drops in the supplying network, and is focused on double-cage motor drives. Obtained results are used for design of a quite new time-current characteristic curve, placed between curves defined in an international standard [4]. This curve allows for a prolonged start, but at the same time, protects the motor windings against small but long-lasting overloads.

Shapes of overcurrent protection characteristics are discussed in References [7–9], but they are based on IEEE standards [10]. Different formulae for the description of these curves are used in the abovementioned papers, while this work uses a formula described in the IEC standards [4]. The problem of a long-term asymmetry of supply voltages is not considered in this paper, as it is not a significant problem in mine networks (due to the very small number and power of single-phase loads).

The detailed modeling of thermal processes in an induction motor is possible [5,6,11] but it needs an extensive knowledge on particular motor design details (materials and internal structure). This knowledge is usually unavailable from the motor manufacturer, so most overcurrent protective relay solutions are based on a simple first-order model [12] and its appropriate mathematical description used in standards [4,10].

2. Materials and Methods

The starting time is related to the starting current, which is a multiple of the nominal current. Therefore, this state must be treated as improper operating conditions, according to Reference [13]. The starting process should be as short as possible, but no longer than the time it would take for the motor winding to heat up (due to the starting current) to the maximum permissible temperature for a given class of insulation. An overload-protective device with a properly chosen time-current characteristic should protect against exceeding the maximum permissible temperature (and thus the thermally induced damage of the motor).

High grid impedance of an underground power network with long cable lines and medium power transformers causes big voltage drops during the flow of the starting current. These voltage drops decrease the voltage level on starting motor terminals and lessen mechanical torque, considerably increasing the starting time. Ventube fan drives specifically are characterized by a large moment of inertia, and in the case of long ventube, they need a long time to fill with air. Thus, a small accelerating torque (at low voltage) causes a long starting time at relatively low starting currents (not detectable by standard overcurrent protection relays set close to locked-rotor currents). This prolonged startup due to the voltage drops problem exists not only in networks with territorially distributed loads, but also in small networks powered from weak (e.g., autonomous) sources [14,15]. A mathematical description of the overload protection device characteristics is included in Reference [4]. The relation between protection delay time, T, and the current value, I/In, of overcurrent protection devices is expressed by the following Equation (1):

$$T = \frac{C}{\left(\frac{I}{I_n}\right)^{\alpha} - 1} \tag{1}$$

The values of parameters α and C included in (1) are shown in Table 1. This table also includes the values of parameters of (1) for the time-current protection device characteristic proposed by the manufacturer of protection relays for Invertim mining equipment [16].

Table 1. Values of parameters α and C in accordance with References [4,16].

Item	Characteristic	α	C
1	Standard Inverse	0.02	0.14
2	Very Inverse	1	13.5
3	Extremely Inverse	2	30
4	Long Inverse	1	120
5	Invertim	1.2	140

Figure 1 below shows the shapes of individual time-current characteristics of overcurrent protections from (1) and values of the parameters included in Table 1.

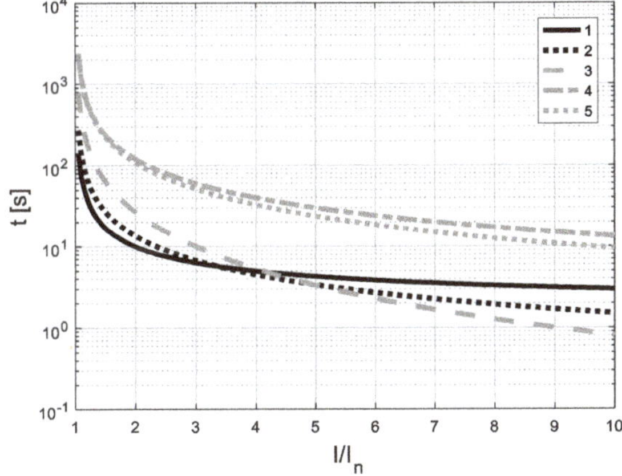

Figure 1. Time-current characteristics of overcurrent protections with characteristics consistent with European Standards [4] and Invertim [16]: 1—Standard Inverse, 2—Very Inverse, 3—Extremely Inverse, 4—Long Inverse, 5—Invertim.

Simulations included the study of the WLE-1013/B/E/1 auxiliary axial flow fan powered by a double-cage dSOKg 200L2B-E electric motor. Table 2 presents the parameters of the motor used in the simulation studies. These motor parameter calculations were performed according to the methodology given in Reference [17].

Table 2. Parameters of the dSOKg 200L2B-E motor [18].

Quantity	Unit	Value	Quantity	Unit	Value
P_{rM}	kW	37.0	J	kg·m^2	0.180
U_{rM}	V	500	m	kg	335
n_{rM}	min^{-1}	2960	R_s	Ω	0.146
M_{rM}	N·m	119	L_s	mH	2.07
η_{rM}	%	83.8	R_{r1}	Ω	0.087
$cos\varphi_r$	–	0.89	L_{r1}	mH	2.611
I_{rM}	A	51	R_{r2}	Ω	0.424
M_{LR}/M_{rM}	–	2.2	L_{r2}	mH	0.14
I_{LR}/I_{rM}	–	6.7	L_m	mH	42
M_{max}/M_{rM}	–	2.0			

The motor parameters have been calculated on the basis of rated data, and as such, they acted as a basis for the design of the speed regulation characteristic, as well as stator current ratio as a function of rotational speed. The obtained speed-torque and speed-current curves indicate sufficient similarity with device characteristics presented by the manufacturer, the Cantoni Group [18]. These motor parameters were used to evaluate the stator voltage influence on the accelerating time characteristics of the dSOKg 200L2B-E motor, M_e = f(n), $I1$ = f(n), and the course of the RMS value of stator current in time at the assumed range of stator voltage variation between 0.7·U_n and 1.0·U_n with a step of 0.1·Un in a situation of powering the WLE-1013/B/E/1 auxiliary jet fan with that motor. Many interesting methods exist to calculate the starting time (such as References [19,20]). However, those are limited to single cage motors, while this paper focuses on the double-cage motors. Double-cage induction motors simultaneously provide a lower value of starting current and a higher value of starting torque, hence they are widely used in mining. This feature is particularly important in relatively weak power supply networks. In a later stage of the study, calculated series of motor characteristics—where the stator voltage is the main parameter—were used to analyze the selection of time-current characteristics of overcurrent protections for the induction motor in the drive of the axial auxiliary jet fan. The motor was supplied by a flameproof mining mobile transformer station IT3Sm-400/6 (Izol-Plast) and a mining certified cable YHKGXSekyn 3 × 16/15 mm^2 (manufactured by TeleFonika). Parameters of the power supply system are presented in Tables 3 and 4. The medium voltage (6kV) network impedance has been neglected because of its relatively small value (as it is divided by the square of the transformer voltage ratio).

Cable length was calculated to obtain selected values of the motor terminals voltage at startup. The selected cable length values together with corresponding voltage levels are presented in Table 5. These calculations were done using a scheme of voltage divider composed of a supply system (transformer and cable) and a load (motor).

Table 3. Parameters of the flameproof mobile transformer station used for the supply of the analyzed ventube fan drive.

Quantity	Unit	Value
Type		IT3Sm = 400/6
Rated power	kVA	400
Primary voltage	V	6000
Secondary voltage	V	525
Short-circuit voltage	%	4.3
Rated load losses	kW	1.4

Table 4. Cable parameters.

Quantity	Unit	Value
Current carrying capacity	A	104
Per-unit length resistance	Ω/km	0.088
Per-unit length inductance	mH/km	0.28

Table 5. Cable length values used in further calculations to obtain the selected motor terminals voltage levels.

Cable Length (km)	Voltage Level at Motor Startups
0.65	0.9 U_n
1.74	0.8 U_n
3.131	0.7 U_n

Figure 2 below shows the relationship between the startup voltage level and cable length in the whole analyzed range. Selected points were marked with circles. The number of these points was limited for the sake of clarity.

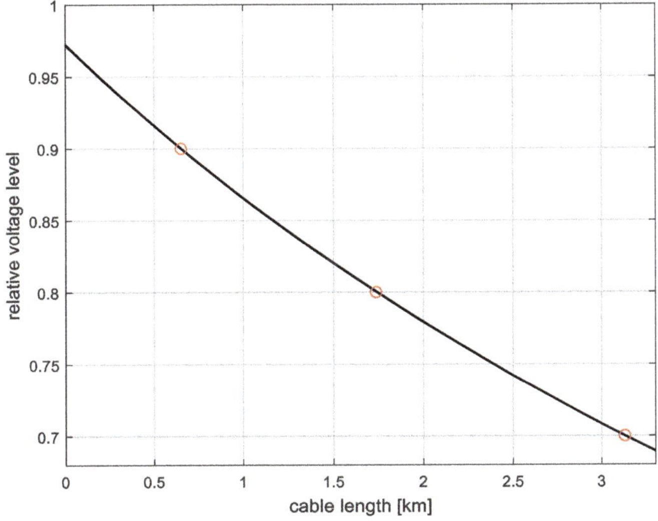

Figure 2. Dependence of the startup voltage level on cable length in the whole analyzed range.

Figures 3–5 show the influence of the stator voltage on the mechanical characteristics of the motor driving the auxiliary jet fan, on the stator current, as well as on the rotational speed during starting of the motor.

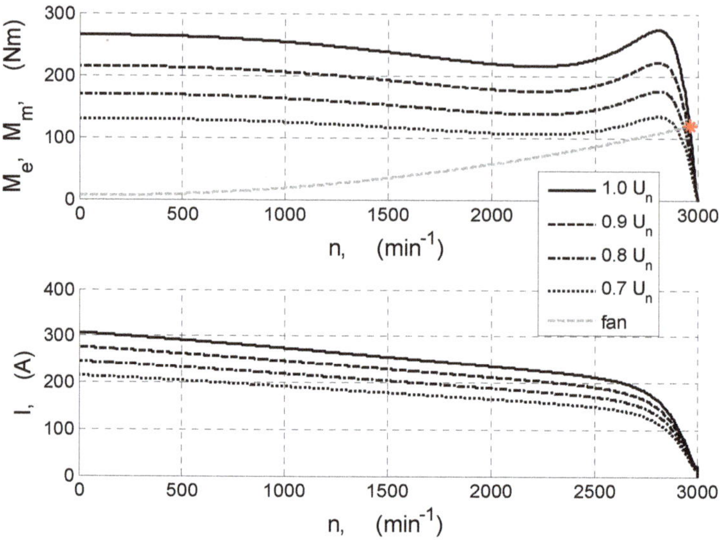

Figure 3. Relation of electromagnetic torque and stator current of the dSOKg 200L2B-E motor as a function of speed for four voltage levels at the motor terminals.

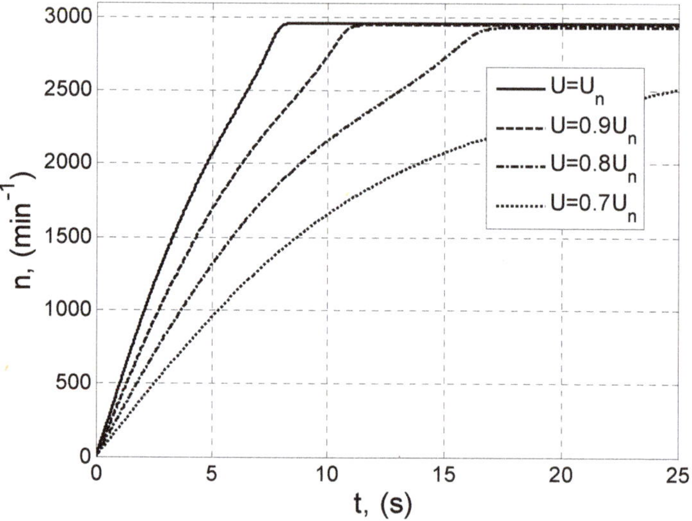

Figure 4. Course of rotational motor speed while powered by different power network voltages in the range between $0.7 \cdot U_n$ and U_n.

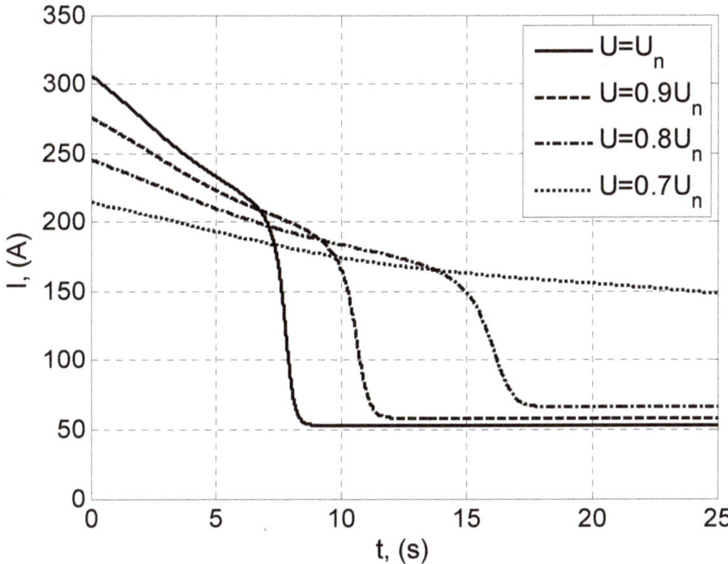

Figure 5. Course of motor current while powered by voltage in the range between $0.7 \cdot U_n$ and U_n.

As can be seen in Figure 4, the lowest momentary value of accelerating torque occurs at the lowest voltage level, i.e., $0.7 \cdot U_n$, which results in the inability to conduct the starting process in a time shorter than 25 s (Figures 4 and 5). In this situation, the motor current reaches a RMS value which is many times higher than the nominal one and in that time, it will not be reduced to a value close to the nominal (Figure 5).

Reaching a stable motor operating point ($dn/dt = 0$) is possible for stator voltage equal to at least $0.8 \cdot U_n$, while the starting time is in a range between 8 (for $U_1 = U_n$) and 17.5 s ($U_1 = 0.8 \cdot U_n$).

Overcurrent protection should protect the motor (and supplying cable) against the effects of overcurrent, allowing for the starting of the device. This is not an easy task in case of drives with prolonged starting times, which may be observed by matching courses of stator currents from Figure 4 with time-current characteristics of the protections (Figure 6).

The point of intersection between the time-current characteristic of the protection and the current course curve of the motor is the time at which the motor shuts off prematurely, which makes it impossible to finish the starting process of the drive.

The selection of overcurrent protection settings in accordance with Condition (1) indicates that the choice to use a protection with characteristics consistent with Reference [4] was formally correct.

But, as can be seen in Figure 6, protections with characteristics of Standard Inverse, Very Inverse and Extremely Inverse, set in accordance with Reference [21] to a value of $1{,}1 \cdot I_n$, will end the starting process after ca. 3.8 to 5 s, regardless of the voltage value on the motor terminals. When using protection with characteristics of Long Inverse or Invertim, starting the motor will be possible for each analyzed voltage. However, the efficiency of the protection might prove insufficient in case of a failed start, which occurs at voltage equal to 0.7 U_n, or in case of small but long-lasting overloads.

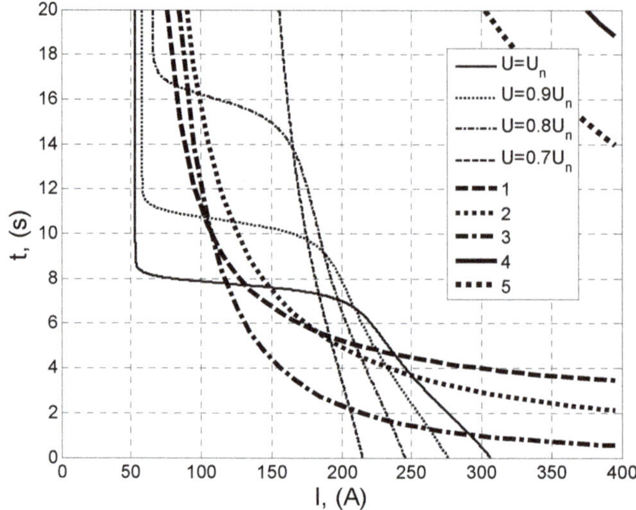

Figure 6. Courses of RMS value of stator current during its starting against the time-current characteristics of overcurrent protections with characteristics in accordance with Reference [4] and Invertim [16].

3. Results

From the results of the simulation studies, it can be concluded that there is a need to design an intermediate protection characteristic situated between Extremely Inverse and Invertim. In relation to the courses of starting currents from Figure 6, depicting a situation where the stator voltage of the motor is in range (0.8 ÷ 1) U_n, the parameters α and C of Equation (1) were identified in order to designate a characteristic of an overcurrent protection with features intermediate between protections with Standard Inverse characteristics and those with Invertim characteristics. It was assumed in a simplistic manner that the winding temperature in induction devices at starting time up to 20 s will not exceed the permissible temperature for a given class of insulation. In such a situation, the investigated characteristic should run above the characteristics of starting currents for the assumed voltage range, and its shape should be as similar as possible to the Extremely Inverse characteristic. The values of designated parameters of Equation (1) describing the investigated time-current characteristic are $\alpha = 2.3$ and $C = 178$, respectively. The course of the characteristic with designated parameters α and C is marked with the number "6" in Figure 7. As can be seen in that figure, the protection with the characteristic marked as "6" allows for the starting of the auxiliary ventube fan motor in the stator voltage range (0.8 ÷ 1)·U_n. At the same time, the protection with this characteristic ensures safety from the effects of a failed starting process, so, in the case where $U = 0.7 \cdot U_n$. It must be noted that while this protection is not as effective as devices with the characteristics of Standard Inverse, Very Inverse and Extremely Inverse, it is better than protection devices with Long Inverse and Invertim characteristics. The detailed methodology is described below.

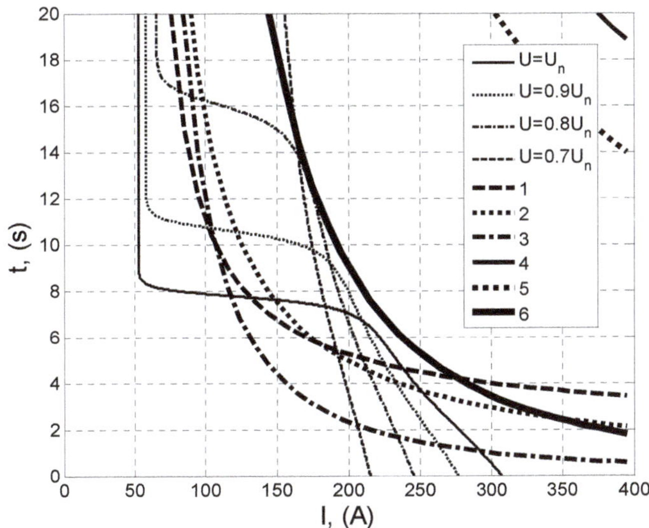

Figure 7. Proposed time-current characteristic of an overcurrent protection for an induction motor powering the auxiliary jet fan against the time-current characteristics of overcurrent protections with characteristics in accordance with References [4,16] and time courses of RMS values of starting currents of this motor for various stator voltages.

We attempted to determine a new time-current characteristic of the overload protection, as none of the standard characteristics of the overload protection given in Reference [4] take into account the specific supply and operation of mining machines with extended start-up times. The following assumptions were made to identify the Equation (1) parameter values:

1. For large current values (above 5 I_n), the characteristic should have a shape similar to the curve "2", i.e., Very Inverse.
2. The course of the characteristic should enable starting the machine in voltage conditions (0.8 ÷ 1.0) U_n, i.e., the protection characteristic should not intersect the motor time-current characteristics, which reach the steady state in a time shorter than the maximum value (20 s).
3. The characteristic should be modeled with regard to the shut-off time of a motor supplied with 0.7 U_n voltage, in a time not longer than the maximum value of 20 s.

The identification of the Equation (1) parameter values was carried out for the data points specified in the characteristics of Figure 8 that fulfill the above assumptions. Data point values for currents above 3.2 I_n are arbitrary so that the third assumption is met.

In order to calculate the values of the Equation (1) α and C parameter values, the Equation was transformed into the following form (2):

$$\hat{I}(k) = I_n \cdot \left(\frac{C}{T(k)} + 1 \right)^{\frac{1}{\alpha}} \tag{2}$$

where

$T(k)$—data points of the response time values,
$\hat{I}(k)$—calculated values of currents for given times $T(k)$ and values of coefficients α and C.
$k = 1, 2, \ldots, N$,
N—number of data points.

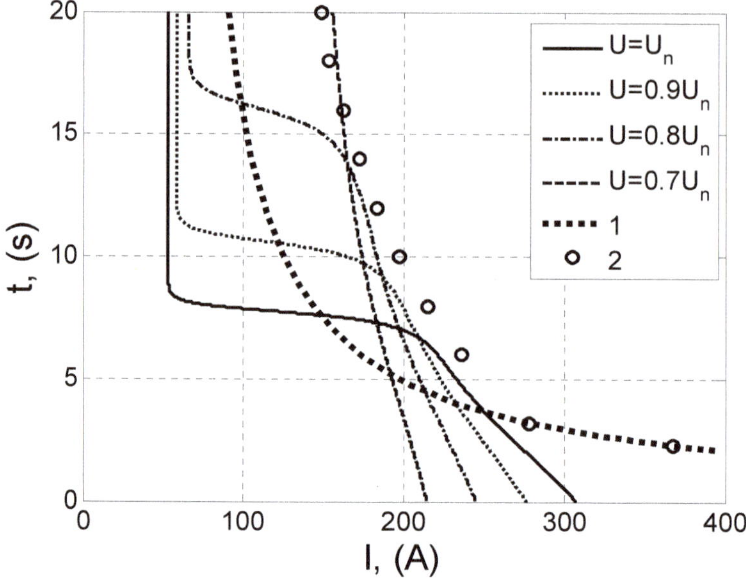

Figure 8. Data points selected for the determination of a new time-current characteristic on the background of simulated current RMS waveforms and Very Inverse characteristics. 1—Very Inverse characteristics, 2—selected data points.

Subsequently, iterative computer calculations were carried out to determine from the Equation (2) the values of the $I(k)$ currents with given values of time $T(k)$ for the parameters α and C, which were selected from the ranges ⟨2.0, 3.0⟩ and ⟨150, 190⟩, respectively. The identification task was therefore reduced to determine such values of parameters α and C that would minimize the accepted criterion in the form of the mean square error (MSE). This condition can be written as follows (3):

$$(\alpha, C) = \min\{MSE\} = min\left\{\frac{1}{N}\sum_{k=1}^{N}\left(I(k) - \hat{I}(k)\right)^2\right\} \tag{3}$$

where:

$I(k)$—data points of current values (from Figure 8) for the corresponding $T(k)$ time values.

The dependence of MSE value surface on the parameters α and C in the whole analyzed area is shown in Figure 9.

In order to verify the results obtained from the identification of the parameters of Equation (1) described above, a second method was used to determine these constants. Equation (1) can be transformed into the following form:

$$y = \left(\frac{C}{T(k)} + 1\right) = \alpha \ln\left(\frac{I(k)}{I_n}\right) = \alpha \cdot x \tag{4}$$

This formula represents a linear function. In such a situation, the α coefficient at a given C (from the range as in Figure 9) can be determined either by linear regression, or by the method of least squares. The value of Criterion (8) is then calculated for the obtained data. Using Equation (4) significantly reduces the calculation time.

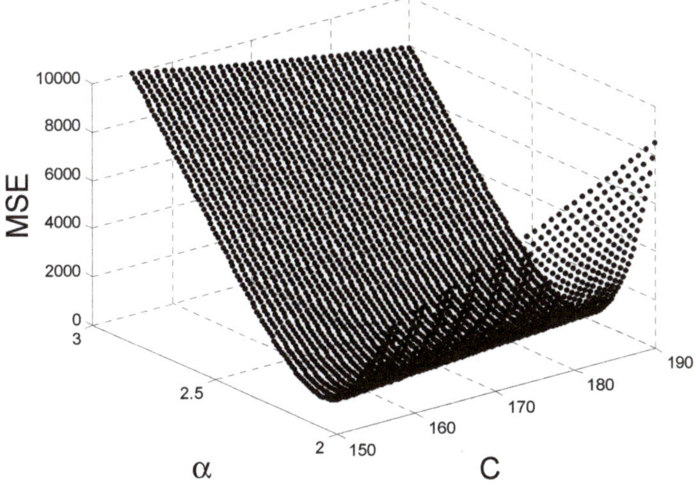

Figure 9. Mean square error (MSE) value surface dependence on α and C parameter values in the whole analyzed area for the studied motor case

If Equation (4) transforms due to α to the form:

$$\alpha(k) = \frac{\ln\left(\frac{C}{T(k)} + 1\right)}{\ln\left(\frac{I(k)}{I_n}\right)} \tag{5}$$

then its value for a given C should be constant for each point k. In fact, there are small differences for different points, so it can be determined as an arithmetic mean for a given value of C.

$$\alpha = \frac{1}{N}\sum_{k=1}^{N} \frac{\ln\left(\frac{C}{T(k)} + 1\right)}{\ln\left(\frac{I(k)}{I_n}\right)} \tag{6}$$

The performed identification calculations using the Formula (6) and Criterion (7) for C in the range as shown in Figure 9 gave identical results as the method described above—using Criterion (3). The dependence of the criterion on the C parameter for different α values obtained in this situation is shown in Figure 10.

$$C = \min\{MSE\} = \min_{\alpha}\left\{\frac{1}{N}\sum_{k=1}^{N}\left(I(k) - \hat{I}(k)\right)^2\right\} \tag{7}$$

Minimum MSE values were obtained for α = 2.3 and C = 178, which is exactly the same as the method based on Formula (2) and Criterion (3). The values of Criteria (3) and (7) are 123 in both cases.

The protection device with the proposed "6" characteristics exhibits features which are intermediate between protections with Very Inverse characteristics and those with Invertim characteristics, while the shape of this characteristic is similar to Extremely Inverse characteristics, which is consistent with the previous assumptions. In more complex operating conditions (e.g., consecutive starts) and using a more complicated model description, characteristics design can be made using more complicated and much more time-consuming evolutionary computations—like Particle Swarm Optimization, as in Reference [22].

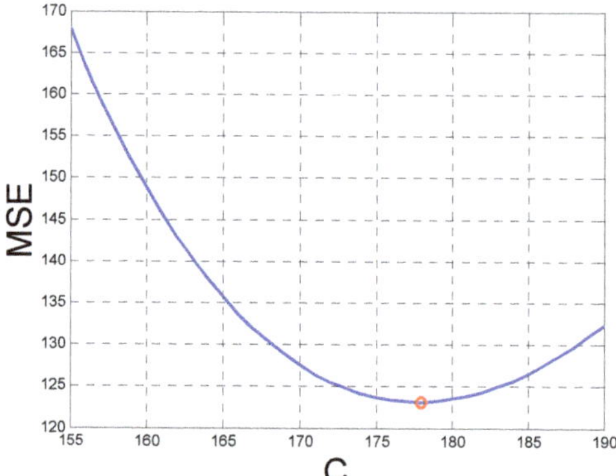

Figure 10. MSE values obtained when Formula (7) is used to identify α and C parameters.

4. Discussion

Because of the time-varying motor torque value as well as the speed-dependent ventube load torque, the starting RMS current and the starting time analysis have to be done using numerical simulation, based on the parameters for the double-cage induction motor equivalent circuit. Because of the specific structure of mine electrical power network, there is a need to analyze the motor starting conditions at different values of network voltage below the rated value. Overcurrent protections with Standard Inverse, Very Inverse and Extremely Inverse characteristics consistent with Reference [4] do not allow for starting of the drive with a high moment of inertia, especially in conditions of low power voltage. Whereas, protections with Long Inverse [4] or Invertim [16] characteristics allow for the proper starting of the motor, but barely protect the motor in case of failed starting and in case of relatively slight overloads—due to the long operating time.

It would be beneficial to introduce protection relays with characteristics that are intermediate between Standard and Extremely Inverse, as well as Long Inverse and Invertim. This paper presented the proposed parameters describing such a characteristic, designed on the basis of a numerical simulation study. These new characteristics can be easily implemented in modern digital protection relays widely used in mining power networks as they need no new equation (and therefore no significant changes in relay algorithm) but only small changes in coefficient values. Nowadays, microprocessor-based protection relays are widely used even in low-voltage mining electrical equipment, as they conveniently integrate many different protective functions (like short-circuit protection, overload protection, ground-fault protection, insulation monitoring, temperature monitoring) in one compact device. Flexible user-defined characteristics are hard to implement in electromechanical protective relays, but their role in modern industrial power networks is diminishing [23].

5. Conclusions

The time-current characteristic of overcurrent relay protection has to be carefully adjusted to the load characteristics, particularly for high-inertia drives. European Standard [4]-defined models are not suitable for some machines, nor for drives powered from weak (relatively high impedance) networks. Ventube fans and their drives are very important as their failure can cause methane explosions, or at least may force an evacuation of miners from the workplace, which leads to large economic losses. Although reliable startup of mining ventube drives is crucial for proper mine operation, no relevant relay protection characteristics exist. Nowadays, they are protected using relay characteristics far from their real startup characteristics. While such a solution enables a reliable start of the motor, it does not protect it from small but long-lasting overloads which can significantly degrade insulation characteristics and therefore shorten the life of the motor. The solution proposed in our paper can help overcome this contradiction. In addition to the selection of several pre-defined standard characteristics, there should be a possibility of continuous change of the time-current characteristic parameters.

Author Contributions: Conceptualization, A.B.; formal analysis, S.B.; methodology, J.J.; software, A.H.; writing—original draft preparation, A.H.; supervision, J.J.; investigation, A.H. and J.J.; validation, J.J.; visualization, A.H.; funding acquisition, S.B. All authors have read and agreed to the published version of the manuscript.

Funding: The APC was funded by the Silesian University of Technology, Faculty of Mining, Safety Engineering and Industrial Automation.

Acknowledgments: This work was supported by the Faculty of Mining, Safety Engineering and Industrial Automation of the Silesian University of Technology, Gliwice, Poland.

Conflicts of Interest: The authors declare no conflict of interest. The funder had no role in the design of the study; in the collection, analyses, or interpretation of data; in the writing of the manuscript, or in the decision to publish the results.

References

1. Minister of Energy ordinance regarding detailed requirements concerning the operation of coalmines. *Journal of Laws*, pos. 1118, 2017, Warsaw. Available online: https://dziennikustaw.gov.pl/DU/rok/2017/pozycja/1118 (accessed on 28 August 2020).
2. Proctor, M. Safety Considerations for AC Motor Thermal Protection. In Proceedings of the 69th Annual Conference for Protection Relay Engineers, College Station, TX, USA, 4–7 April 2016.
3. Smith, K.; Sajal, J. The necessity and challenges of modelling and coordinating microprocessor based thermal overload functions for device protection. In Proceedings of the 70th Annual Conference for Protective Relays Engineers (CPRE), College Station, TX, USA, 3–6 April 2017.
4. *IEC 60255-151:2010 Measuring Relays and Protection Equipment—Part 151: Functional Requirements for over/under Current Protection*; International Electrotechnical Commission: Geneva, Switzerland, 2010.
5. Mróz, J. The model of double-cage induction motor for the analysis of thermal fields in transient operations. *Arch. Electr. Eng.* **2017**, *66*, 397–408. [CrossRef]
6. Mróz, J.; Poprawski, W. Improvement of the Thermal and Mechanical Strength of the Starting Cage of Double-Cage Induction Motors. *Energies* **2019**, *12*, 4551. [CrossRef]
7. Gonzalez-Cordoba, J.L.; Osornio-Rios, R.A.; Granados-Lieberman, D. Thermal-Impact-Based Protection of Induction Motors Under Voltage Unbalance Conditions. *IEEE Trans. Energy Convers.* **2018**, *33*, 1748–1756. [CrossRef]
8. Ernst, T.; Farison, K. Motor Thermal Capacity Used—How Does the Relay Know When I've Reached 100%? In Proceedings of the 71st Annual Conference for Protective Relay Engineers, College Station, TX, USA, 26–29 March 2018.
9. Venkataraman, B.; Godsey, B.; Premerlani, W.; Shulman, E. Fundamentals of a Motor Thermal Model and its Applications in Motor Protection. In Proceedings of the Conference Record of 2005 Annual Pulp and Paper Industry Technical Conference, Jacksonville, FL, USA, 20–23 June 2005.
10. *Guide for AC Motor Protection, IEEE Standard C37.96*; IEEE: New York, NY, USA, 2012.
11. Krok, R. Influence of work environment on thermal state of electric mine motors. *Arch. Elect. Eng.* **2011**, *60*, 357–370. [CrossRef]
12. Boglietti, A.; Carpaneto, E.; Cossale, M.; Vaschetto, S. Stator Winding Thermal Models for Short-Time Thermal Transients: Definition and Validation. *IEEE Trans. Ind. Electr.* **2016**, *63*, 2713–2721. [CrossRef]
13. Gawor, P. *Electrical Power Networks in Mining*; Silesian University of Technology: Gliwice, Poland, 2011.
14. Haesen, E.; Minne, F.; Driesen, J.; Bollen, M. Hosting Capacity for Motor Starting in Weak Grids. In Proceedings of the IEEE International Conference on Future Power Systems, Amsterdam, The Netherlands, 18 November 2005.
15. Saiprasad, S.; Soni, N.; Doolla, S. Analysis of motor starting in a weak microgrid. In Proceedings of the IEEE International Conference on Power Electronics, Drives and Energy Systems (PEDES), Mumbai, India, 16–19 December 2014.
16. *Control and Protection Microprocessor-Based Relay Type PM-2. User Manual Ex-DTR-901.01.02*; Invertim Ltd.: Otwock Maly, Poland, 2015.

17. Pedra, J.; Corcoles, P. Double-cage induction motor parameters estimation from manufacturer data. *IEEE Trans. Energy Convers.* **2004**, *19*, 310–317. [CrossRef]
18. Cantoni Group. *Catalogue of Explosion-Proof Motors for Mine Auxiliary Jets*; Cantoni Group: Cieszyn, Poland, 2018.
19. Natarajan, R.; Misra, V.K.; Oommen, M. Time domain analysis of induction motor starting transients. In Proceedings of the 21st North American Power Symposium, Rolla, MI, USA, 9–10 October 1989.
20. Aree, P. Starting time calculation of large induction motors using their manufacturer technical data. In Proceedings of the 19th International Conference on Electrical Machines and Systems (ICEMS), Chiba, Japan, 13–16 February 2016.
21. *PN-G-42042:1998 Preventive and Protective Measures in Mine Electrical Power Engineering—Short-Circuit and Overload Protection—Requirements and Selection Rules*; Polish Committee for Standardization: Warsaw, Poland, 1998.
22. El-Amary, N.H.; Ezzat, F.A.; Mostafa, Y.G. Thermal Protection for Successively Starting Three Phase Induction Motors Using Particle Swarm Optimization Technique. In Proceedings of the 11th International Conference on Environment and Electrical Engineering, Venice, Italy, 18–25 May 2012.
23. Kullkarni, A.S.; Ugale, S.P. Advanced Thermal Overload Protection for High Tension Motors Using Digital Protection Relays. In Proceedings of the IEEE International Conference on Computer, Communication and Control, Indore, India, 10–12 September 2015.

Article

Modelling of Distributed Resource Aggregation for the Provision of Ancillary Services

Adam Lesniak *, Dawid Chudy and Rafal Dzikowski

Institute of Electrical Power Engineering, Lodz University of Technology, Stefanowskiego Str. 18/22, PL 90-924 Lodz, Poland; dawid.chudy@dokt.p.lodz.pl (D.C.); rafal.dzikowski@dokt.p.lodz.pl (R.D.)

* Correspondence: adam.lesniak@dokt.p.lodz.pl; Tel.: +48-501-699-454

Received: 26 July 2020; Accepted: 1 September 2020; Published: 4 September 2020

Abstract: Nowadays, ancillary services (ASs) are usually provided by large power generating units located in transmission networks, while smaller assets connected to distribution systems remain passive. It is expected that active distribution systems will start to play an important role due to numerous issues related to power system operation caused mainly by developing renewable generation and restrictions imposed on conventional power generating units by climate policies. The future development of the power system management will also lead to the establishment of new market agents such as distributed resource aggregators (DRAs). The article presents the concept of the DRA as part of an active distribution system enabling small resources to participate in wholesale markets, provide ASs and indicates the functions of the DRA coordinator in the modern power system. The proposed method of the DRA structure modelling with the use of the mixed-integer linear programming (MILP) is aimed at evaluating the optimal operation pattern of participating resources, the desired shape of the load profile at the point of common coupling (PCC) and the AS provision. The performed simulations of the DRA's operation show that various types of aggregated resources located in distribution networks are able to provide different services effectively to support the power system in terms of load–generation balancing and allow for further development of renewables.

Keywords: aggregation; ancillary services; distributed energy resources; optimization; power system operation

---

1. Introduction

A structure of the current power systems is based on active transmission systems where centrally dispatched generating units (CDGUs) provide different services such as balancing, frequency regulation and operating reserves for maintaining proper power system operation. However, the existing distribution systems from the point of view of the transmission system operator (TSO) still remain passive with fixed demand profiles—the left side of Figure 1.

Distributed energy resources (DERs) located in distribution networks are developing rapidly at present. Many of them, such as gas and biogas power plants and energy storages (ESs), might be controllable in order to perform services similar to those provided by CDGUs. Furthermore, active loads (ALs) and curtailment of renewable energy resources (RESs) can be employed to support the management of the power system.

The above-mentioned entities by being financially attractive or forced by legal obligation could be aggregated into the distributed resource aggregator (DRA) structure to be controlled by the DRA coordinator for the provision of ancillary services (ASs). The scope of those potentially provided ASs should be therefore identified for a given DRA.

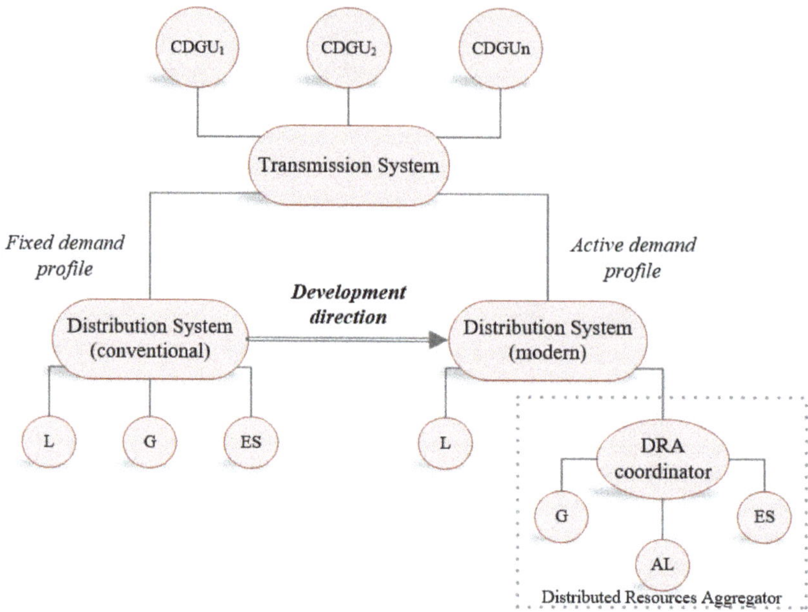

Figure 1. The development of modern power systems and relations between their entities.

The small size of distributed assets is a problem causing a lack of ability to participate in wholesale markets such as balancing markets and AS markets; moreover, legal obligations constitute a barrier for some entities. Therefore, an additional agent, that is, an aggregator, is required [1–4]. Figure 1 depicts the DRA coordinator as an entity responsible for distributed asset coordination and control enabling active participation of DERs in the management of power system operation. Nowadays, distribution system operators (DSOs) have passive operation profiles which in the near future could be modified by the DRA operation depending on the current system needs to provide different services—an active demand profile.

The concept of active distribution networks is widely discussed in the literature. For example, [5–8] propose the use of active distribution for the provision of services such as balancing and congestion management. Possible relations and details of cooperation between different markets and technical entities, including aggregators, are indicated, thus forming a good foundation for future research in the field of active distribution networks.

Other publications state that the development of power systems and increasing level of energy source diversification are mainly driven by the growth of renewables which are installed mostly in the distribution networks; hence, the electrical energy supply's security and the proper power system operation is harder to be maintained [9,10]. In order to keep the balance between demand and generation, the corresponding level of the power system's flexibility is required [11]. Subsequently, to ensure this appropriate level of flexibility, it is necessary to coordinate the operation of distributed resources [12]. Such an operation, provided by aggregators located inside distribution networks, creates an opportunity to reinforce the power system's ability to react to the rapid changes of the demand and supply [13,14]. In order to obtain the desired results, next to DERs and ESSs, the demand-side coordination was employed to provide chosen ASs at the point of common coupling (PCC). The provided ASs cover i.e., load profile shaping, load levelling and congestion management [15–19].

The possibility for the participation of distributed resources coordinated by an aggregator in competitive energy markets and their positive impact on the power system's operation were also considered. Small ESSs for household application, due to their growing number, can be aggregated for long-term cooperation in order to maximize aggregators' profit and the system welfare [3]. Another study describes demand-side resources utilized for load scheduling to minimize the total cost of electricity procurement [20]. The high potential of aggregated flexibility, especially loads located in the residential sector, should be developed as a replacement for fossil fuel power plants' contribution in the continuous demand and supply balancing [4,21]. The need for peak-to-average demand reduction is addressed in [22], where the authors by the use of demand-side response reduce electricity charges for end-users and improve the shape of the load profile.

Nevertheless, the majority of the above-presented studies contain an optimized aggregation of one type of a distributed resource, mainly the demand-side entities and storages. It should be underlined that in modern distribution networks, different types of resources, including small generators and renewables, are installed and may be properly utilized—not only for cost reduction but also for AS provision. The described articles focused mostly on market aspects of the aggregation, analyzing the offering strategies and the competitiveness between different agents modelled in particular by game theories.

In order to propose remedies for the challenges presented, the purpose of this article is to introduce a new DRA structure and its management as a development of formerly proposed concepts of the active distribution system and aggregation approaches. The functions of the DRA coordinator in the modern power system and relations with other entities are also defined. The novel methodology proposes the modelling of the DRA structure with the use of the mixed-integer linear programming (MILP) which aims at the evaluation of the optimal operation pattern of different types of participating resources, the desired shape of the load profile at the PCC and AS provision. The performed research examines whether the proposed solution could be a step toward improvements in power system operation, and by the use of its flexibility, whether it can facilitate load-generation balancing and maintain a system's proper operation during continuous RES development.

The article is organized as follows: The second section describes services which can be provided by assets located in distribution networks. Section 3 presents a proposed structure of a DRA and a way to implement its operation into the MILP optimization model. Section 4 shows the main assumptions. Section 5 discusses the results of simulations as examples of ASs provided by the DRA, while the last section concludes the article.

2. Background: Ancillary Services Portfolio

The AS portfolio comprises services which may be provided by the aggregated resources, taking into account their distinctive features and the composition of the DRA structure. These services are described thoroughly below.

2.1. Peak Shaving and Valley Filling

Peak shaving and valley filling, also known as load levelling, is an AS comprising increased consumption of electrical energy during periods of low demand, storing it and then returning it to the grid when high demand occurs. During the later periods, the energy injected back to the power system reduces peaks of demand to be covered by conventional power plants and therefore decreases overall system operation costs, as production from more expensive peak-generating power units may be limited. Therefore, it is clear that load levelling can be provided mainly by ESSs but also by ALs, as consumption, during peak demand, can be limited or shifted to lower consumption periods [23].

Load levelling may be desirable not only because the reduction of overall system operation costs but also due to an opportunity to limit investments in new power generating units and grid upgrades (load levelling can extend the set of tools for the congestion management) [5]. Units providing this type of services could be additionally remunerated; however, shifting from low demand to high demand periods is inherent in price arbitrage which generates a basic income. A visualization of peak shaving and valley filling services is presented in Figure 2.

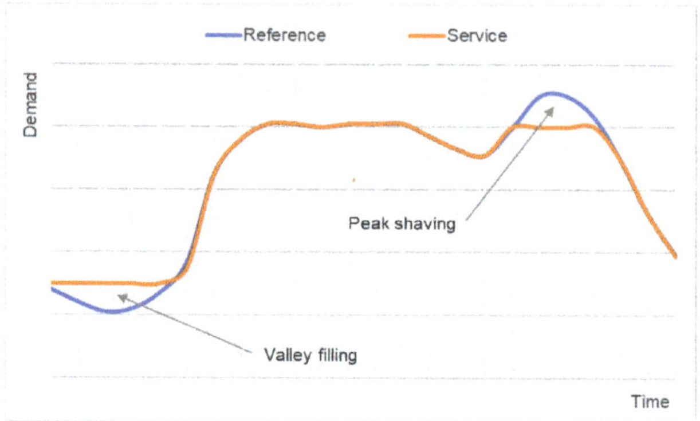

Figure 2. Peak shaving and valley filling services.

2.2. Load Profile Smoothing

Due to the growing share of RESs and their variable generation, the power system operation can be disturbed. It could happen especially for power systems with a significant share of renewables where difficulties in the balancing of demand and supply occur. Therefore, smoothing, which embraces reduction of rapid changes in the load profile, is another example of the AS that can be provided by the DRAs to assist the power system operation [24].

At the PCC, the load profile can be shaped by controlled operation of active resources incorporated into the DRA structure. The load profile smoothing could improve balancing, currently provided by the transmission system connected CDGUs because the smoothed profile is characterized by a smaller variability and milder up and down ramps. A visualization of load profile smoothing is presented in Figure 3.

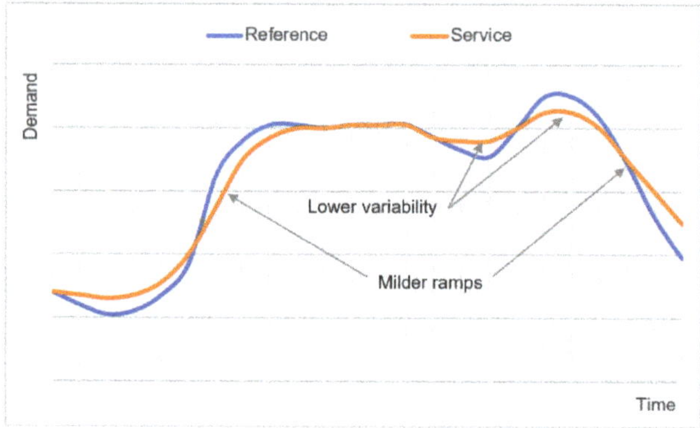

Figure 3. Load profile smoothing.

2.3. Balancing and Reserves

Electrical energy consumption must be equal to its generation at all times. The frequency of the grid is the best indicator of this balance. DERs managed by an aggregator are able to support the maintenance of the frequency within acceptable limits by quickly responding to its deviations. Such services are currently provided mainly by large CDGUs located in transmission networks, but properly aggregated structures may also be present on the balancing markets [25].

The balancing services may be also provided by compensation of balance deviations inside an aggregated structure in order to obtain a determined generation/load profile for the whole cluster. Some resources may be able to respond to other entities' deviations caused, for example, by variable weather conditions or rapidly changing demand.

The reserves are required to maintain the power system's balance in both the short and the long term. In this context, DRAs may additionally share a part of its capacity to cover TSOs' needs and be remunerated for the willingness to provide reserve services and for their activation.

3. Structure of the Distributed Resource Aggregator

3.1. Roles and Structure of an Aggregator

The main roles of the DRA coordinator are to select suitable assets located in distribution networks and to group them into a cluster in order to strengthen the significance of resources distributed on a small scale and thereby allow them to participate in wholesale markets, such as the balancing and AS markets. The operation of the aggregated assets is managed by the DRA coordinator in order to optimize the overall profit of the whole cluster. The structure of the DRA comprises assets located in the distribution level: passive loads, ALs, ESs, controllable and noncontrollable power generating resources. Figure 4 presents the proposed structure of the DRA and its cooperation with other entities located in the distribution level.

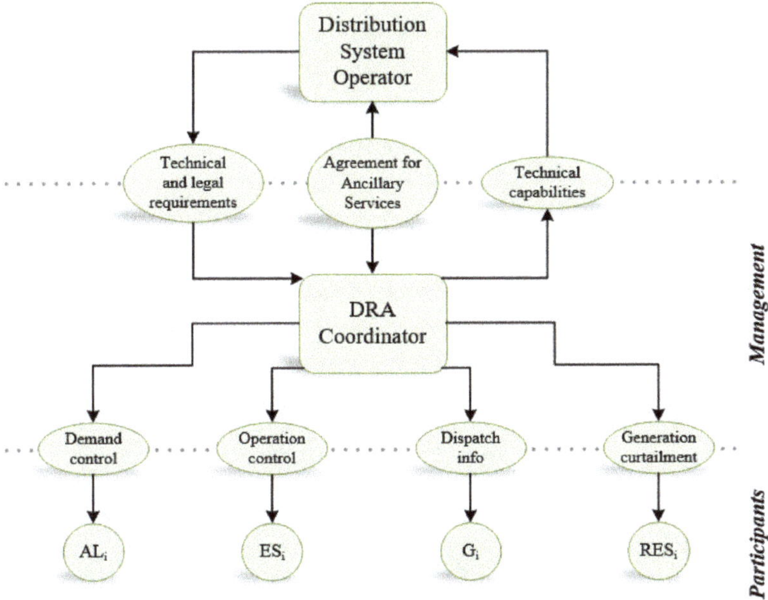

Figure 4. The proposed structure of the distributed resource aggregator (DRA) and its cooperation with other entities.

New technical and legal requirements should be introduced to allow the DRA to operate in the power system. These requirements should consist of basic technical parameters (power, response time, regulatory range, etc.) and legal conditions relating to inter alia, resource ownership, aggregation areas and financial settlements. After defining the aggregated structure, the DRA coordinator submits its technical capabilities to the DSO and makes agreements for the provision of the ASs due to the power system conditions and demands.

The DRA coordinator is responsible for supervision, coordination and control of the operation of resources. The coordinator has the ability to select suitable resources and create an optimal structure oriented towards the desired operation pattern which aims at a maximum profit while assisting the power system's operator.

In order to improve the power system flexibility, the DRA operation may prevent the negative impact of intermittent RES generation (left side of Figure 5) and stabilize operation of base load and combined heat and power (CHP) generation units (right side of Figure 5). The DRA can adjust output power from P_{MIN} to P_{MAX} to provide ASs and indirectly assist the TSO in the balancing of demand and supply.

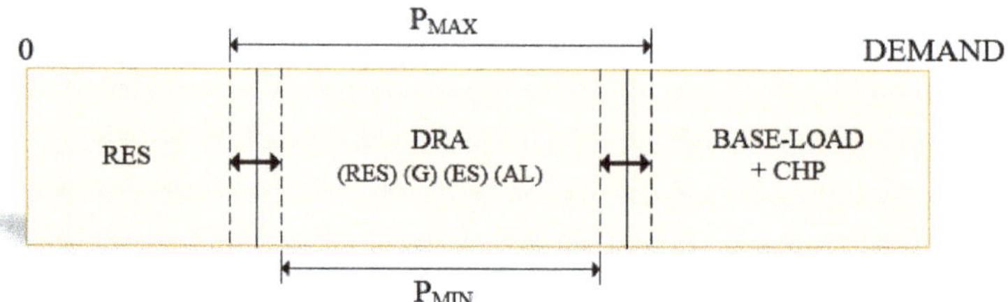

Figure 5. The proposed position of the DRA operation in the power system.

3.2. Modelling of Distributed Resources

In the presented concept, four groups of distributed assets were modelled and can be incorporated into the aggregator's structure: intermittent RESs, dispatchable generating units, ESs, ALs. The proposed model is based on mixed integer linear programming (MILP). For further simulation purposes, 15 min intervals of a 24 h time span were assumed.

3.2.1. Renewable Energy Resources

RESs represent a group of noncontrollable generating units whose production is dependent only on weather conditions. For the purpose of further simulations, wind and solar generators were implemented in the simulation model. Figure 6 depicts adopted generation profiles for those RESs.

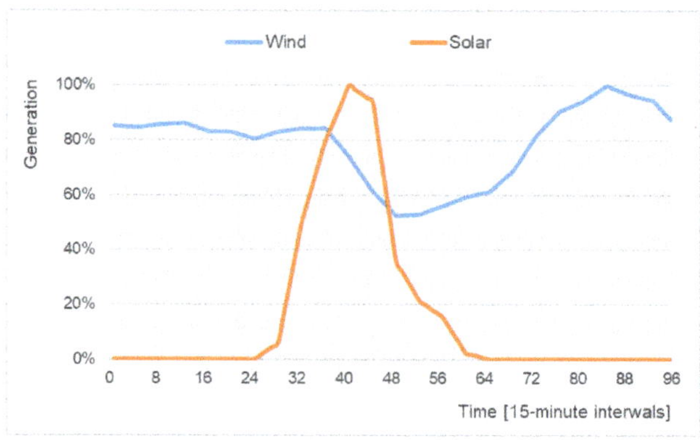

Figure 6. Generation profiles for wind and solar sources assumed for the simulation purposes (own development based on [26,27]).

The possibility of RES generation curtailment was also assumed, and is expressed by Equation (1).

$$\forall r \in R, \forall t \in T : P_{\min r} \leq P^{t}_{cur\ r} \leq P^{t}_{r} \tag{1}$$

3.2.2. Dispatchable Generating Units

The electricity production from units like gas, biogas, biomass, etc., is dependent on the amount of provided fuel; therefore, their output power can be controlled within their operational limits (Equation (2)).

$$\forall g \in G,\ \forall t \in T : P_{\min g} \leq P_g^t \leq P_{\max g} \tag{2}$$

3.2.3. Energy Storages

The adopted model of energy storage allows for the implementation of any type of ESS by defining appropriate parameters. Equation (3) reflects capacity limits.

$$\forall s \in S,\ \forall t \in T : E_{\min s} \leq E_s^t \leq E_{\max s} \tag{3}$$

A given ESS charges and discharges taking into account actual capacity and storage efficiency—Equation (4).

$$\forall s \in S,\ \forall t \in 2, \ldots, 96 : E_s^t = E_s^{t-1} + E_{\text{imp}s}^t \cdot \eta_s - \frac{E_{\text{exp}s}^t}{\eta_s} \tag{4}$$

Operational constraints given by Equations (5)–(7) contain binary variables to ensure switching between charging, discharging and idle mode of a given ESS.

$$\forall s \in S,\ \forall t \in T : P_{imp\ s}^t \leq s_{imp\ s}^t \cdot P_{\max s} \tag{5}$$

$$\forall s \in S,\ \forall t \in T : P_{exp\ s}^t \leq s_{\exp s}^t \cdot P_{\max s} \tag{6}$$

$$\forall s \in S,\ \forall t \in T :\ s_{imp\ s}^t + s_{exp\ s}^t \leq 1 \tag{7}$$

3.2.4. Active Loads

Due to a significant impact on the network's load profile, heavy industrial loads can also be a member of the DRA. Optimized operation of these resources allows them to provide a load levelling service. The peak demand can be shifted to the valley within the set limits (Equation (8)). The assumed constraint (Equation (9)) states that the sum of energy imported by the active load (AL) during the day period has to be equal to the daily electricity consumption, while the load remains passive.

$$\forall a \in A,\ \forall t \in T : 0.8 \cdot E_{L\ a}^t \leq E_{AL\ a}^t \leq 1.1 \cdot E_{L\ a}^t \tag{8}$$

$$\forall a \in A : \sum_{t=1}^{T} E_{AL\ a}^t = \sum_{t=1}^{T} E_{L\ a}^t \tag{9}$$

3.3. *Optimization*

The aim of the proposed model is the maximization of the aggregator's total profit which consists of income obtained from the operation of renewable resources (wind and solar), DERs, ALs and ESSs, all managed by the aggregator. The total income is divided among the DRA participants due to their contribution.

For the purposes of the simulations, only the market price of electrical energy was taken under consideration. Income resulting from RES subsidies were neglected, as they are the topic of separate research.

The resulting objective function is given by Equation (10).

$$obj = max\left\{\sum_{t=1}^{T}\left[\sum_{r=1}^{R}(E_{RES\,r}^{t}\cdot p^{t}) + \sum_{g=1}^{G}(E_{g}^{t}\cdot p^{t}) - \sum_{a=1}^{A}(E_{AL\,a}^{t}\cdot p^{t}) + \sum_{s=1}^{S}\left((E_{\exp s}^{t}\cdot p^{t}) - (E_{imp\,s}^{t}\cdot p^{t})\right)\right]\right\} \tag{10}$$

The provision of the ASs is enforced by the DRA through proper constraints. The proposed model comprises three types of services: load profile shaping, load levelling and a combined service (simultaneous smoothing and load levelling). Changes in operational points for the provision of ASs may be formulated by the DRA coordinator as an offer submitted to the market. Previously described balancing and reserves remain outside the scope of this article.

Both Equations (11) and (12) describe constraints that correspond to maximum ramps (load profile smoothing) when Equation (13) results in the provision of peak shaving and valley filling (load levelling service).

$$\Delta PCC_{\%} = \frac{PCC^{t-1} - PCC^{t}}{PCC_{maxref}}\cdot 100\% \tag{11}$$

$$-\Delta PCC_{\%} \le \Delta PCC_{des\%} \le \Delta PCC_{\%} \tag{12}$$

$$PCC_{des\ min} \le PCC^{t} \le PCC_{des\ max} \tag{13}$$

All parameters with the indication des have to be treated as values desired by the system operator as a part of the provision of ASs.

The demand change at the PCC is given by Equation (14) and described as the percentage ratio of the difference between the maximum and the minimum load regarding the maximum load during the simulations' time horizon.

$$\Delta Dem = \frac{PCC_{max} - PCC_{min}}{PCC_{max}}\cdot 100\% \tag{14}$$

4. Simulations

4.1. Assumptions

4.1.1. Parameters of the Simulation Model

Tables 1 and 2 present the assumed parameters of the simulation model expressed as a percentage of the peak demand.

Table 1. Parameters of the model—demand.

Parameter	Description
Share of commercial loads	31 $\frac{2}{3}$% of the peak demand
Share of residential loads	31$\frac{2}{3}$% of the peak demand
Share of industrial loads	31$\frac{2}{3}$% of the peak demand
Share of ALs	5% of the peak demand

Table 2. Parameters of the model—generation and storage.

Parameter	Description
Wind resource power	15% of the peak demand
Solar resource power	10% of the peak demand
Non-renewable DER power	8% of the peak demand
Storage maximum charge/discharge power	5% of the peak demand
Storage maximum capacity	4 h · maximum charge power
Storage efficiency	90%

4.1.2. Passive Load Patterns

Beyond the DRA participants, the passive loads are also located inside the modelled distribution network. The information of their demand is gathered by the DSO in order to set the desired operation points for the DRA to shape the profile at the PCC. The presented concept assumed three types of passive loads: commercial, residential and industrial, which were modelled by the fixed demand profiles presented in Figure 7.

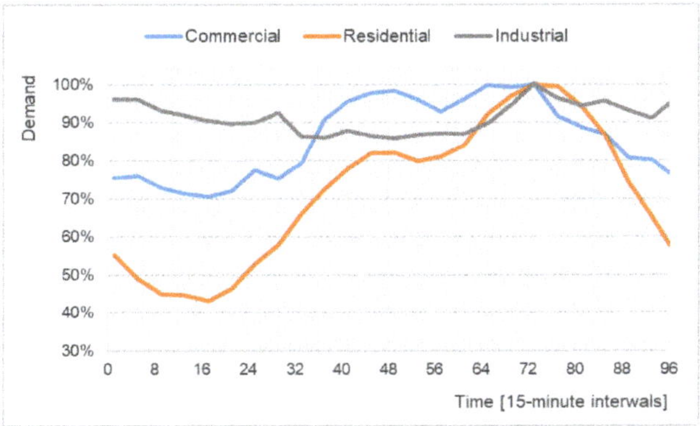

Figure 7. Demand profiles for commercial, residential and industrial passive loads (own development based on [28]).

4.1.3. Market Price Pattern

The market price of electricity is implemented as the pattern presented in Figure 8. The assumed profile is based on real market data and is treated as predictions of market price used by the aggregator.

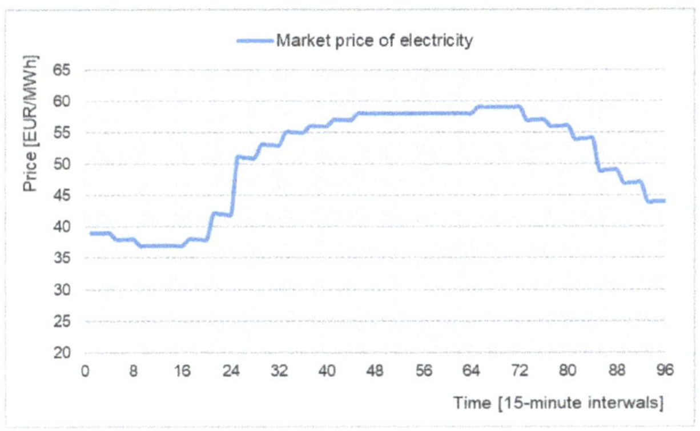

Figure 8. Market price pattern (own development based on [29]).

4.2. Scenarios

In order to examine the impact of the DRA on the operation of the active distribution system, seven simulation scenarios were proposed (Table 3). Each scenario presents different services provided by the maximum utilization of the resources coordinated by the DRA depending on various assumptions.

Table 3. Description of the simulation scenarios.

Scenario	Description
Reference Scenario	No DRA influence
Scenario 1	Smoothing of the load profile; RES curtailment not allowed
Scenario 2	Smoothing of the load profile; RES curtailment allowed
Scenario 3	Load levelling; RES curtailment not allowed
Scenario 4	Load levelling; RES curtailment allowed
Scenario 5	Combined service (simultaneous smoothing and load levelling); RES curtailment not allowed
Scenario 6	Combined service (simultaneous smoothing and load levelling); RES curtailment allowed

5. Results and Discussion

The reference scenario assumes a lack of influence of the DRA on the operation of subordinate resources such as RESs, DERs, ESs and ALs. All the resources aim for maximum profit without performing any ASs. The simulation output describes power flow at the PCC formed by all entities located inside the distribution network and is presented as *PCC_%_ref* in Figure 9. The results of this scenario form a basis for comparisons with further scenarios.

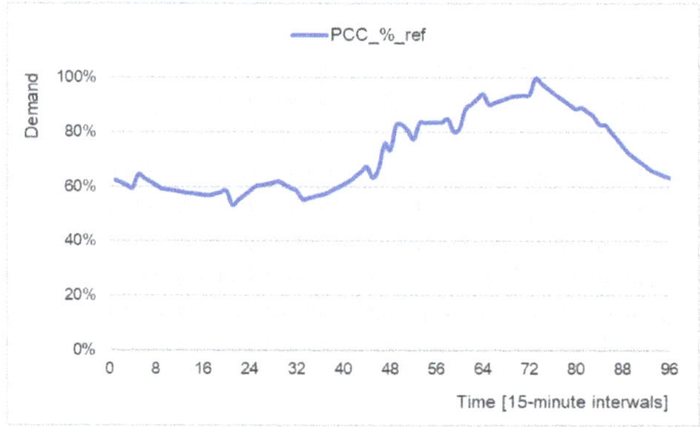

Figure 9. Results of the reference scenario—power flow at the point of common coupling (PCC).

The obtained value resulting from the objective function in the reference scenario is denoted as 100% for the purpose of future comparisons. The maximum percentage change in the power flow at the PCC between two adjacent hours is 19%/h when the demand change (Equation (14)) stands at 46.7%. The shape of the profile results from the assumptions presented earlier in Tables 1 and 2 and from the uncontrolled operation of distributed resources.

The shapes of the load profiles at the PCC for Scenarios 1–6 compared with reference simulation are presented in Figure 10.

The presented results are characterized by the different impact on the shape of the load profile at the PCC. The first column of Figure 10 (Scenarios 1, 3 and 5) corresponds to services provided without renewable curtailment, while in the second column (Scenarios 2, 4 and 6), curtailment is allowed. *PCC_%_ref* denotes reference load profile, while *PCC_%_S1-6* correspond to modified profiles according to the simulation scenarios.

In Scenarios 1, 3 and 5, ASs are provided mainly through ESs, ALs and non-renewable DERs; hence, the modifications of the load profile are visibly lower than in Scenarios 2, 4 and 6 where the services are provided also by wind and solar renewables. The analyzed differences are related also to the assumed share of different resources presented previously in Tables 1 and 2. Following the assumptions, the share of generating units to be curtailed (wind and solar) corresponds to 25% of the peak demand.

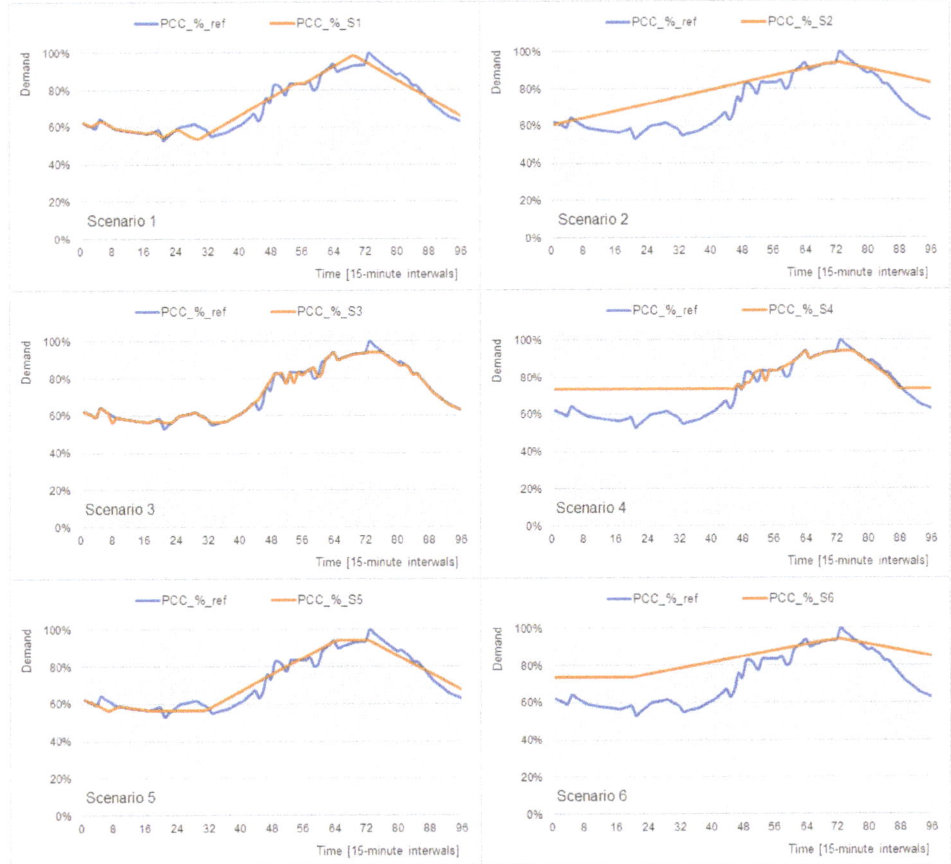

Figure 10. Results of Scenarios 1–6—power flow at the PCC.

From the scenarios assuming a lack of renewable curtailment, the best results are visible in Scenario 5. The load profile is significantly smoother, resulting in milder up and down ramps. Nevertheless, the impact of the RES curtailment and the combination of both services (smoothing of the load profile and load levelling) gives the best results: the profile is smooth, peak demand is slightly reduced and night valleys of the demand are visibly filled (Scenario 6). The slight reduction of the peak demand in all scenarios is associated with a relatively low share of entities that are able to shift the demand from night valleys to the peaks: ESSs and ALs.

Table 4 presents a summary of the obtained results, where:

- Objective function denotes the value of the objective function (profit) obtained in a given scenario expressed as a percentage of profit obtained in the reference scenario as given by Equation (10);
- Maximum ramp denotes a maximum percentage change in the power flow at the PCC between two adjacent hours included in the simulations' time horizon as given by Equation (11);

- Demand change denotes the percentage ratio of the difference between the maximum and the minimum demand regarding the maximum demand during the simulations' time horizon (Equation (14)).

Table 4. Summary of the results.

Scenario	Objective Function	Maximum Ramp	Demand Change
Reference Scenario	100%	19.3%/h	46.7%
Scenario 1	99.1%	4.9%/h	45.7%
Scenario 2	78.5%	1.9%/h	35.4%
Scenario 3	99.8%	13.5%/h	40.0%
Scenario 4	83.4%	7.1%/h	22.1%
Scenario 5	98.6%	4.6%/h	40.0%
Scenario 6	71.9%	1.6%/h	21.9%

The provision of the ASs caused deviation from the reference operation points of aggregated resources, hence, the values of the objective function for all scenarios were lower compared to the reference case. The difference between the obtained value of the objective function (for Scenarios 1–6) and reference value should be treated as the minimum price of the offer for AS provision (Figure 11). The lowest profits were obtained in Scenarios 2, 4 and 6 in which RES curtailment was allowed, and a significant decrease in incomes due to lost generation occurred. For this reason, the AS provision has to be properly valued in order to ensure the desired income for DRAs and therefore encourage them to participate in different markets.

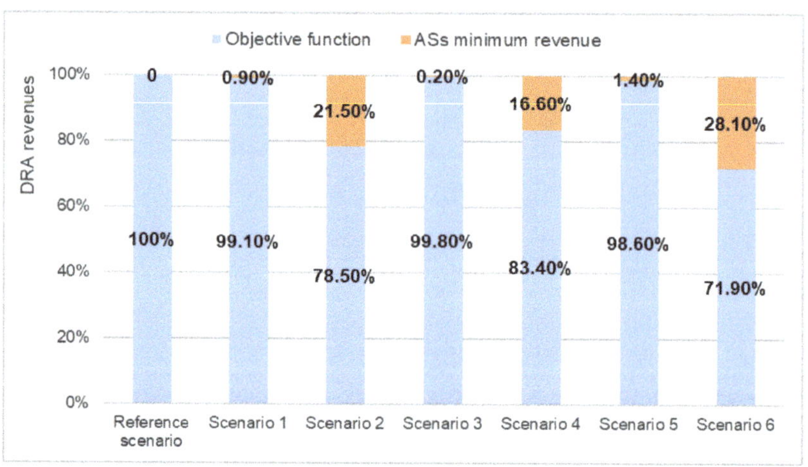

Figure 11. Minimum required revenues from the ancillary service (AS) provision for each scenario.

In summary, Scenarios 2, 4 and 6 were characterized by the best technical performance. RES curtailment caused significant smoothing of the load profile at the PCC and visibly reduced the difference between the maximum and the minimum demand. Such actions can have a positive impact on the operation of the whole power system, as load levelling and smoothing of the demand profile facilitate the real-time balancing of the load and generation due to the reduction of the profile's variability, and they may allow further RES development.

In Scenarios 1, 3 and 5 where RES curtailment was not allowed, ASs were provided mainly by active resources like ESs and ALs. The assumed share of such entities (Tables 1 and 2) was lower than the share of RESs; therefore, the quality of the services' provision was strongly dependent on the contribution of different distributed resources inside the aggregator's structure.

6. Conclusions

The implemented model of the DRA and the performed simulations showed that distributed resources aggregated into the DRA structure are able to provide ASs together with CDGUs and therefore contribute to the improvement of the power system's operation. The establishment of the DRA structures, as part of the modern power system, enables small resources to participate in wholesale markets. The flexibility of the DRA resources facilitates load-generation balancing and allows for further RES development due to the mitigation of their intermittent operation. The variety of the provided services is strictly dependent on the DRA composition and legal regulations (e.g., the possibility of renewables curtailment).

The performed studies indicate that further research should consider how to determine the optimal set of DRA participants. The effectiveness of the service provided has an impact on the reduction of the volume of produced electricity and therefore reduces the basic market income for the aggregated resources' owners. The properly designated revenues for participants should encourage them to share their capabilities for the AS provision. Additionally, other types of services, e.g., operating reserves, could be considered.

The MILP formulation of the optimization model resulted in a short computation time.

Author Contributions: All authors developed the initial idea of providing ancillary services (ASs) by the distributed resource aggregator (DRA). A.L. and D.C. designed the DRA structure, relations with other entities and the optimization model. R.D. provided critical comments and research suggestions. All authors have read and agreed to the published version of the manuscript.

Funding: This research received no external funding.

Acknowledgments: The authors would like to express the gratitude to the FICO® corporation for programming support and provision of academic licenses for Xpress Optimization Suite to Institute of Electrical Power Engineering at Lodz University of Technology.

Conflicts of Interest: The authors declare no conflict of interest.

Nomenclature

Indices and Sets	
$t \in T$	Time span (24 h with 15-min intervals)
$g \in G$	Dispatchable generating unit
$r \in R$	Renewable Energy Source (wind and solar)
$s \in S$	Energy Storage System
$a \in A$	Active Load
Parameters	
p^t	Market price of electricity in time period t
$P_{\text{min}g}$	Minimum output power of dispatchable generating unit g
$P_{\text{max}g}$	Maximum output power of dispatchable generating unit g
$P_{\text{min}r}$	Minimum required output power of Renewable Energy Source r in time period t during generation curtailment
P_r^t	Output power of Renewable Energy Source r in time period t without generation curtailment
$E_{RES\ r}^t$	Energy generated by Renewable Energy Source r in time period t
$E_{\text{min}s}$	Minimum capacity of Energy Storage System s
$E_{\text{max}s}$	Maximum capacity of Energy Storage System s
η_s	Efficiency of Energy Storage System s
$P_{\text{max}s}$	Maximum charge/discharge power of Energy Storage System s
$E_{L\ a}^t$	Energy consumed by Load a in time period t
$\Delta PCC_{des\%}$	Percentage change of the demand at the Point of Common Coupling between two adjacent hours desired by the system operator
$PCC_{\text{max}ref}$	Maximum demand at the Point of Common Coupling in reference scenario
$PCC_{des\ min}$	Minimum demand at the Point of Common Coupling desired by the system operator
$PCC_{des\ max}$	Maximum demand at the Point of Common Coupling desired by the system operator

Variables

P_g^t	Output power of dispatchable generating unit g in time period t
$P_{cur\ r}^t$	Output power of Renewable Energy Source r in time period t during generation curtailment
E_g^t	Energy generated by dispatchable generating unit g in time period t
E_s^t	Capacity of Energy Storage System s in time period t
E_{imps}^t	Energy imported (charged) by Energy Storage System s in time period t
$E_{\mathrm{exp}\,s}^t$	Energy exported (discharged) by Energy Storage System s in time period t
P_{imps}^t	Charge power of Energy Storage System s in time period t
$P_{\mathrm{exp}\,s}^t$	Discharge power of Energy Storage System s in time period t
$E_{AL\ a}^t$	Energy consumed by Active Load a in time period t
$\Delta PCC_{\%}$	Percentage change of the demand at the Point of Common Coupling between two adjacent hours
PCC^t	Demand at the Point of Common Coupling in time period t
ΔDem	Maximum demand change at the Point of Common Coupling during simulation period
PCC_{max}	Maximum demand at the Point of Common Coupling during simulation period
PCC_{min}	Minimum demand at the Point of Common Coupling during simulation period
Binary Variables	
$s_{imp\ s}^t$	Charging status of Energy Storage System s in time period t
$s_{exp\ s}^t$	Discharging status of Energy Storage System s in time period t

References

1. Calvillo, C.F.; Sánchez-Miralles, A.; Villar, J.; Martín, F. Optimal planning and operation of aggregated distributed energy resources with market participation. *Appl. Energy* **2016**, *182*, 340–357. [CrossRef]
2. Poplavskaya, K.; Vries, L. Distributed energy resources and the organized balancing market: A symbiosis yet? Case of three European balancing markets. *Energy Policy* **2019**, *126*, 264–276. [CrossRef]
3. Contreras-Ocana, J.; Ortega-Vazquez, M.; Zhang, B. Participation of and Energy Storage Aggregator in Electricity Markets. In Proceedings of the 2018 IEEE Power & Energy Society General Meeting (PESGM), Portland, OR, USA, 5–9 August 2018. [CrossRef]
4. Bruninx, K.; Pandžić, H.; Le Cadre, H.; Delarue, E. On the Interaction between Aggregators, Electricity Markets and Residential Demand Response Providers. *IEEE Trans. Power Syst.* **2020**, *35*, 840–853. [CrossRef]
5. TSO-DSO REPORT. An Integrated Approach to Active System Management with the Focus on TSO-DSO Coordination in Congestion Management and Balancing. April 2019. Available online: https://eepublicdownloads.blob.core.windows.net/public-cdn-container/clean-documents/Publications/Position%20papers%20and%20reports/TSO-DSO_ASM_2019_190416.pdf (accessed on 18 June 2020).
6. EURELECTRIC. Active Distribution System Management: A Key Tool for the Smooth Integration of Distributed Generation, The Union of the Electricity Industry. February 2013. Available online: https://www.eurelectric.org/media/1781/asm_full_report_discussion_paper_final-2013-030-0117-01-e.pdf (accessed on 1 July 2020).
7. Trebolle, D.; Hallberg, P.; Lorenz, G.; Mandatova, P.; Guijarro, J.T. Active distribution system management. In Proceedings of the 22nd International Conference and Exhibition on Electricity Distribution (CIRED 2013), Stockholm, Sweden, 10–13 June 2013. [CrossRef]
8. Li, R.; Wang, W.; Chen, Z.; Jiang, J.; Zhang, W. A Review of Optimal Planning Active Distribution System: Models, Methods, and Future Researches. *Energies* **2017**, *10*, 1715. [CrossRef]
9. Chalvatzis, K.J.; Ioannidis, A. Energy supply security in the EU: Benchmarking diversity and dependence of primary energy. *Appl. Energy* **2017**, *207*, 465–476. [CrossRef]
10. European Union. Energy Research Knowledge Centre Report. In *Research Challenges to Increase the Flexibility of Power Systems*; European Union: Brussels, Belgium, 2014; pp. 1–38.
11. Danish Energy Agency. Flexibility in the Power System-Danish and European Experiences. October 2015. Available online: https://ens.dk/sites/ens.dk/files/Globalcooperation/flexibility_in_the_power_system_v23-lri.pdf (accessed on 10 July 2020).
12. Ulbig, A.; Goran, A. Role of Power System Flexibility. In *Renewable Energy Integration: Practical Management of Variability, Uncertainty and Flexibility in Power Grids*; Elsevier: San Diego, CA, USA, 2014; pp. 227–238.

13. National Renewable Energy Laboratory. Flexibility in 21st Century Power Systems. May 2014. Available online: https://www.nrel.gov/docs/fy14osti/61721.pdf (accessed on 4 June 2020).
14. Electric Power Research Institute (EPRI). Electric Power System Flexibility: Challenges and Opportunities. February 2016. Available online: https://www.naseo.org/Data/Sites/1/flexibility-white-paper.pdf (accessed on 6 June 2020).
15. Rosso, A.; Ma, J.; Kirschen, D.S.; Ochoa, L.F. Assessing the Contribution of Demand Side Management to Power System Flexibility. In Proceedings of the 50th IEEE Conference on Decision and Control and European Control Conference (CDC-ECC), Orlando, FL, USA, 12–15 December 2011. [CrossRef]
16. Nursimulu, A.; Florin, M.-V.; Vuille, F. Demand-Side Flexibility for Energy Transitions: Policy Recommendations for Developing Demand Response, Lausanne, International Risk Governance Council (IRGC). Switzerland, 2016. Available online: https://irgc.org/wp-content/uploads/2018/09/Demand-side-Flexibility-for-Energy-Transitions-Policy-Brief-2016.pdf (accessed on 1 July 2020).
17. Lesniak, A. Optimization of the Flexible Virtual Power Plant Operation in Modern Power System. In Proceedings of the 15th International Conference on the European Energy Market (EEM18), Łódź, Poland, 27–29 June 2018. [CrossRef]
18. Mielczarski, W. HANDBOOK: Energy Systems & Markets, Edition I—June 2018. Available online: http://www.eem18.eu/gfx/eem-network/userfiles/_public/handbook_energy_systems___markets.pdf (accessed on 6 June 2020).
19. Ni, L.; Wen, F.; Liu, W.; Meng, J.; Lin, G.; Dang, S. Congestion management with demand response considering uncertainties of distributed generation outputs and market prices. *MPCE J. Mod. Power Syst. Clean Energy* **2016**, *5*, 66–78. [CrossRef]
20. Kumar Panwar, L.; Reddy Konda, S.; Verma, A.K.; Panigrahi, B.; Kumar, R. Demand response aggregator coordinated two-stage responsive load scheduling in distribution system considering customer behaviour. *IET Gener. Transm. Distrib.* **2016**, *11*, 1023–1032. [CrossRef]
21. Lucas, A.; Jansen, L.; Andreadou, N.; Kotsakis, E.; Masera, M. Load Flexibility Forecast for DR Using Non-Intrusive Load Monitoring in the Residential Sector. *Energies* **2019**, *12*, 2725. [CrossRef]
22. Mohsenian-Rad, A.-H.; Wong, V.W.; Jatskevich, J.; Schober, R.; Leon-Garcia, A. Autonomous Demand-Side Management Based on Game-Theoretic Energy Consumption Scheduling for the Future Smart Grid. *IEEE Trans. Smart Grid* **2010**, *1*, 320–331. [CrossRef]
23. DSO Committee on Flexibility Markets. Flexibility in the Energy Transition—A Tool for Electricity DSOs, Belgium. February 2018. Available online: https://www.edsoforsmartgrids.eu/wp-content/uploads/Flexibility-in-the-energy-transition-A-tool-for-electricity-DSOs-2018-HD.pdf (accessed on 8 June 2020).
24. Wang, M.; Mu, Y.; Jiang, T.; Jia, H.; Li, X.; Hou, K.; Wang, T. Load curve smoothing strategy based on unified state model of different demand side resources. *MPCE J. Mod. Power Syst. Clean Energy* **2018**, *6*, 540–554. [CrossRef]
25. Flex4RES Flexible Nordic Energy Systems. Framework Conditions for Flexibility in the Electricity Sector in the Nordic and Baltic Countries. December 2016. Available online: https://backend.orbit.dtu.dk/ws/portalfiles/portal/128130121/Flex4RES_Electricity_Report_final.pdf (accessed on 2 July 2020).
26. Polish Power System Operator—PSE S.A. Polish Power System Operation—Generation of Wind Farms and Solar Farms. Available online: https://www.pse.pl/web/pse-eng/data/polish-power-system-operation/generation-in-wind-farms (accessed on 2 July 2020).
27. Olek, B. Optimization of Energy Balancing and Ancillary Services in Low Voltage Networks. Ph.D. Thesis, Lodz University of Technology, Lodz, Poland, June 2013.
28. ENEA Operator, Sp. z o.o. Instrukcja Ruchu i Eksploatacji Sieci Dystrybucyjnej - IRiESD, January 2014. Available online: https://www.enea.pl/operator/dla-firmy/iriesd/iriesd_enea-operator_tj_od-20160201.pdf (accessed on 10 June 2020).
29. Polish Power Exchange—TGE S.A. Statistical Data. Available online: https://tge.pl/statistic-data (accessed on 24 August 2020).

Article

Dynamic-Model-Based AGC Frequency Control Simulation Method for Korean Power System

Han Na Gwon and Kyung Soo Kook *

Department of Electrical Engineering, Jeonbuk National University, Deokjin-gu, Jeonju 54896, Korea; canna08@jbnu.ac.kr

* Correspondence: kskook@jbnu.ac.kr; Tel.: +82-63-270-2368

Received: 19 August 2020; Accepted: 22 September 2020; Published: 25 September 2020

Abstract: To fulfill the need of operating power systems more effectively through diverse resources, frequency control conditions for maintaining a balance between generators and loads need to be provided accurately. As frequency control is generally achieved via the governor responses from local generators and the automatic generation control (AGC) frequency control of the central energy management system, it is important to coordinate these two mechanisms of frequency control efficiently. This paper proposes a dynamic-model-based AGC frequency control simulation method that can be designed and analyzed using the governor responses of generators, which are represented through dynamic models in the planning stage. In the proposed simulation model, the mechanism of the AGC frequency control is implemented based on the dynamic models of the power system, including governors and generators; hence, frequency responses from the governors and AGC can be sequentially simulated to coordinate and operate these two mechanisms efficiently. The effectiveness of the proposed model is verified by simulating the AGC frequency control of the Korean power system and analyzing the coordination effect of the frequency responses from the governors and AGC.

Keywords: automatic generation control (AGC); frequency control; dynamic deadband; energy management system (EMS); governor response

1. Introduction

The Korean power system has undergone fundamental changes due to the shift towards new energy sources such as wind, solar, and battery [1]. Korea intends to obtain 20% of its total electric energy from renewable energy sources (RES) by 2030; accordingly, more than 50 GW of RES is expected to be integrated into the Korean power system. This would require a new and challenging operation of frequency control as the characteristics of RES are different from those of the conventional generators. Moreover, as the Korean power system has been operated based on an electricity market with diverse participants, efficiency and transparency would be the most important aspects of its operation. In particular, this becomes more critical when the system should be operated under challenging conditions such as a high penetration level of RES.

The frequency of the power system has been controlled to maintain the balance between generators and loads. The corresponding framework is configured by ancillary services that provide various frequency control methods depending on the performance of the power system. In particular, primary and secondary frequency control methods that utilize the governor responses from generators and the automatic generation control (AGC) of the energy management system (EMS) respectively, are the most commonly used frameworks for the control of the frequency of the power system. The two mechanisms for providing frequency control services differ in terms of their performance and objective. However, they need to coordinate with each other to minimize the frequency drop caused by a disturbance and restore the frequency to the nominal value, as the secondary frequency control method is designed to take over the primary frequency control method during a transient period. For days on which the resources for providing the frequency control methods were sufficient, the coordination between the primary and secondary frequency control methods could also be considered to secure the controlled performance of the frequency using a sufficient margin rather than to maximize the efficiency of the operation of these methods.

However, the efficiency of power system operations such as frequency control is expected to increase based on the electricity-market-based circumstances and the constraints of power system operations are expected to be more stringent with a higher penetration level of RES. Hence, the coordination of the primary and secondary frequency control methods needs to be designed and operated more efficiently. Moreover, while the primary frequency control method is mainly characterized by the individual dynamics of generators, including governors, the secondary frequency control method is focused on the central control of all the generators in a normal condition. As it is typical to analyze these two mechanisms in different domains, a common framework for practically designing and analyzing their coordination is required to consider both the dynamics and AGC operation of the power system.

Although several synchronous generators providing frequency control in power systems have been proposed, limited studies have been conducted on the coordination of primary and secondary methods. Among them, Reference [2] proposed that the governor deadband should be wider to prevent an overlap between the primary and secondary control methods in normal operating conditions; however, there is a possibility of ineffective usage because of the overlap between the two services in transient operating conditions. In Reference [3], as the AGC and governor response from the generator are in different domains, the governor response from the generator causes performance degradation by the AGC signal. Thus, Reference [3] proposed that the AGC target of the plant-level controller (PLC) should be modified to sustain the governor response from the generator. However, when the AGC system in an EMS does not consider the processing on the PLC, it would cause wear and tear leading to high operation and maintenance costs. Thus, the AGC system in the EMS should also be reviewed to compensate for this problem.

Therefore, this paper proposes a dynamic-model-based AGC frequency control simulation method for the Korean power system, implemented using Python and the power system simulator for engineering (PSS/E) program. The proposed method not only analyzes the coordination between the governor responses from the generators and the AGC frequency control under normal and transient operating conditions, but also reviews the AGC algorithm in the EMS. The effectiveness of the proposed model is verified by simulating the AGC frequency control of the Korean power system and analyzing the coordination between the frequency responses from the governors and AGC.

2. AGC Frequency Control in Korean Power System

As the Korean power system has no interconnection with neighboring countries and the maximum capacity of the generators (which is now 1.5 GW) continues to increase, frequency instabilities resulting from a disturbance have become increasingly important. The Korean power system uses both the AGC and governor responses as ancillary methods for controlling the frequency. Its operation strategy has been implemented more efficiently after the new domestic EMS called K-EMS came into operation in late 2014. This section analyzes the major characteristics of frequency control using AGC in the Korean power system for implementing a simulation model based on the dynamic model of the power system.

2.1. AGC Operation

The purpose of the AGC is to maintain: (1) the system frequency at or very close to a nominal frequency value, (2) the correct value of the interchange power between the control areas, and (3) the power output of each generator at the most economic value, as depicted in Reference [4]. In this context, as noted above, as the Korean power system is an isolated system, its AGC frequency control is operated only in the constant-frequency control mode, as depicted in Reference [5]. Figure 1 shows an overview of the AGC frequency control of the Korean power system, as illustrated in Reference [6].

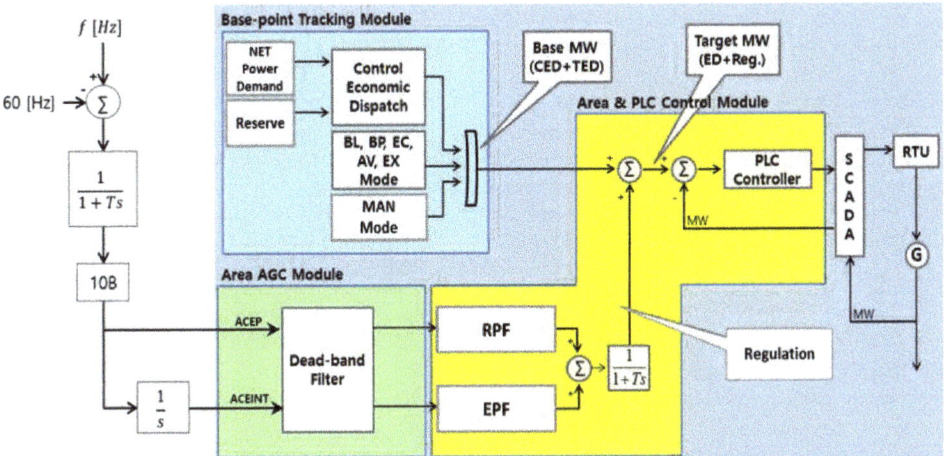

Figure 1. Automatic generation control (AGC) operation in the Korean power system.

As shown in Figure 1, the AGC frequency control of the Korean power system is performed by the EMS. The system frequency is sampled every two seconds via supervisory control and data acquisition (SCADA) and is filtered through a low-pass filter (LPF), which softens out rapid frequency variations. Then, the area control error (ACE) is calculated using the filtered frequency according to Equation (1):

$$\text{ACE} = -10 \times B \times (F_A - F_S - TE) \quad (1)$$

where F_A and F_S are the measured system frequency and nominal system frequency (60 Hz), respectively. B is the frequency bias in MW/0.1 Hz, and it was set to 913 MW/0.1 Hz in 2019. TE is the time error.

Thus, the ACE is the index of imbalance between the generators and loads in the power system. In this context, as the frequent activation of AGC frequency control under normal operating conditions affects the wear and tear of the generator, the AGC frequency control should not respond to instantaneous random variation. Moreover, as AGC should not conflict with the governor response under transient operating conditions, the AGC frequency control should respond slowly [7]. Thus, AGC must be filtered or delayed using the LPF or ACE processing algorithm to protect equipment, and it should be coordinated with the governor response. The AGC should be able to make this coordinating operation by tuning its control parameters, and such a required characteristic of AGC is defined as the flexibility of AGC frequency control in this section. In order to do this, the ACE is generally processed in various ways depending on the control philosophy of the utility or country [8]. For the Korean power system, the Korean control philosophy was reflected in the deadband filter to operate the AGC frequency control flexibly [9]. The dynamic deadband filter is described in the next section in detail.

Then, the ACE adjusted using the dynamic deadband filter distributes the required MW power to each generator. In this context, the adjusted ACE is derived from the proportional and integral components. The proportional component of the ACE is ACEP and is calculated by using Equation (1). The integral component of the ACE is ACEINT and accumulates the system frequency deviation over time. Its calculation can be described by Equation (2):

$$\text{ACEINT} = \text{ACEINT}_{\text{pre}} + \frac{\text{ACEINT}_{\text{pro}} \times \text{AGC Cycle}}{3600} \quad (2)$$

$\text{ACEINT}_{\text{pre}}$ and $\text{ACEINT}_{\text{pro}}$ are the previous value of ACE integrated value and the current value of ACE integrated value, respectively. The AGC cycle is the calculating time for ACE and it was set to two seconds in the Korean power system.

That is, the proportional component of the ACE is distributed to each generator considering its ramp rate (ramping power factor, RPF), and the integral component of the ACE is distributed considering its generating unit price (economic power factor, EPF). The required MW power of each generator is filtered through an LPF.

The AGC signal sends the economic dispatch (ED) and the MW power required for each generator to maintain the desired frequency through SCADA to the PLC of the generators every four seconds. The ED is calculated every five minutes.

2.2. Deadband Filter of AGC Frequency Control

In the Korean power system, the deadband filter has been used to implement the flexibility of AGC frequency control [9]. The main function of the deadband filter is to delay ACE activation by using the dynamic deadband and regulate its amount by applying different gains depending on system operating condition when the frequency drops in the power system. Figure 2 depicts a detailed overview of the dynamic deadband filter.

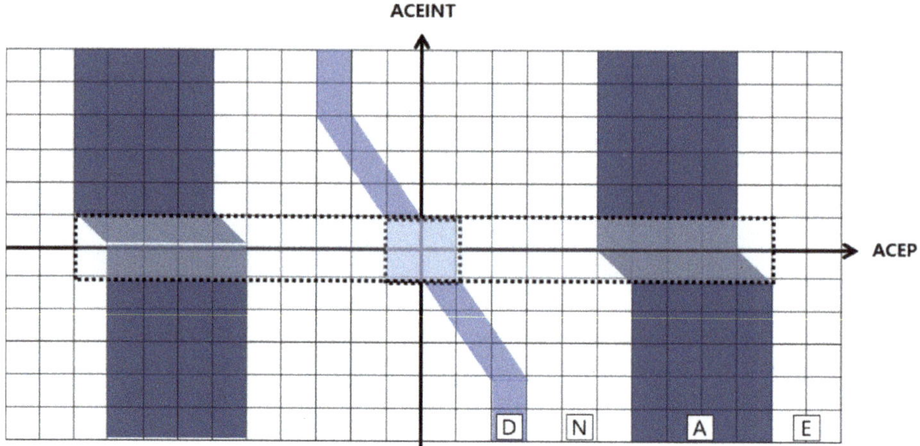

Figure 2. Concept of dynamic deadband in deadband filter.

As shown in Figure 2, the control mode of the dynamic deadband filter is classified into deadbands (D), normal (N), assist (A), and emergency (E), depending on the magnitude of the calculated ACE. The first of the above categories consists of static and dynamic deadbands [10]. As the calculated value of ACE increases, the control mode is changed to N, A, and E mode, and accordingly, the response of the generators according to the AGC frequency control is increased by increasing the size of the control gain multiplied by the ACE. The control mode is determined at every time-step depending on the amount of calculated ACE.

When the calculated ACE is smaller than the static deadband, ACE is not activated. However, ACE does not get activated due to the effect of the dynamic deadband even though the calculated ACE exceeds the static deadband. Thus, the gains of either the static or dynamic deadband make the value of ACE zero.

If the frequency is still not recovered to the expected frequency, the dynamic deadband starts decreasing in order to approach the static deadband at a rate of time. When the ACE exceeds the decreasing dynamic deadband, it is activated by applying the gain of the control mode. At this instant, the gains are applied differently depending on the control mode. Once the frequency is recovered and its ACE is smaller than the static deadband, the dynamic deadband is initialized and the ACE is not activated.

2.3. Coordination of Governor Response and AGC

The ancillary service of the Korean power system has recently amended the market rules to ensure that the power system operates with safety and security with the increasing penetration level of RES, as depicted in References [4,11]. Figure 3 shows the concept of the amended ancillary service.

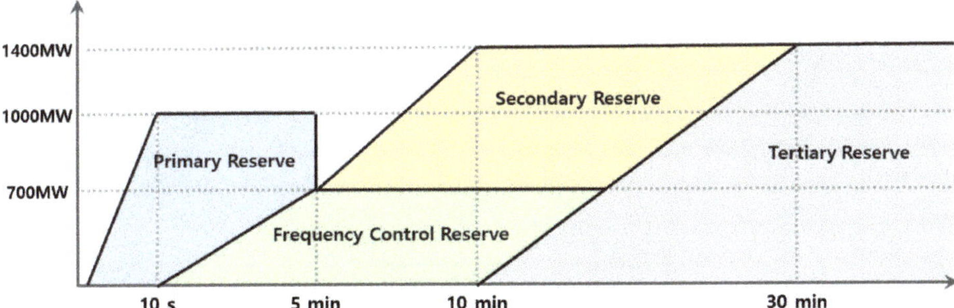

Figure 3. Concept of amended ancillary service in the Korean power system.

In Figure 3, the frequency regulation reserve, which was obtained by integrating the governor response and AGC frequency control, is classified as a frequency control reserve of 700 MW for normal operating conditions and a frequency recovery reserve for transient operating conditions. The frequency recovery reserve is subdivided into a primary reserve of 1000 MW, a secondary reserve of 1400 MW, and a tertiary reserve of 1400 MW.

The primary reserve of the Korean power system is activated by the governor of the generator, except for the nuclear power plant. When a large-scale generator trips in the Korean power system, the primary reserve secured by 1000 MW must be activated within 10 s and last for at least 5 min to maintain the frequency decrease. In this case, the system frequency is maintained at a frequency of 59.7 Hz and is restored above 59.8 Hz within 1 min.

The frequency control reserve generally called "AGC" activates the system frequency to smooth out the variability of load under normal operating conditions. In addition, the frequency control reserve secured by 700 MW must be activated within 5 min and last for at least 30 min. The secondary reserve is also called "AGC" but is used only for contingency. This is responsible for backing up the primary reserve. It must be activated within 10 min and last for at least 30 min following the event of a generator trip in the Korean power system. Furthermore, the secondary reserve of the Korean power system must be secured by 1400 MW.

Thus, the interaction between the governor response and the AGC frequency control was clarified by the amended ancillary service concept. Consequently, it becomes important to coordinate these services with each other.

3. Dynamic Simulation Model of AGC

The dynamic-model-based simulation frameworks, which are typically used for power system analysis by the utilities, are adapted to include the AGC frequency control to implement the simulation model for both the AGC and governor responses. As the database of the Korean power system has been implemented in the framework of PSS/E, Python is used as its application program interface (API) for implementing the AGC frequency control and interfacing its responses to the dynamic-model-based simulation using PSS/E. Accordingly, the proposed simulation model can flexibly implement the AGC frequency control and effectively use the existing models of power systems. This section describes the details of the proposed simulation model by dividing it into a system aspect implementing the AGC operation and a generator aspect linking it to dynamic simulation.

3.1. Dynamic-Model-Based Implementation of Load Frequency Control in the Korean Power System

As AGC controls the power system frequency by adjusting the generation targets of centrally dispatched generators based on the frequency deviations, its simulation model should be able to calculate the corresponding amount of LFC required by the disturbance and allocate it to each generator. The operational mechanism of the AGC frequency control is implemented in Python, which is an API program of PSS/E where the dynamic data of the Korean power system have been managed, for the flexibility and compatibility of the proposed simulation model [12–15].

In addition, the proposed model uses an existing program, such as PSS/E, as a solver of the power system with the responses of the generators assigned by the AGC frequency control. As the dynamic characteristics of the power system are simulated in solving the power system using the dynamic-model-based program PSS/E, the proposed model can determine the response of the power system to the AGC frequency control considering the dynamics of the power system.

Figure 4 shows the simulation framework of the dynamic-model-based AGC frequency control.

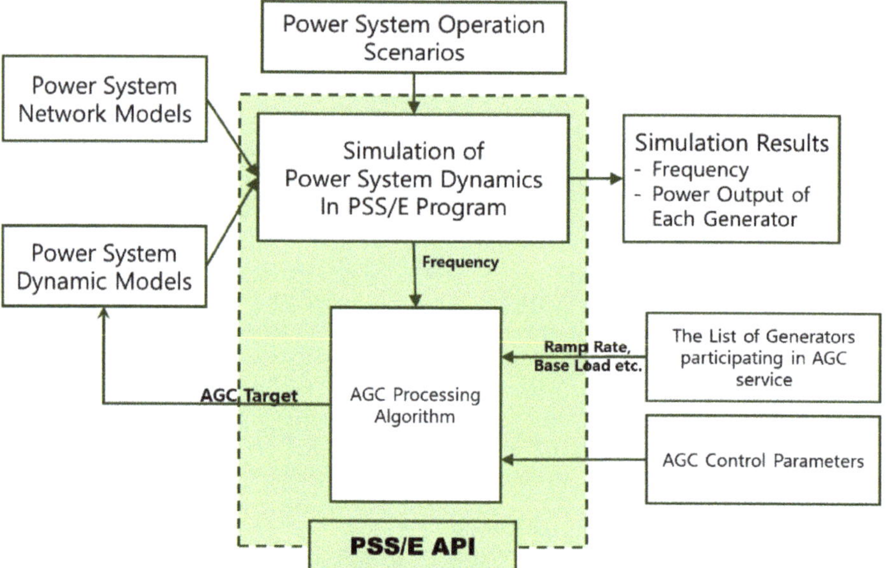

Figure 4. Simulation framework for AGC in the Korean power system.

As shown in Figure 4, in every AGC cycle, Python API runs the dynamic simulation of the power system with the given data by calling the PSS/E program and performs AGC frequency control using the simulated operation results. By repeating this process, the proposed model can simulate the AGC frequency control based on the dynamic model of the power system.

Figure 5 shows the process of simulation for the AGC frequency control using Python.

As shown in Figure 5, the proposed model calculates ACEP and ACEI using the dynamic-model-based simulation method at every AGC cycle. The calculated ACEP and ACEI are adjusted by applying the dynamic deadband filter and then assigning the response to the generators. Finally, the dynamics of the power system are simulated using the responses of the generators to the assigned AGC targets.

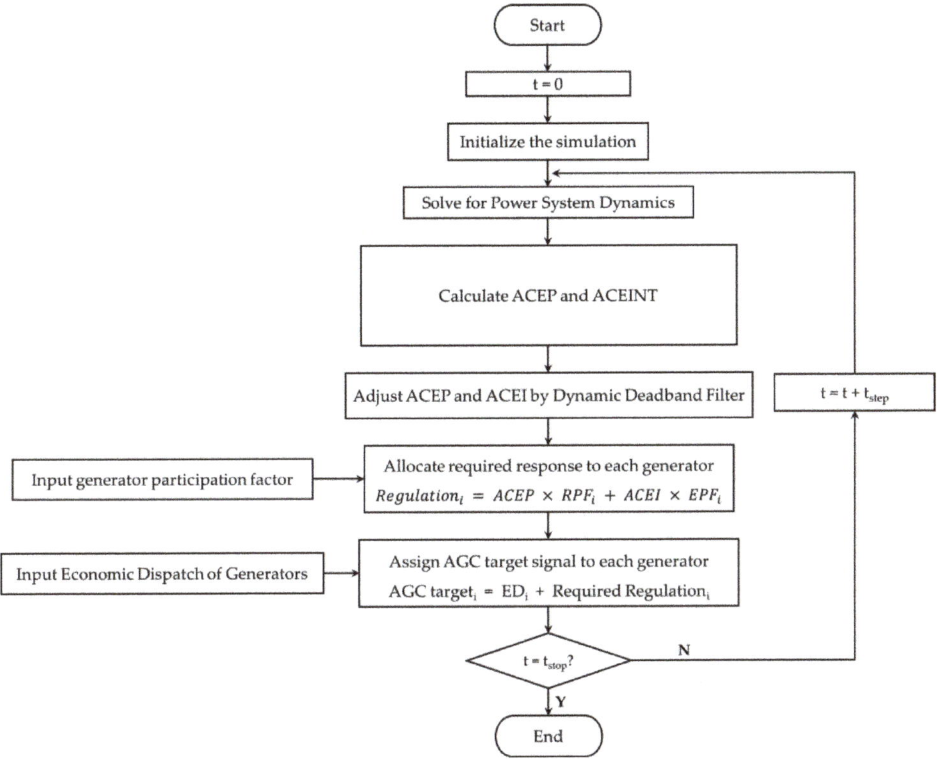

Figure 5. Simulation process of the AGC frequency control.

3.2. Modeling of Generators for Simulating AGC

As the existing dynamic models of generators have been implemented to simulate the dynamic responses of the generators themselves, these models need to be modified to receive and process the AGC command for those outputs in the proposed simulation framework. In the operating system in the field, each generator takes the AGC signal via PLC, as illustrated in Reference [16]. The role of PLC is to adjust the AGC target for coordination with the governor response [3]. For instance, consider a case where the governor of each generator orders an increase in its power output immediately after the frequency drop caused by a contingency, but the AGC target is not changed from the previous command at this time due to the AGC updating cycle. Consequently, each generator would not be able to provide the primary frequency response. Therefore, the PLC is used to adjust the AGC target when the governor responds to the disturbance so that the response can be sustained.

In the proposed simulation framework, the PLC functions are implemented in the existing dynamic models of generators. In the Korean power system, the mainly used models for governors are GGOV1, GAST TGOV1, and IEEEG1. In the case of the GGOV1 model, the interfacing AGC command with the matched input of the GGOV1 model would be able to implement the PLC function, as shown in Figure 6.

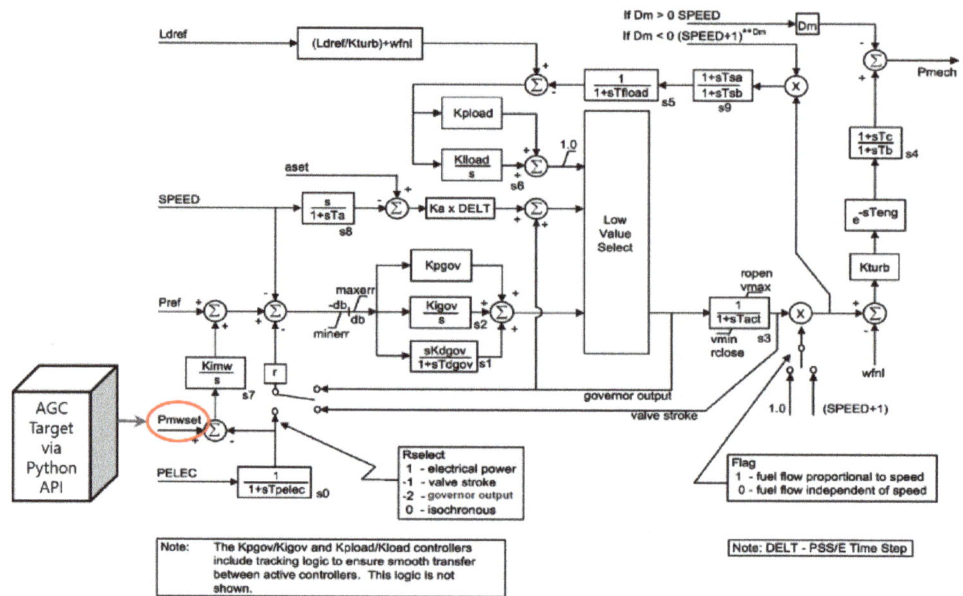

Figure 6. AGC interfacing with the GGOV1 model.

As shown in Figure 6, the implementation of the PLC function of the GGOV1 model is indicated by a red circle. Upon sending the AGC target to the Pmwset of the GGOV1 model through Python-API, the GGOV1 model responds to the AGC frequency control in coordination with the governor response.

In the case of the remaining models, the LCFB1 model which is a standard turbine load controller model for the governor in PSS/E program is added to implement the PLC function for linking the AGC command to the governor, as shown in Figures 7–9.

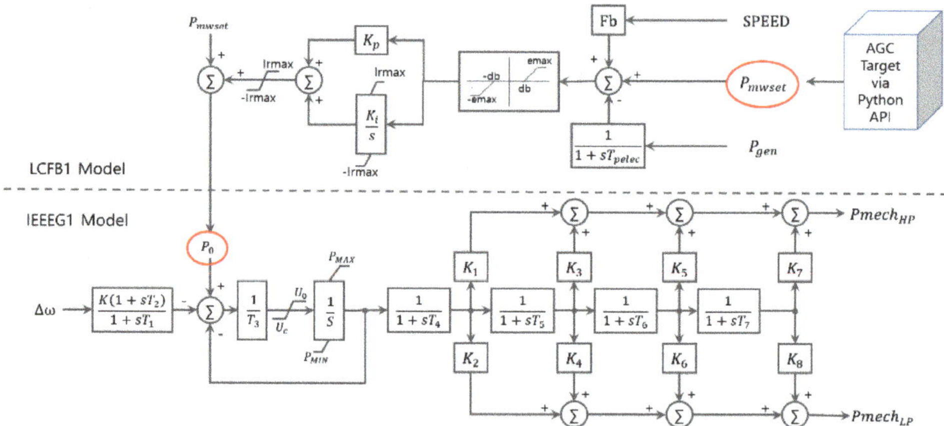

Figure 7. IEEEG1 model interfacing with the LCFB1 model for simulating AGC.

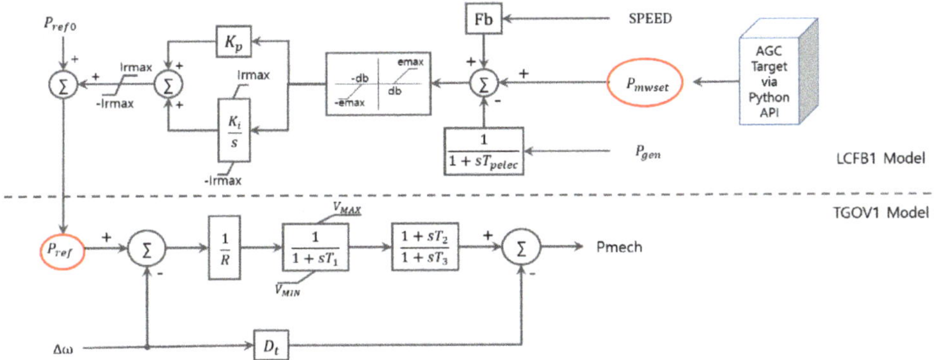

Figure 8. TGOV1 model interfacing with the LCFB1 model for simulating AGC.

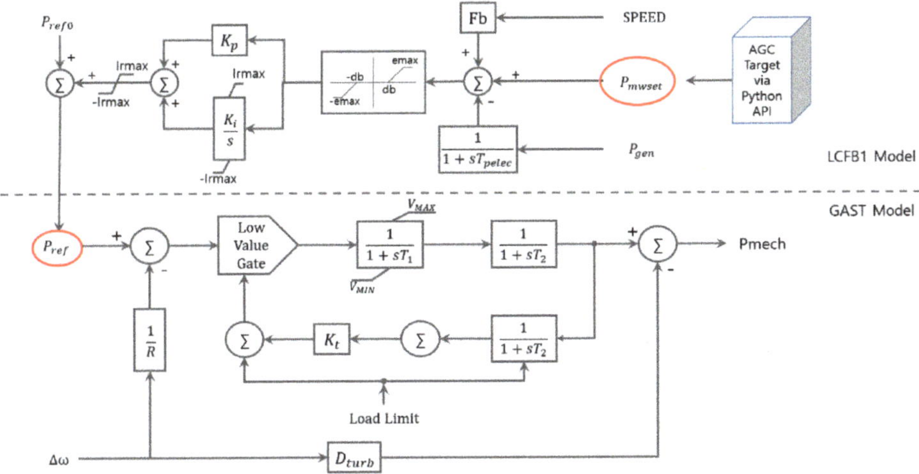

Figure 9. GAST model interfacing with the LCFB1 model for simulating AGC.

As shown in Figures 7–9, the added LCFB1 model revises the AGC target with the amount of difference between the received AGC target and the actual power output of the generator.

In the proposed simulation model, by feeding the revised AGC target from the LCFB1 model to the governor model as its output reference, the PLC function for the coordination between the AGC frequency control and the governor response immediately after the disturbance is implemented.

4. Case Study

In this section, the effectiveness of the proposed method is verified through case studies employing the Korean power system. As the Korean power system does not have any interconnection with neighboring countries, its frequency control is controlled only by its own operating reserves, including the governor responses and AGC frequency controls. However, their effective operation has increasingly become a concern with the increasing penetration level of RES.

The dynamic data of the Korean power system are obtained from the planning database of the power utility [17], which is maintained in the format of PSS/E. The operating conditions of the power system are assumed to be a set of typical values used in the planning database, where the total electricity demand is 94 GW, provided by 276 generators. Among the 276 generators, all the generators except for 27 nuclear generators provided the governor response, and 143 generators provided the AGC frequency control.

4.1. Verifocation of the Simulation Model for AGC Frequency Control

This case study reviewed each process function of the simulation for the AGC frequency control according to Figure 5 to verify the effectiveness of the proposed simulation model. Figure 10 shows the filtered frequency, which is used to calculate the ACE in the proposed model assuming a generator trip.

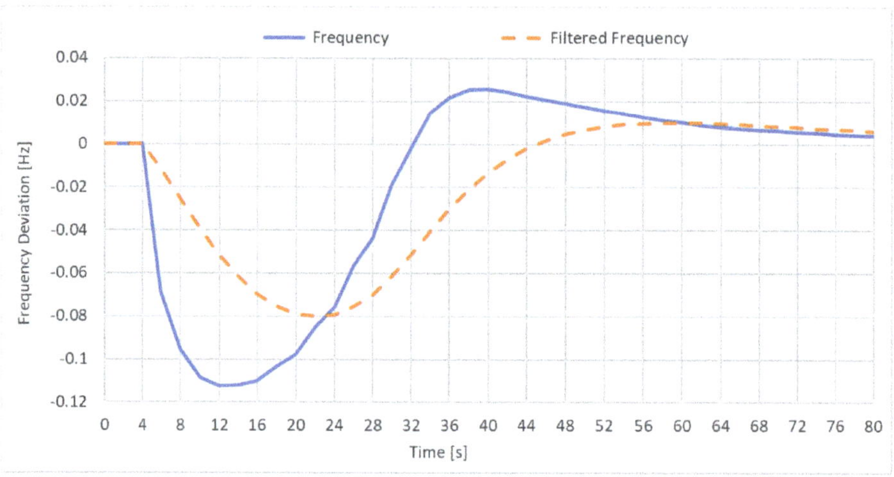

Figure 10. Comparison of frequency and filtered frequency using low-pass filter.

An efficient and stable response of the AGC frequency control can be achieved by using the filtered frequency for calculating the ACE.

The function of the dynamic deadband is verified by comparing the operating results of the AGC frequency control with and without it. Figure 11 shows the simulation results for the comparison in terms of the ACE and controlled frequency.

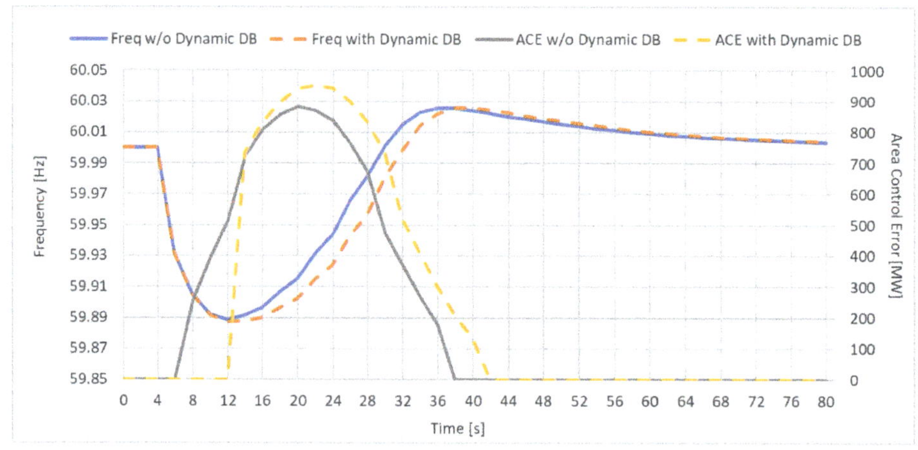

Figure 11. AGC frequency control with and without dynamic deadband.

As shown in Figure 11, the dynamic deadband delays the activation of the ACE when the frequency drops. However, the performance of the frequency control is not deteriorated by the delayed activation of the ACE because the nadir of the controlled frequency is the same as that in the case where the ACE is instantly activated without the dynamic deadband. Therefore, it would be efficient to operate the AGC frequency control with some degree of delay by using the dynamic deadband during a transient period, as even an instant response of the AGC frequency control would not be able to contribute to the frequency control owing to the overlap with the governor responses immediately after the disturbance. The detailed simulation results of the proposed model are shown in Figure 12.

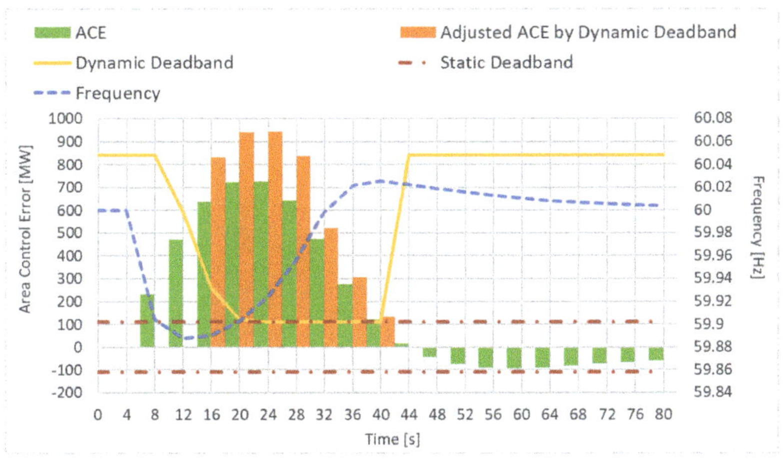

Figure 12. Verification of dynamic deadband activation.

As illustrated in Figure 12, the ACE is not activated immediately after the frequency drop because of the dynamic deadband. However, if the frequency is not recovered until its ACE exceeds the decreasing dynamic deadband, the ACE is activated and its amount is adjusted by the control gain depending on the control mode. Once the frequency is recovered and its ACE is smaller than the static deadband, the dynamic deadband is initialized. The activated ACE is allocated to each generator as the AGC frequency control target. Figure 13 shows an example of ACE allocation using the proposed simulation model.

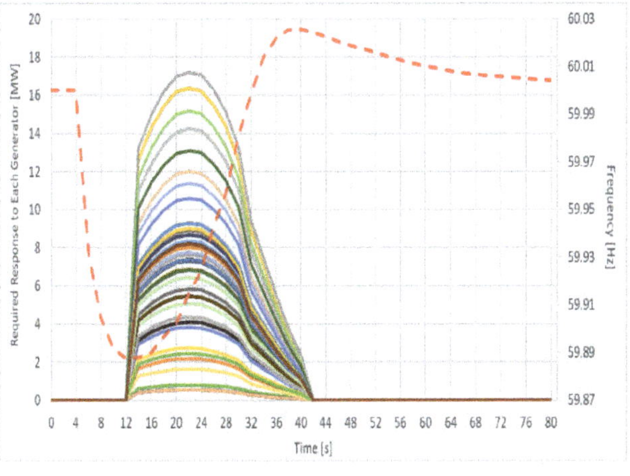

Figure 13. *Cont.*

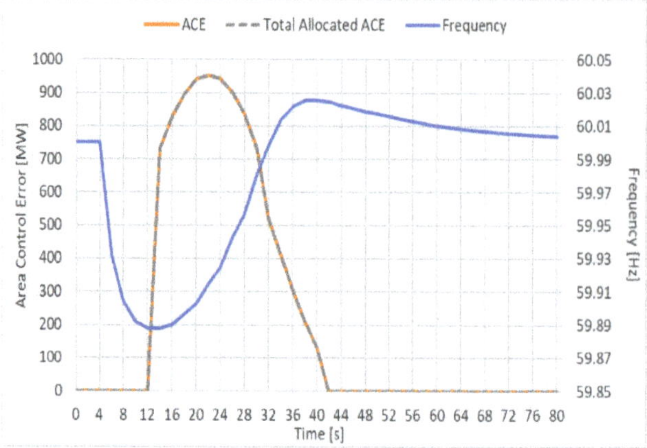

Figure 13. Example of area control error allocation using the proposed simulation model (**upper**) and comparison of between the total allocated area control error and pre-allocated area control error (**lower**).

As shown in Figure 13, the amount of activated ACE depends on the frequency drop, and it is allocated to each generator using the proposed simulation model.

In the AGC frequency control operation, the allocated ACE is received by each generator through its PLC so that the governor response could be prioritized in the frequency control. In this study, the PLC of the proposed model is verified by applying the contingency of the generator trip. Figure 14 illustrates the simulation results of the PLC function.

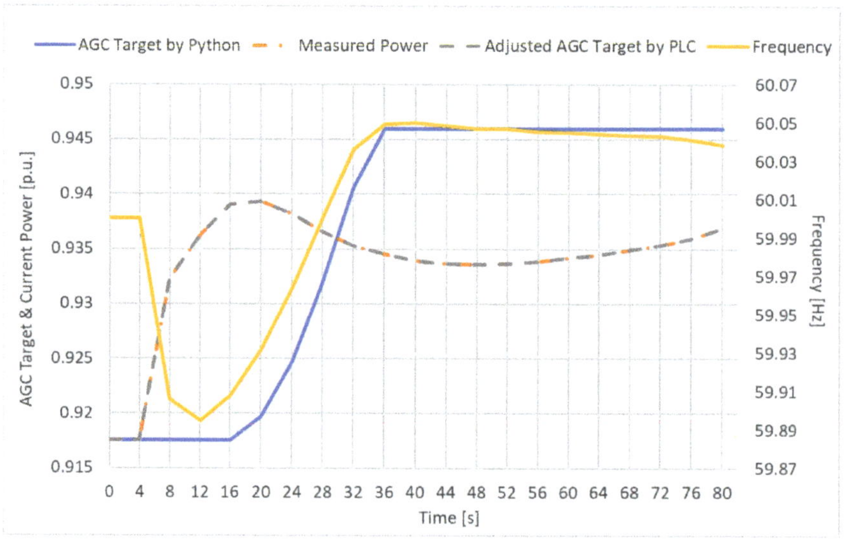

Figure 14. Revised AGC target by the plant-level controller function.

As illustrated in Figure 14, once the measured output of the generator is different from the received AGC target, the PLC is activated to revise the received AGC target by adding a governor response to it. Therefore, the output of the generator follows the revised AGC target rather than the received target.

Once the AGC target is received by each generator, it should be met with a response from the generator even considering its PLC function. The performance of the proposed simulation model is verified by comparing the output of the generator and the AGC target in terms of the appropriateness of the AGC allocation and generator dynamics. Figure 15 shows the AGC response of the generators with the highest and lowest amounts of ACE allocation.

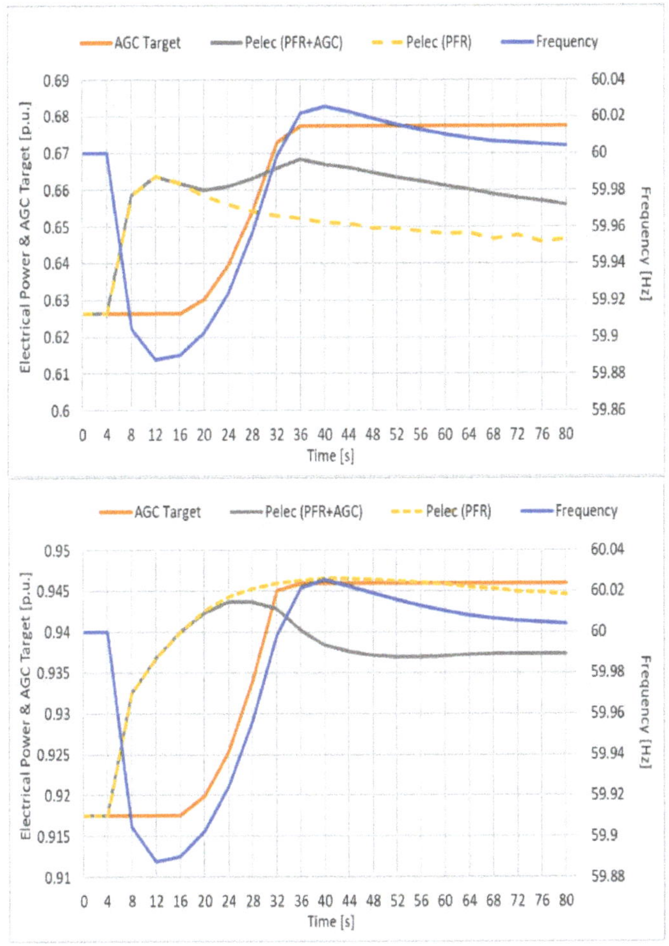

Figure 15. AGC frequency control response of GEN#1 (**upper**) and GEN#2 (**lower**).

As shown in Figure 15, the amount of ACE allocated to GEN#1 is smaller than that allocated to GEN#2. This is because GEN#1, a coal generator, has a lower performance than GEN#2, a gas generator. As the output of the generators could satisfy the revised AGC target, the appropriateness of the AGC allocation could be verified by considering the dynamics of the two generators.

To validate the proposed model, the simulated results are also compared by the selected actual frequency responses in the Korean power system. The real-event data was recorded when a nuclear generator of 1 GW tripped in 2013. Since there must be differences in the detailed operating conditions of the real event and the simulated case, those should be compared only in terms of AGC frequency control. Figure 16 compares ACE calculations of the measured and simulated case to the same frequency recorded in the real case.

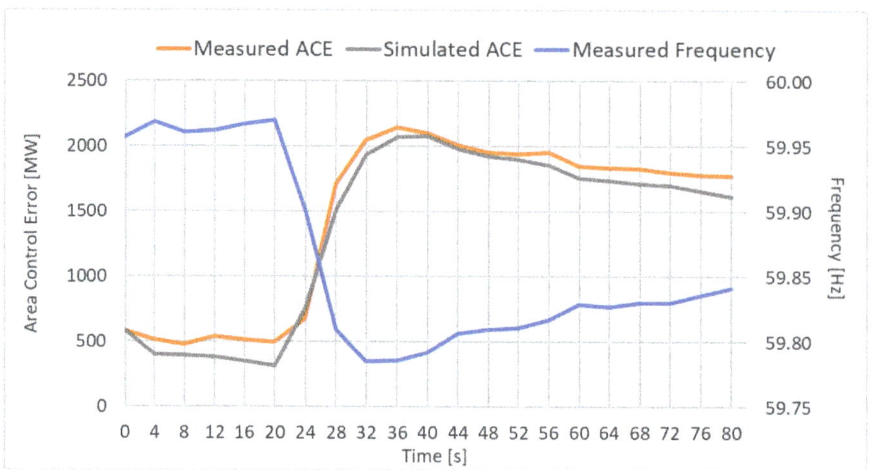

Figure 16. Comparison of the measured and the simulated ACE.

As shown in Figure 16, the simulated ACE was matched with the measured ACE and this validates that the proposed model could be used to simulate real ACE calculations by appropriately setting the control parameters.

In addition, AGC command dispatched to a specific generator and its response is compared using the same event as shown in Figure 17.

As shown in Figure 17, the time delay in the AGC activation that occurred in the measured data could be also found in the simulated results. This validates that the proposed model could simulate the time response of the real AGC operation.

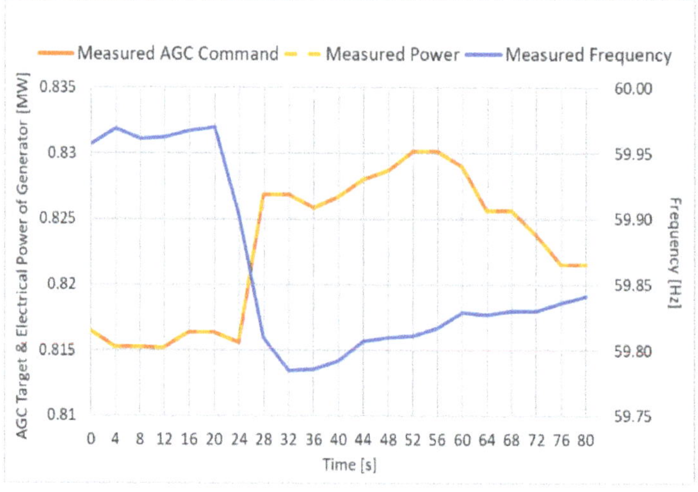

Figure 17. *Cont.*

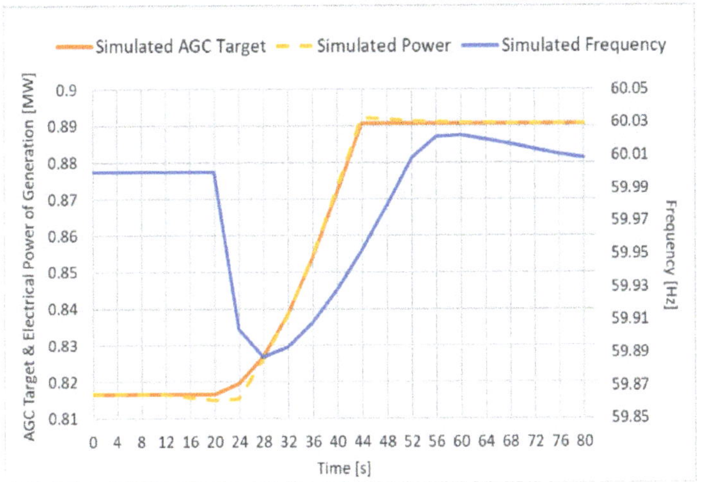

Figure 17. Comparison of the AGC activation time between the measured (**upper**) and simulated (**lower**).

4.2. Analysis of Tuning AGC Frequency Control

In this section, case studies using the proposed simulation model for analyzing the effects of setting the deadband filter of the AGC frequency control, which should consider both the dynamics and AGC operation of the power system, are described. As the proposed model simulates the frequency performance of the power system synthesizing the governor response and AGC operation, the effect of parameter setting can be analyzed by considering not only the effect of the parameter itself but also the coordination with the governor response.

The deadband filter has three control parameters: variable time, the boundary of the control mode, and the control gain at each mode. The variable time delays the activation of the AGC frequency control for coordination with the governor response. Figure 18 illustrates the simulation results of the frequency control with different variable times of the deadband filter in the AGC operation.

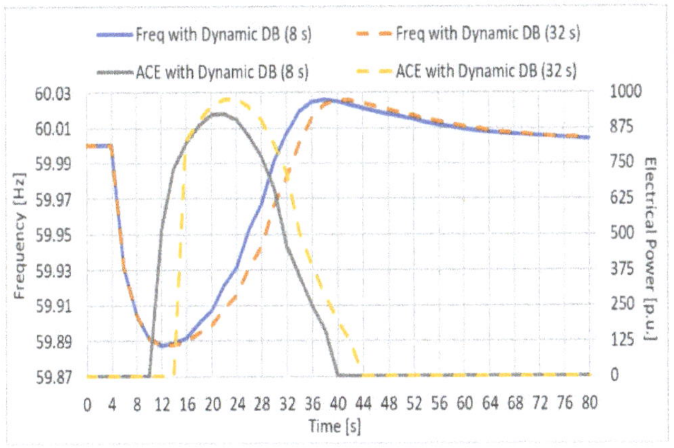

Figure 18. *Cont.*

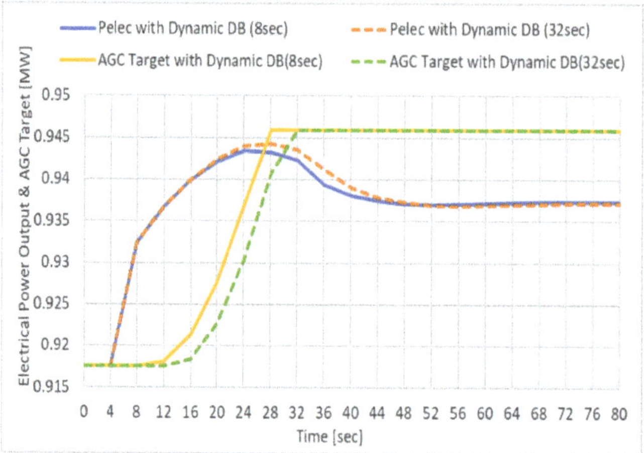

Figure 18. The area control error activation time (**upper**) and the frequency response of generator (**lower**) depending on variable times of the dynamic deadband.

As shown in Figure 18, the ACE activation time is controlled by the variable time of the dynamic deadband, and it affects the frequency response of the generators and the frequency of the power systems. Although the frequency recovered faster when the variable time of the dynamic deadband was reduced from 32 to 8 s, the AGC frequency control could not improve the nadir frequency. This appears to be because the generator power was saturated with the governor response, as shown in the lower figure in Figure 18. Accordingly, the proposed model can be used to set the variable time of the dynamic deadband appropriately in AGC operations.

The boundary conditions of the control modes are used to determine the status of the power system depending on the amount of ACE. In this case study, three cases are assumed for the proposed simulation model with different sets of boundary conditions, as summarized in Table 1. As one goes from cases 1 to 3, the range of each mode becomes wider and the emergency mode appears later. For analyzing the effectiveness of boundary conditions of the control modes only, the variable time of the dynamic deadband is set as 32 s, and the control gains are set as 1.0 in all three cases.

Table 1. Bounds of the control mode (ACE: Area Control Error; Static DB: Static Deadband).

Cases	Static DB	Normal	Assist	Emergency
Case 1	ACE < 110	110 ≤ ACE < 390	390 ≤ ACE < 500	500 ≤ ACE
Case 2	ACE < 110	110 ≤ ACE < 500	500 ≤ ACE < 840	840 ≤ ACE
Case 3	ACE < 110	110 ≤ ACE < 700	700 ≤ ACE < 2100	2100 ≤ ACE

Figure 19 shows the simulation results using the proposed model for the three cases and the distribution of the control mode determined every 2 s during the simulation.

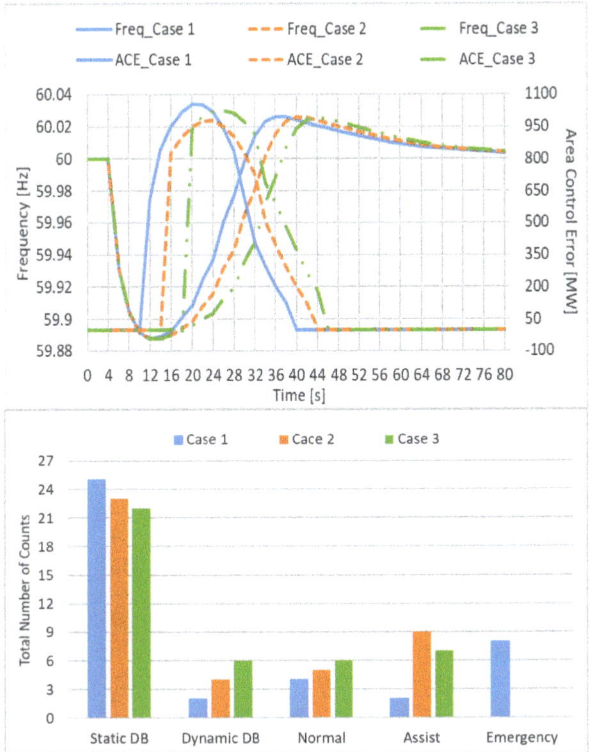

Figure 19. Comparison of frequency response with different bounds of control mode (**upper**) and the distribution of the determined control mode (**lower**).

As shown in Figure 19, the activation of the AGC frequency control is also delayed by setting the ranges of the control modes wider. Among the three cases, the boundary condition of case 2 appears to be relatively more reasonable considering the frequency drop because the frequency could recover earlier without the emergency mode. In the case of 1, the nadir frequency could not be improved even by the more frequent emergency modes.

Accordingly, the proposed model can be used to set the boundary conditions of the control modes appropriately in the AGC operations. It can also be observed from Figure 19 that the nadir frequency is the same in all three sets of boundary conditions, but the frequency recovers earlier with narrower ranges of the control modes.

The control gain of each mode determines the strength of the AGC frequency control. In this study, three cases are assumed with different sets of proportional control gains, as summarized in Table 2 where the boundary conditions of the control modes are assumed to be the same as those in case 2 of Table 1. As one goes from cases 1 to 3 in Table 2, the proportional control gain of each mode increases. The boundary conditions of the control modes are assumed to be the same as those in case 2 of Table 1.

Table 2. Gain setting of each control mode.

Cases	Normal	Assist	Emergency
Case 2.1	0.1	0.3	0.5
Case 2.2	1.1	1.3	1.5
Case 2.3	2.1	2.3	2.5

Figure 20 shows the simulation results using the proposed model for these three cases.

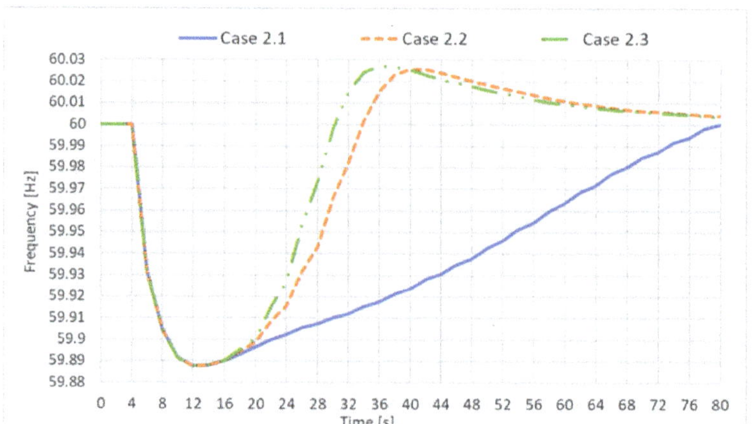

Figure 20. Comparison of frequency responses with different control gains.

As shown in Figure 20, the frequency recovers faster with higher control gains of the AGC frequency control. However, the high control gain in the AGC frequency operation can cause an overshoot of the power system control. Accordingly, the proposed model can be used to set the control gains of each mode appropriately in the AGC operations.

As shown in the results of the case studies, the proposed simulation model can be used to tune the control parameters of the AGC frequency control considering both the dynamics and AGC operations of the power system. This is one of the benefits of developing a simulation model for the AGC frequency control based on the dynamic models of the power systems.

5. Conclusions

This paper proposed a dynamic-model-based AGC frequency control simulation method employing the Korean power system that can be designed and analyzed coordinately with the governor responses of the generators. The existing dynamic-model-based simulation frameworks typically used for power system analysis were adapted to include the AGC frequency control to implement the simulation model for both the AGC frequency control and governor responses. As the database of the Korean power system has been implemented in the framework of PSS/E, Python was used as the API for implementing the AGC frequency control and interfacing its responses to the dynamic-model-based simulation model through additional PLC functions of the governor models using PSS/E. Accordingly, the proposed simulation model could flexibly implement the AGC frequency control and effectively use the existing models of power systems.

The effectiveness and applicability of the proposed simulation model for AGC frequency control were verified using various case studies. These case studies examined each function of the AGC frequency control and analyzed the effects of variation of the control parameters related to the deadband filter of the AGC frequency control.

Owing to the need to operate the power system more competitively with more variable resources, the proposed simulation model can be used to tune the control parameters of the AGC frequency control to achieve a highly efficient coordination of operations with the governor responses from local generators.

Author Contributions: Conceptualization, H.N.G. and K.S.K.; methodology, H.N.G. and K.S.K.; software, H.N.G.; validation, H.N.G., and K.S.K.; writing-original draft preparation, H.N.G. and K.S.K.; writing-review and editing, K.S.K.; supervision, K.S.K. All authors have read and agreed to the published version of the manuscript.

Acknowledgments: This research was partially supported by the Korea Electric Power Corporation (grant number: R18XA04).

Conflicts of Interest: The authors declare no conflict of interest.

References

1. Korean Ministry of Trade, Industry and Energy. *The 8th Basic Plan for Long-Term Electricity Supply and Demand (2017–2031)*; Korean Misnistry of Trade, Industry and Energy: Sejong, Korea, 2017.
2. Hector, C.; Ross, B.; Julija, M. The joint adequacy of AGC and primary frequency response in single balancing authority systems. *IEEE Trans. Sustain. Energy* **2015**, *6*, 959–966.
3. NERC. *Reliability Guideline: Primary Frequency Control*; NERC: Washington, DC, USA, 2015.
4. Wood, A.J.; Wollenberg, B.F. *Power Generation Operation and Control*, 2nd ed.; John. Wiley & Sons: New York, NY, USA, 1996.
5. Korea Power Exchange (KPX). *Electricity Market and Power System Operating Guide*; Korea Power Exchange (KPX): Naju, Korea, 2019; p. 278.
6. Choi, Y.M.; Lee, G.W. Introduction of EMS AGC Group Control. In Proceedings of the KIEE 2008 Power Engineering Society Fall Conference and General Meeting; pp. 133–135. Available online: http://www.dbpia.co.kr/journal/articleDetail?nodeId=NODE01336617 (accessed on 25 September 2020).
7. Choi, S.H.; Hwang, K.I.; Chun, Y. Evaluation of AGC characteristics and a novel AGC control strategy for independent power systems. *Trans. Korean Inst. Electr. Eng. A* **2005**, *54*, 540–547.
8. Unbenhauen, H.D. Automation and Control of Electrical Power Generation and Transmission Systems. In *Control Systems, Robotics and Automation–Volume: Industrial Applications of Control Systems*; EOLSS Publishers Co. Ltd.: Oxford, UK, 2009; Volume 18, p. 213.
9. Korea Power Exchange (KPX) & Jeonbuk National University (JBNU). *A Study on Optimal Perform Method of AGC and Criteria Establishment for Control Variables*; Korea Power Exchange (KPX): Naju, Korea, 2014.
10. Chang Soo, O.; Suk Ha, S.; Woon Hee, L. Optimal AGC Control Parameter Tuning. In Proceedings of the KIEE 2008 Power Engineering Society Fall Conference and General Meeting; pp. 321–322. Available online: https://www.koreascience.or.kr/article/CFKO200814835315828.pdf (accessed on 25 September 2020).
11. Korea Power Exchange (KPX). *Decree on Reliability Standard and Electricity Quality*; Korea Power Exchange (KPX): Naju, Korea, 2019.
12. Siemens PTI. *PSS/E 34 Application Program Interface (API)*; SIEMENS: New York, NY, USA, 2015.
13. Wang, L.; Chen, D. Experience with Automatic Generation Control (AGC) Dynamic Simulation in PSS/E. Power Technol. 2012. Available online: https://static.dc.siemens.com/datapool/us/SmartGrid/docs/pti/2012August/PDFs/Experience_on_AGC_Dynamic_Simulation_in_PSSE.pdf (accessed on 25 September 2020).
14. Wang, L.; Chen, D. Automatic Generation Control (AGC) Dynamic Simulation in PSS/E. Power Technol. 2011. Available online: https://static.dc.siemens.com/datapool/us/SmartGrid/docs/pti/2011February/PDFs/AGC%20Dynamic%20Simulation%20in%20PSSE.pdf (accessed on 25 September 2020).
15. Siemens PTI. *PSS/E 34 Model Library*; SIEMENS: New York, NY, USA, 2015.
16. NERC. *Reliability Guideline; Application Guide for Modeling Turbine-Governor and Active Power-Frequency Controls in Interconnection-Wide Stability Studies*; NERC: Atlanta, GA, USA, 2019.
17. Korea Power Exchange (KPX). *Annual Operating Results of Korean Electricity Market in 2019*; Korea Power Exchange: Seoul, Korea, 2020; p. 2.

Article

Adaptive Algorithm of a Tap-Changer Controller of the Power Transformer Supplying the Radial Network Reducing the Risk of Voltage Collapse

Robert Małkowski [1,*], Michał Izdebski [2] and Piotr Miller [3]

[1] Department of Electrical Power Engineering, Faculty of Electrical and Control Engineering, Gdańsk University of Technology, 80-233 Gdańsk, Poland

[2] Gdansk Division, Institute of Power Engineering, 80-870 Gdańsk, Poland; m.izdebski@ien.gda.pl

[3] Department of Power Engineering, Faculty of Electrical Engineering and Computer Science, Lublin University of Technology, 20-618 Lublin, Poland; p.miller@pollub.pl

* Correspondence: robert.malkowski@pg.edu.pl

Received: 15 September 2020; Accepted: 10 October 2020; Published: 16 October 2020

Abstract: The development of renewable energy, including wind farms, photovoltaic farms as well as prosumer installations, and the development of electromobility pose new challenges for network operators. The results of these changes are, among others, the change of network load profiles and load flows determining greater volatility of voltages. Most of the proposed solutions do not assume a change of the transformer regulator algorithm. The possibilities of improving the quality of regulation, which can be found in the literature, most often include various methods of coordination of the operation of the transformer regulator with various devices operating in the Medium-Voltage (MV) network. This coordination can be decentralized or centralized. Unfortunately, the proposed solutions often require costly technical resources and/or large amounts of real-time data monitoring. The goal of the authors was to create an algorithm that extends the functionality of typical transformer control algorithms. The proposed solution allows for reducing the risk of voltage collapse. The performance of the proposed algorithm was validated using multivariate computer simulations and tests with the use of a physical model of the distribution network. The DIgSILENT PowerFactory environment was used to develop the simulation model of the proposed algorithm. Then, tests were conducted on real devices installed in the LINTE^2 Laboratory at the Gdańsk University of Technology, Poland. Selected test results are included in this paper. All results have shown that the proposed algorithm makes it possible to increase the reserve of the voltage stability of the node, in which it is applied, thus mitigating the risk of a voltage collapse occurring. The proposed algorithm does not require complex and costly technical solutions. Owing to its simplicity, it has a high potential for practical application, as confirmed by the real-time control experiment in the laboratory.

Keywords: voltage collapse; voltage control; transformer controller; adaptive algorithm; voltage regulation; distribution system; power distribution network; distributed generation; OLTC

---

1. Introduction

The interface between the High-Voltage (HV) transmission grid and the Medium-Voltage (MV) distribution network are HV/MV substations. An example of the structure of such a station is shown in Figure 1. In some HV/MV substations, stations shunt compensators are installed. Most often they are Capacitor Banks (CB), less often reactors.

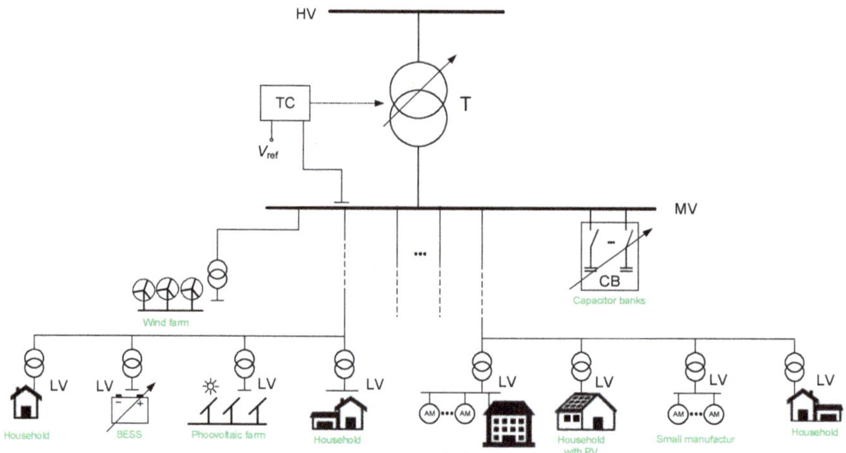

Figure 1. Simple radial Medium-Voltage (MV) grid with High-Voltage (HV)/MV substation.

The variability of active power and reactive power of sources creates the need for voltage regulation. Voltage regulation is done primarily by changing the power transformer voltage ratio in the HV/MV substation and by controlling shunt capacitors (if any). A criterion for the operation of a transformer regulator in a HV/MV substation is to maintain the preset voltage value on the MV side V_{ref}. The range of voltage variations is determined by the transformer parameters and the network characteristic. The regulation capacities of a transformer are usually a dozen or so taps that enable voltage change within a range of several dozen percent of the rated voltage. Preset values along with a schedule of changes are defined offline by the operation staff. The voltage regulation is continued either until an extreme maximum or minimum tap position is achieved, or until the voltage value attains one of the limiting values: of over-voltage V_{Tmax} or under-voltage V_{Tmin}.

An item of the equipment of some HV/MV substations are also capacitor banks. Their regulation involves switching individual sections on or off. Switching processes are determined by, inter alia, the time schedule and are done by manual switch-on, either remotely or locally, or by an individual controller operating in voltage regulation mode.

To minimize the influence of load variability on the voltage levels in the network, requirements to improve the reactive power management are imposed on both the generators and consumers [1]. These are generally limited to the requirement to maintain a constant preset magnitude of reactive power or not to exceed the preset value of the coefficient tg.

Usually, when the volume of distributed generation and daily variability of loads are low, the efficiency of voltage regulation by changing the power transformer ratio is satisfactory. However, in networks with a high saturation of distributed generation, the efficiency of such a method of voltage regulation is increasingly often insufficient. High generation variability (wind, photovoltaic generation) deteriorates the voltage profiles on the grid [2–9]. An equally important problem is periodically too low or too high voltage in the network [10,11]. Too high voltage occurs as a result of a change in the typical direction of power flow (flow—from HV/MV substations to consumers) caused by a large generation volume, e.g., as a result of favorable weather conditions.

The possibilities of improving the quality of regulation, which can be found in the literature, e.g., [2–6,10,12], most often include various methods of coordination of the operation of the transformer regulator with various devices operating in the MV network. This coordination can be decentralized or centralized.

Most of the proposed solutions do not assume a change of the transformer regulator algorithm.

The analysis of the operation algorithms of the transformer regulators supplying the distribution network showed that no solution would take into account the change like the dQ/dV coefficient. There is, however, a commercial device called Collapse Prediction Relay (CPR-D) offered by A-Eberle [13–15] but its algorithm is very complex. To determine the need to block or properly control the transformer tap changer, the following are used: bifurcation theory [16–18] in combination with neural network elements, determination of Lyapunov exponents, identification of voltage drop and, finally, damping coefficients.

Bearing the above in mind, an attempt was made to develop an effective and simple method of controlling the operation of the transformer supplying the distribution network. The main difference between the aforementioned CPR-D system and the proposed co-authoring algorithm lies in the specific purpose of operation. In the case of the CPR-D system, the aim is to identify the state of emergency, while the purpose of the proposed adaptive control system for transformers supplying the distribution network is to prevent the emergence of a voltage avalanche hazard.

The paper will present an algorithm for the operation of the regulator of a distribution network-supply transformer, which is protected by the patent [19]. The solution presented in the paper extends the functionality of the typical algorithms of the power transformer regulator. In the described solution, in the normal operating state of the network, the power transformer regulator works according to a typical manufacturer-defined algorithm. The proposed algorithm is activated only when there is a risk of voltage collapse, possibly leading to a blackout.

Moreover, in this paper, theoretical fundamentals concerning the effect of the regulation of the transformer voltage ratio on the risk of an occurrence of a voltage collapse, and the results of the simulation studies and laboratory tests of the proposed algorithm will be described.

2. Voltage Stability Phenomenon

Voltage stability is the ability of a power system to maintain steady acceptable voltages at all buses in the system under normal operating conditions and after being subjected to a disturbance. The main factor causing instability is the inability of the power system to meet the demand for reactive power [20–25].

A system is voltage unstable if, for at least one bus in the system, the bus voltage magnitude (V) decreases as the reactive power injection (Q) is increased at the same bus. In other words, a system is voltage stable if V-Q sensitivity is positive for every bus and voltage unstable if V-Q sensitivity is negative for at least one bus.

Voltage instability is essentially a local phenomenon; however, its consequences may have a widespread impact. Voltage collapse is more complex than simple voltage instability and is usually the result of a sequence of events accompanying voltage instability leading to a low-voltage profile in a significant part of the power system.

The physical cause of a voltage collapse occurring are primarily some phenomena taking place within the complex load. A reduction in load node voltage causes a reduction of the driving torque of asynchronous motors making up the load. The reduction of asynchronous motor torques is accompanied by an increase of slips and thereby a substantial increase in reactive power taken up from the network, and, as a consequence, a further reduction in load node voltage. As the voltage goes down, subsequent motors within the complex load halt. The consequences of this course of events are obvious to the consumers. For the electric power system, the consequences generally do not end up with a voltage collapse in only one load node. At the same time, the voltages of adjacent nodes will go down, and the phenomenon may spread to other loads.

Voltage instability may occur in several different ways. In its simple form, it can be illustrated by considering the two-terminal network of Figure 2 [25]. It consists of a constant voltage source (E_s) supplying a load (P_l + iQ_l) through the equivalent reactance of the power system (X).

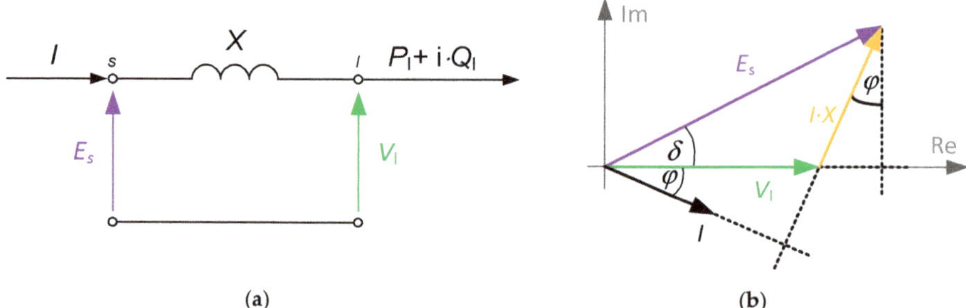

Figure 2. Power system as a source of reactive power: (**a**) equivalent circuit; (**b**) a diagram of the nodal currents and voltages of the substitute source. s—generating node; l—load node; E_s, X—fictitious electromotive force and system equivalent reactance.

The properties of the power system as a source supplying a selected load node will be described using the reactive power generation characteristic, hereinafter in abbreviation called the generation characteristic and denoted by $Q_s(V)$. The generation characteristic defines, as a function of voltage, the reactive power given up by the power system to a given load node, with that node being loaded with the preset load active power, $P_l(V)$.

On the right-hand side in Figure 2b, the complex load with a given characteristic is cut off, and only the sole equivalent source is separated. For this source, formulae can be written, which define the active power and reactive power supplied to the load node:

$$P_s = V_l \cdot I \cdot \cos\varphi = V_l \cdot I \cdot \frac{E_s \cdot \sin\delta}{I \cdot X} = \frac{E_s \cdot V_l \cdot \sin\delta}{X} \tag{1a}$$

$$Q_s = V_l \cdot I \cdot \sin\varphi = V_l \cdot I \cdot \frac{E_s \cdot \cos\delta - V_l}{I \cdot X} = \frac{E_s \cdot V_l}{X} \cdot \cos\delta - \frac{V_l^2}{X} \tag{1b}$$

These formulae can be transformed into the following form:

$$P_s^2 + \left(Q_s + \frac{V_l^2}{X}\right)^2 = \left(\frac{E_s \cdot V_l}{X}\right)^2 \tag{2}$$

According to the definition of the generation characteristic, the source should be loaded with the active power, that is to substitute $P_s(V) = P_l(V)$. With this assumption, the following will be obtained from Formula (2) after making simple transformations:

$$Q_s = \sqrt{\left(\frac{E_s \cdot V_l}{X}\right)^2 - P_s^2} - \frac{V_l^2}{X} \tag{3}$$

Formula (3) defines the generation characteristic, or the relationship of the reactive power $Q_s(V)$ supplied to the load node by the source, with the source being loaded with active power with the preset voltage characteristic $P_l(V)$. The generation characteristic corresponding to Formula (3) is a curve similar to an inverted parabola. The smaller the equivalent reactance of the source X and the smaller active power loading the source P_l, the more the V^2/X parabola shifts towards the center of the coordinating system.

The source and the load may only operate with each other at such a voltage, where the power supplied by the source is equal to the power taken up by the load $Q_s(V) = Q_l(V)$. In Figure 3, this corresponds to the points of intersection of both characteristics. Obviously, the point of the stable operation of the system may only be a locally stable point of equilibrium. The condition of equilibrium is as follows:

$$\frac{dQ_s}{dV} < \frac{dQ_l}{dV} \quad \text{or} \quad \frac{d\Delta Q}{dV} < 0 \tag{4}$$

whereas $\Delta Q = Q_s - Q_l$ is the excess of the source's reactive power over the load demand.

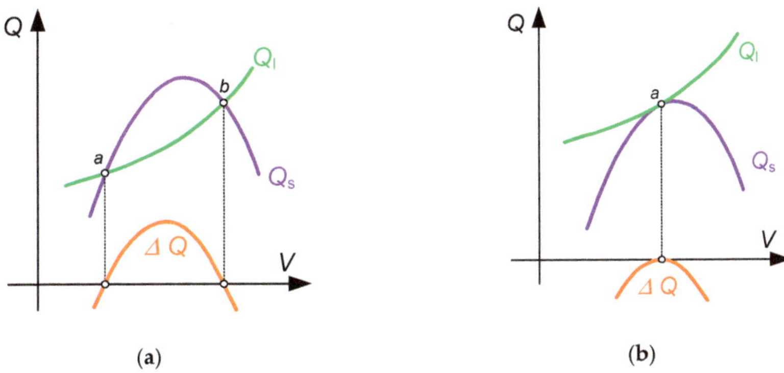

Figure 3. Illustration of the mutual location of the production and reception characteristics: (**a**) two points of equilibrium; (**b**) one point of equilibrium.

In the case shown in Figure 3a, the stability condition (4) is met at point "b" only. At that point, the system can operate in a stable manner. Electrically, the behavior of the system is as follows. A momentary increase in voltage by a small magnitude of ΔV is accompanied by an excess of received power over the generated power, which entails a drop of voltage and the return to the point of equilibrium. Similarly, a momentary decrease in voltage is accompanied by an excess of generated power over the received power, which results in an increase in voltage and the return to the point of equilibrium. A special case of equilibrium point (the point "a") is the situation illustrated in Figure 3b. At point "a", the stability condition (3) is not satisfied. At this point, the system could not operate in a stable manner. Any change in voltage causes an excess of receiver power over the generated power, which results in a further decrease in voltage. The system departs from the point of equilibrium with continuously decreasing voltage. The loss of the system's stability, manifesting itself in a voltage breakdown due to a small disturbance, is called the voltage collapse.

3. The Voltage Dependency of Loads

The effect of operation and some regulation properties of any electric device taking up active and/or reactive power can be described using, e.g., static characteristics. Typical characteristics describe the variability of active P_L and reactive Q_L power take-up as a function of variations in voltage and frequency. Simulation studies concerning the analyses of the operation of electric power systems rarely consider single loads. More often, aggregated load models are used, in which the load power is the sum of the powers of electric loads of different characteristics [26–34]. This sum includes the proportionality factors k_P and k_Q, defined as the percentage fraction of individual groups of loads of the total load power, that are determined at the point of common connection (PCC) (e.g., a specific outlet field in a HV/MV substation), which, assuming a constant frequency level, can be described in a general form using the following relationships:

$$P_L = P_{Ln} \cdot \sum_i k_{P,i} \cdot \left(\frac{V_i}{V_n}\right)^{\alpha_{P,i}} \tag{5a}$$

$$Q_L = Q_{Ln} \cdot \sum_j k_{Q,j} \cdot \left(\frac{V_j}{V_n}\right)^{\alpha_{Q,j}} \tag{5b}$$

where:

- P_{Ln}, Q_{Ln}—total value of, respectively, the active and reactive power of a load group, corresponding to the rated voltage magnitude;
- k_P, k_Q—factors defining the percentage fractions of individual load groups of the total power;
- α_P, α_Q—exponents defining the voltage dependence of a group of loads;
- i, j—parameters defining the number of different voltage dependency groups.

From the point of view of the regulation of voltage in HV/MV substation, power variations are the most important, as they determine the greatest voltage variability. With small voltage deviations from the rated voltage of ±10%, it can be assumed that, in the vast majority of instances, the coefficient of voltage sensitivity of the reactive power of loads, defined as dQ/dV, has a positive value. This is an advantageous situation since at that case one has the so-called self-regulation of the power system—where a decrease in voltage results in a reduction in reactive power take-up by the loads. It should be noted, however, that the value of the dQ/dV coefficient is variable in time. This is caused by the variation of network voltages, the variation of the fractions of individual groups of loads on the total node power (the variation of the coefficients k_P and k_Q), etc. The variation of the dQ/dV coefficient applies not only to the absolute value but also to the sign. In a situation where the dQ/dV coefficient is less than 0, lowering the voltage at the load's terminals will result in an increase in reactive power uptake. Such behavior of loads is undesirable in a situation where a voltage drop is due to a deficit of reactive power in the system, as it will contribute to a deepening of that deficit. A negative value of the dQ/dV coefficient may occur, e.g., in networks with large participation on asynchronous motors [26]. Examples of static characteristics V-Q for selected cases are shown in Figures 4 and 5.

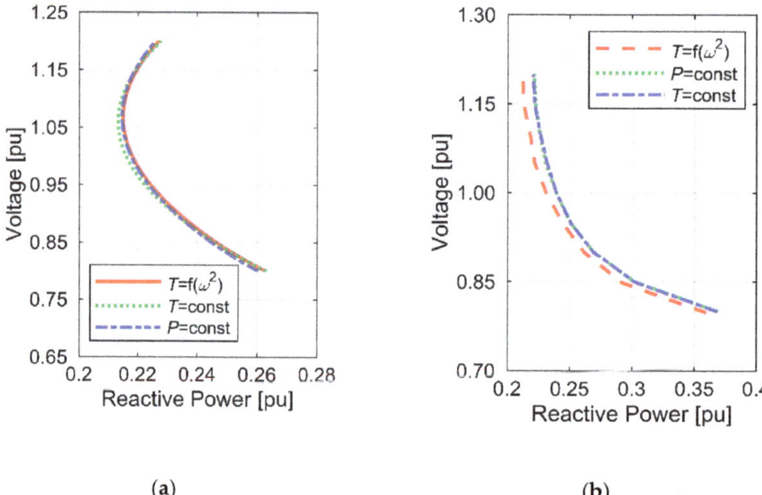

Figure 4. Characteristics of a drive-controlled (elaborated base on [26,27]): (**a**) closed-loop operated three-phase motor for nominal speed; (**b**) closed-loop operated single-phase motor.

It should be noted that the value of the dQ/dV coefficient depends not only on the voltage level but also on the machine loading. The examples shown (Figures 4 and 5) include the cases of loading the machine with a constant moment, T = const.; a constant power, P = const.; and a moment dependent on the square of velocity, T = f(ω^2).

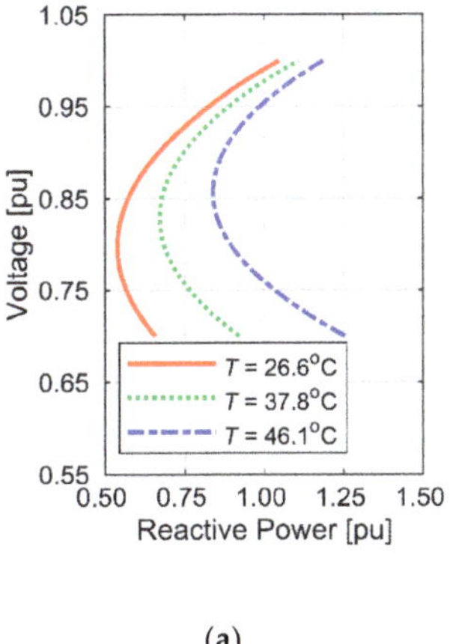

(a)

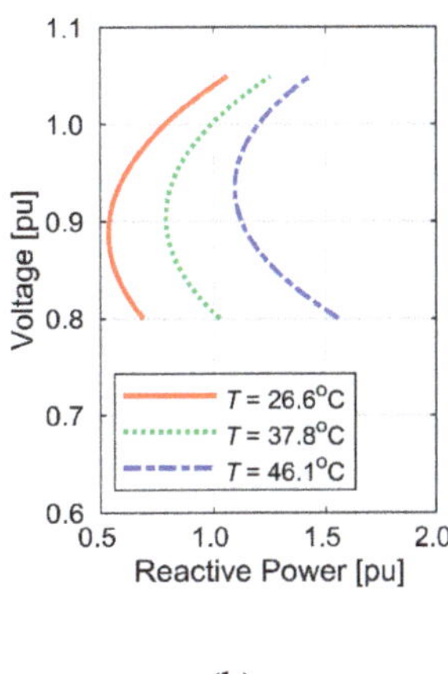

(b)

Figure 5. Reactive power characteristics of a residential air conditioner (RAC) during normal conditions (elaborated base on [29]): (**a**) research by the Electric Power Research Institute (EPRI); (**b**) research by Southern California Edison.

Even though the results of the investigation [35] carried out by the author in the years 2007–2008 show that instances where the dQ/dV is negative or close to 0 are not frequent, this problem should be expected to exacerbate in the nearest future. One of the main causes may be the increased number of refrigerating equipment units—air conditioners. The International Energy Agency (IEA) has published a report entitled The Future of Cooling [36] which forecasts that the growing use of air conditioners in households and offices worldwide will become, over the nearest three decades, one of the main factors of the global demand for electric energy. This problem is addressed also in other publications [29,37]. As has been shown by the investigation results reported in [29,33,38–40], the coefficient of reactive power voltage susceptibility, dQ/dV, of the above-mentioned equipment is negative at lower voltages, as shown in Figure 5. It is also worth noting that the demand for reactive power increases with increasing ambient temperature.

The characteristic of voltage sensitivity dQ/dV, as seen from the terminals of the distribution network supply transformer, is a resulting characteristic. In a situation, where a large number of loads with an "unfavourable" characteristic of voltage susceptibility are present in the network under analysis (Figures 4 and 5), this may pose a threat to the voltage stability in a specific area. The problems may deepen in a situation, where capacitor banks are installed in a large number of network nodes, or the network has a large number of cable lines.

4. The Influence of Transformer Tap Position Changes on the V-Q Characteristics

The characteristics shown in Figures 4 and 5 apply to the situation where the voltage ratio of a transformer is constant. At a constant voltage ratio, the reduction of voltage on the primary side V_1 results in a corresponding reduction of voltage on the secondary side V_2. In the situation where the transformer voltage controller would maintain a constant voltage value on the secondary side, the reactive power Q_2 taken up by the loads would not change (Figure 6).

The discrete nature of voltage regulation by changing the voltage ratio, associated with the stepwise change of the number of coils and the dead zone, also causes a discrete change in the reactive power of the loads.

The process of voltage regulation by changing the transformer voltage ratio is slow. In real power systems, the aim is to reduce the number of tap switchovers during 24 h. It is achieved, e.g., by increasing the settings of switchover delay times or by increasing the dead zone. This is an intentional action, dictated, e.g., by economical and operational reasons. The characteristic presented in Figure 6 proves itself excellently in the analysis of steady states or in the case of slow voltage changes—slower than the action of transformer control systems.

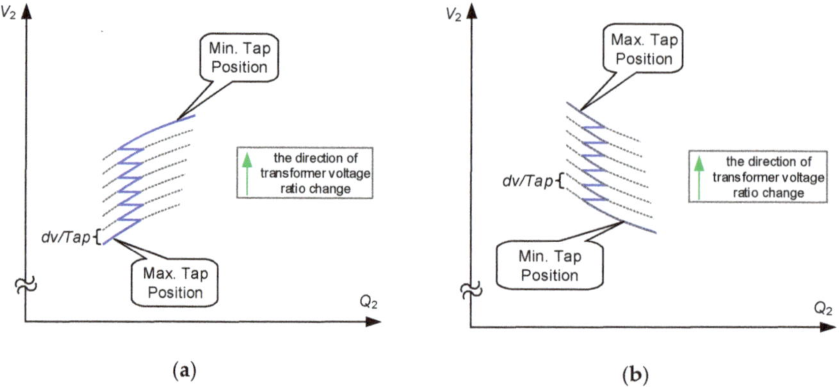

(a) (b)

Figure 6. Influence of transformer tap position changes on the *V*-*Q* characteristics: (**a**) the situation corresponds to work in point "b" shown in Figure 3a; (**b**) the situation corresponds to work in point "a" shown in Figure 3a.

In the case of large overloads, the rate of voltage changes might be greater than the transformer voltage ratio change rate. In such instances, load characteristics should be taken for consideration, which do not allow for the transformer regulation action.

Depending on the phenomena under consideration, the voltage ratio value should remain either unchanged or equal to the extreme voltage ratio values corresponding to the outermost position of the tap changer.

5. The Effect of Transformer Voltage Ratio Regulation on the Risk of a Voltage Collapse Occurring

The mechanism of a voltage collapse occurrence is very complex and comprises many components of the electric power system. Regardless of the causes of its occurrence, the loss of voltage stability is closely related to a disturbance of the reactive power balance in the system. It can either be a local phenomenon or cover a larger part of the electric power system. The causes of the overloading of an electric power system with reactive power and the methods of analysis of the voltage stability of electric power subsystems were addressed in numerous publications, including [20,21,41–45]. This section discusses phenomena that pose a threat to voltage stability and result from transformer voltage ratio regulation. The following two extreme situations are taken for analysis: the first one, in Figure 7a, in which the coefficient of voltage susceptibility of the power of loads is positive ($dQ/dV > 0$); and the second one, in Figure 7b, where the coefficient is negative ($dQ/dV < 0$). It has also been assumed that the voltage change rate is greater than the speed of action of transformer control systems.

The effect of voltage ratio changes with a reactive power deficit in a situation, where ($dQ/dV > 0$), is illustrated in Figure 7a. As a result of a lowering of voltage in the MV network, the transformer controller changes the voltage ratio magnitude to maintain the preset voltage value on the transformer secondary side. The change of the voltage ratio entails a change in the reactive power uptake characteristic. The fixed working point OP_1 moves toward the increased reactive power uptake up to the point OP_2, in which, in an extremely unfavorable case, limiters in the generator controller might be activated. The activation of any current limiter causes a reduction of the magnitude of excitation voltage and, as a consequence, of the generated reactive power, contributing thereby to an aggravation of the reactive power deficit in the electric power system. In the case under discussion, from the point of view of voltage stability, it would be the most favorable to halt the voltage regulation (a constant transformer voltage ratio).

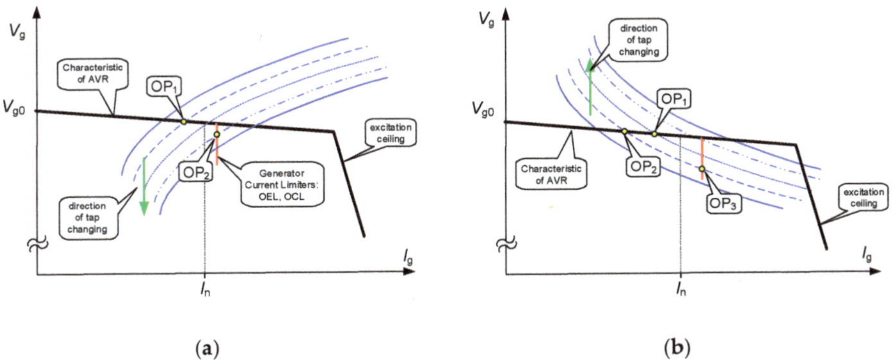

Figure 7. Impact of the transformer tap-ratio control on the risk of voltage collapse: (**a**) positive effect of blocking transformer ratio changes; (**b**) unfavorable effect of blocking transformer ratio changes. Explanation of abbreviations: V_g—generator voltage (Vg0—reference Generator Voltage), —generator current, AVR—Automatic Voltage Regulator, OEL—Over Excitation Limiter, OCL—Over Current Limiter.

The constant transformer voltage ratio will, however, not always bring about desirable effects. As has been mentioned in the introduction, in distribution networks with a large number of asynchronous drives, including air conditioning devices, and a considerable power of capacitors, the dQ/dV coefficient may take on negative values. This means that—with decreasing voltage—the input power increases. In that case, the natural characteristic of loads is distinctly less advantageous than the characteristic determined by the action of the transformer voltage controller that maintains, in a certain range, a constant voltage and the resulting constant reactive power uptake Figure 7b.

When the coefficient $dQ/dV < 0$ of the characteristics of reactive power, operating transformer voltage controller leads to achieving a stable, new working point OP_2. Blocking the controller operation could, in this case, result in a permanent deficit in reactive power (the working point OP_3), causing aperiodic instability—a voltage collapse.

To sum up, if at a given voltage, the voltage susceptibility coefficient dQ/dV determined in the HV/MV substation is negative, while the voltage is lowering, stopping the regulation by changing the power transformer voltage ratio is unfavourable. A stop in transformer regulation may occur, e.g., in a situation when the transformer controller's under-voltage interlock is activated. A similar effect will be observed when the voltage change rate is greater than the rate of voltage correction as a result of transformer voltage ratio regulation (e.g., a large time delay value).

6. New Algorithm of the Transformer Controller Reducing the Risk of Voltage Collapse

6.1. Idea

This section will present the algorithm designed for counteracting the aggravation of the reactive power deficit in a voltage stability risk situation. The algorithm can be described in the following manner: if the voltage level in a HV network is lower than the accepted threshold value, V_{HVmin}, and its decrease rate exceeds the preset threshold value of $\dot{V}_{HVmax}$, then:

- If the coefficient $dQ/dV > 0$, then the transformer controller will operate following the constant lower voltage value criterion (V_{MVref} = const.), with the simultaneous correction of the preset value to the lowest permissible value of $V_{MVref} = V_{MVmin}$. Thus, the controller will reduce the MV side voltage to V_{MVmin}, and after exhausting the regulation capacities (the outermost tap) or reaching the preset value, it will maintain the constant voltage ratio.

- If $dQ/dV \cong 0$, then the transformer controller will operate following the constant lower voltage value criterion with the preset voltage value unchanged.
- If the $dQ/dV < 0$, then the transformer controller will operate following the constant lower voltage value criterion at the unchanged preset voltage value, with the simultaneous correction of the preset value to the highest permissible value of $V_{\text{MVref}} = V_{\text{MVmax}}$.

An exemplary diagram of the operation algorithm of an adaptive HV/MV transformer regulator is shown in Figures 8 and 9.

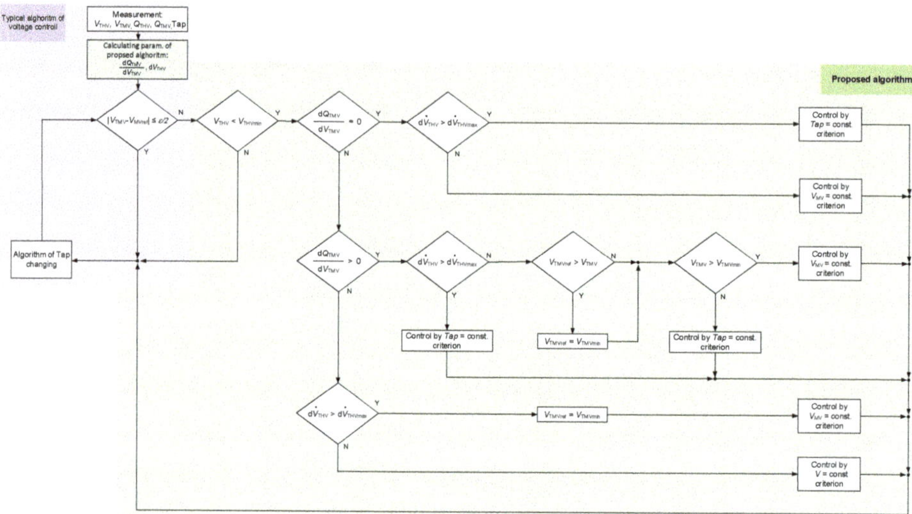

Figure 8. Block diagram of the proposed algorithm of the HV/MV substation transformer controller.

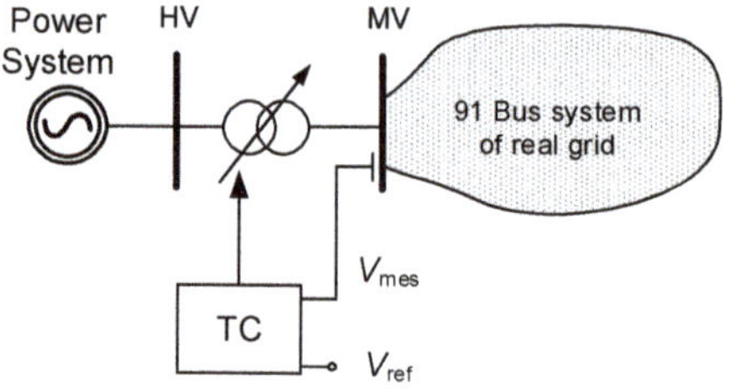

Figure 9. Simplified diagram of the network used in simulation studies.

An important parameter that activates the operation of the proposed algorithm is the maximum rate of reduction of the HV side voltage $\dot{V}_{\text{HVmax}}$. Assuming that the voltage ration value is constant, the voltage decrease rate at either side of the transformer will be identical. If the voltage decrease rate on the primary side is lower than the secondary side voltage change rate caused by the voltage ratio change, then the secondary side voltage will be maintained at a constant level. In that case, the secondary side voltage drop will be compensated for by the change of the voltage ratio. With a view to the above, the maximum voltage decrease rate can be determined from the relationship:

$$\dot{V}_{\text{HVmax}} = \frac{-dv_{\%}}{t_{\Sigma}} \tag{6}$$

while:

$$t_{\Sigma} = t_{\text{op}} + t_{\text{db}} \tag{7}$$

where:

- $dv_{\%}$—tapping step;
- t_{Σ}—total tap change time;
- t_{op}—switch opening time (for contemporary switches it is of the order of 2–10 s);
- t_{db}—dead time associated with the decay of transient processes caused by the switchover (e.g., for TCS—Transformer Control System) this time is of the order of 1 s for the case, where there is a signal confirming the current tap position, or 3 s in the absence of such confirmation).

Relationship (6) describes the maximum theoretical voltage drop value. Considering the measurement accuracy, both the statistical and dynamic errors, and the fact that the value of $dv_{\%}$ is a certain average value, the criterion value of $\dot{V}_{HVmax}$ should be adopted as considerably smaller, as confirmed by the simulation study results reported further in this section.

In cases when $dQ/dV > 0$, a change (reduction) of the preset voltage value occurs in the proposed algorithm. The purpose of this is to reduce the uptake of reactive power. The preset voltage value should be as small as possible and, at the same time, should satisfy the following two conditions:

- Maintaining the preset voltage value should enable the permissible minimum voltages within the power supplied network to be met;
- It should not be lower than the limiting $V_{ext.}$ voltage value, at which the coefficient $dQ/dV \cong 0$.

Bearing the above in mind, and also to estimate the risk of occurring a dramatic reactive power uptake in situations, where the coefficient $dQ/dV < 0$, it would be advisable to determine the voltage susceptibility characteristic, $V = f(Q)$, for each HV/MV substation. The above-mentioned characteristic can be determined through: simulation studies, a network experiment, or, preferably, by the analysis of the results of both investigation types. Regardless of the adopted method, a revision of the current state of the network should be made, including the parameters of network elements, voltage effects in vulnerable network nodes, caused by a change in transformer voltage ration, or, last but not least, the applied types and settings of the safeguards of induction machines in individual network nodes. The accuracy of the model of particular network elements, including safeguards, will determine the reliability of results obtained by means of simulation studies. In the case of a network experiment, a good knowledge of the network under examination will enable the experiment to be safely carried out and reliable results to be obtained. This effect will only be obtained, when a sufficiently low voltage level is successfully obtained, and no emergency trips of the drives occur due to carrying out the network experiment. As indicated by the characteristics shown in Figures 4 and 5, an extremum of the function occurs at voltages lower than the permissible long-lasting voltage; hence, it is essential that the network experiment is carried out at a possibly low voltage. A low voltage level in HV/MV substations increases the risk of emergency trips of asynchronous drives inside the network, caused by an excessive current increase or too low voltage. In practice, two types of safeguards are used for protecting asynchronous drives connected to an MV network. The primary type includes overcurrent protections with a time-dependent characteristic. A secondary, less often used safeguard are undervoltage protections—with typical settings being (0.7–0.8) V_n.

6.2. Methodology for dQ/dV Coefficient Determination

In order to determine the $V = f(Q)$ characteristic and the value of $V_{ext.}$, a change of supply voltage should be induced. The simplest and most effective method of changing voltage in an HV/MV substation is by changing the voltage ratio of the power transformer.

The conditions for conducting the experiment can be defined as follows:

- The measurements are to be made during the normal operation of the medium voltage network;
- The measurements should be made using measuring instruments (network parameter recorders) of the appropriate class;

- The measurements should be conducted in the periods of characteristic loadings, such as a peak or off-peak period, winter, summer, etc.;
- The measurements should be conducted in the period of a limited load variation;
- Voltage changes are to be made within limits permissible in a given network node;
- The test should be started at a maximally high voltage to obtain a possibly large number of measurement samples.

The proposed course of the test:

- Preparation of the measuring apparatus (connection, parametrization, turning on the recording);
- Switching over the transformer voltage controller to the manual operation mode;
- Change of the tap towards the reduction of voltage;
- Successive tap changes should proceed as quickly as possible to minimize the influence of the typical power variation of the loads;
- The switchovers should be done in one direction.

In a situation where a bank of capacitors is installed in the HV/MV substation. The tests should be carried out with the capacitors bank turned on and off, respectively. It should be borne in mind that in the conditions of normal operation, in the situation of lowered voltage, the capacitor bank shall be turned on. Its voltage susceptibility characteristic (negative value of the dQ/dV coefficient) may noticeably change the Q_{subst} = f(V_{subst}) curve. If turning the capacitor bank on–off during the experiment could cause the maximum voltage to be exceeded, the bank should stay turned off.

To emulate the course of the test to determine the Q_{subst} = f(V_{subst}) characteristic, simulation studies were carried out using a model of a 91-node real MV network. The model included 58 MV nodes (interconnected with 66 MV cable lines) and 33 MV/LV substations. A simplified diagram of the network is shown in Figure 9.

Two types of load models are used in each node. The constant admittance model and the dynamic asynchronous motor model were both taken from the library of the PowerFactory program. The motor power was selected to obtain the assumed percentage of asynchronous machines while retaining the reference power value in a node. As the effect of the voltage level at which measurements are made is significant, two variants of the initial supply voltage value were considered: the first, where $V_{MV(t=0)}$ = 1 pu, and the second one, where $V_{MV(t=0)}$ = 0.95 pu. The taps were changed until a voltage collapse was triggered. The obtained characteristics for a network with a 15-percent proportion of induction machines are shown in Figure 10.

These figures show the obtained voltage characteristics, where the bold black solid line represents the trajectory of voltage variation for the linear variation of MV voltage. Two characteristic points are highlighted in Figure 10a,b. They indicate the levels of voltages, at which the network experiment should be aborted. A further change in the voltage ratio of the power transformer would threaten with an emergency trip of the drives due to the exceeding of the rated current value of the machine (denoted as "Crit. I", assuming I_{max} = 1 pu), or because of too low voltage (denoted as "Crit. V", assuming $V \leq 0.85\ V_n$). Figure 10c shows voltage levels at individual load nodes, including the maximum number of the tap, which would not cause the activation of the undervoltage protection. Similarly, Figure 10d shows the drive load level, including the maximum number of the tap, which would not cause the activation of overcurrent protection.

Based on the obtained graphs of V and Q, points for plotting the static characteristics were determined (Figure 11).

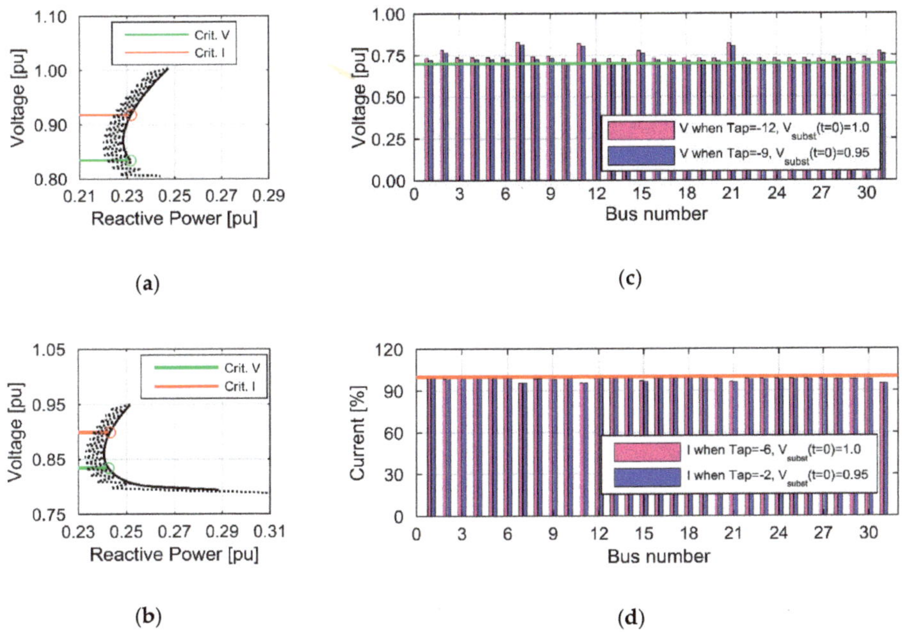

Figure 10. Selected results for the case of voltage change in the HV/MV substation caused by the change of the transformer ratio ("Tap" corresponds to the tap number): (**a**) characteristic $V = f(Q)$—case $V_{MV(t=0)} = 1$; (**b**) characteristic $V = f(Q)$—case $V_{MV(t=0)} = 0.95$; (**c**) nodes voltage on distribution grid; (**d**) values of the load currents of induction machines.

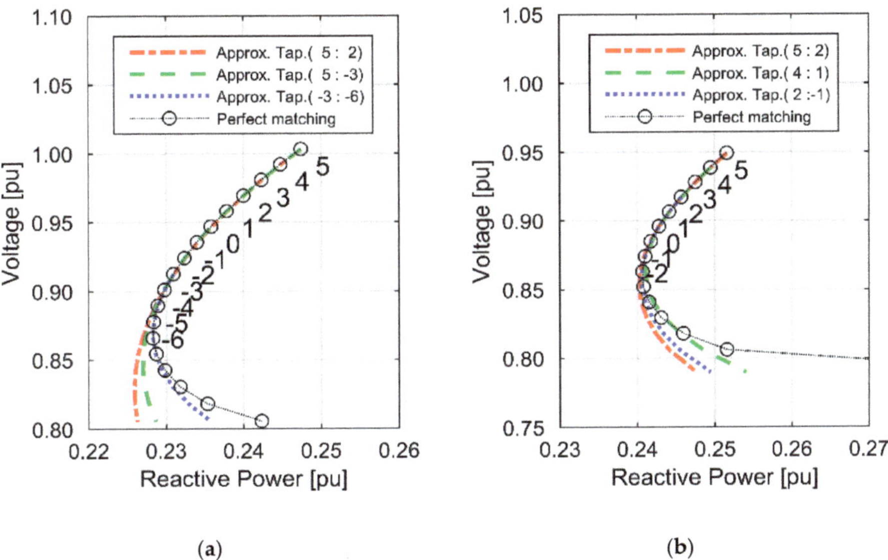

Figure 11. Characteristic $V = f(Q)$ approximation based on measurement points (numerical values correspond to the number of taps position): (**a**) case $V_{MV(t=0)} = 1$; (**b**) case $V_{MV(t=0)} = 0.95$.

The numerical values correspond to successive tap numbers, while the range of data (defined as the tap number) used for determining a specific curve is given in parentheses. The characteristics were obtained by approximation with a third-order polynomial ($i = 3$), which limits the number of required measurement points while maintaining the acceptable accuracy of determining the value of V_{ext}. If the number of measurement points "j" is equal to the degree of the polynomial, "$j = i$", then one has an approximated function defined at "$i + 1$" points and an approximating polynomial of degree "i" is searched for. In this case, one deals with an interpolation, in which the mean-square error equals 0. It is obvious that measurement data obtained from tests on a real object can be burdened with large errors; therefore, in order to reduce them, a possibly large number of measurement points should be used. If a limited number of measurement points is available, it is preferable to adopt points of the lowest voltage. Having the above in mind and also comparing the characteristics shown in Figure 11a,b, a conclusion can be drawn that the test under discussion is best carried out in the conditions of the lowest possible voltage.

6.3. Simulations Studies to Verify the Effectiveness of the Proposed Algorithm

To verify the effectiveness of action of the proposed algorithm (Figure 8), a series of simulation studies were carried out using the network model described above (Figure 9). A disturbance to be modelled was the linear variation of electric power system voltage. Sample results are illustrated in Figures 12 and 13.

At the first step, the parametrization of the model had to be made. The following settings were adopted: tap changer switchover opening time, $t_{op} = 7$ s; time lag associated with the decay of transient processes and the establishment of measurements values, $t_{db} = 1$ s; independent lag delay value, $\tau = 180$ s (which corresponds to the actual switch-off time at a deviation of $\Delta V = 0.2\% \cdot V_n$). Three variants of the transformer controller algorithm were considered, namely:

- "TCS", meaning that the transformer controller algorithm is identical with the algorithm of the commercial controller, manufactured by a renowned manufacturer of power systems automation;
- "New (V_{ref} = var)", or the proposed algorithm (Figure 8);
- "New(ϑ_{freeze})", or the reference variant, which uses an algorithm compatible with the proposed algorithm, except that the voltage ration process was withheld until the condition $V_2(t) \leq V_{ext.}$ had been met. The adopted assumption results, in the first phase of the process, in a natural reduction of reactive load power following from the characteristic $V = f(Q)$.

The driving force was a linear variation in supply voltage on the HV side. The rate and depth of the voltage drop were selected to, on the one hand, bring about a situation that would cause the risk of occurring a voltage collapse, and, on the other hand, to enable the analysis of voltage ratio control effects. In the first case, the HV side voltage drop rate was approx. 2.5 kV/min, which corresponds to the adopted limiting value of $\dot{V}_{HVmax}$—it is about three times smaller than the one determined from Relationship (6). In the second case, the HV side voltage drop rate was at a level of 10 kV/min, corresponding to the value that occurred during the system failure describe in the study [46].

The analysis of the simulation study results (Figures 12 and 13) shows that the proposed authors' solution brings about expected positive results. It should also be noted that the correction (reduction) of the preset voltage value, when the coefficient $dQ/dV > 0$, is an advantageous solution. Because of this operation, a reduction in reactive power demand by the HV/MV substations occurs, which is faster than resulting from the voltage decrease rate, and the time of operation with lower power increases (Figures 12b and 13b). The time increase is associated with the increase in the transformer control range due to the opposite directions of tap change in individual operation phases. In the first phase, the reduction of the preset voltage value is followed by a tap change towards negative tap numbers; in the second phase, when $dQ/dV < 0$, the taps are switched over in the direction opposite to the initial direction (Figures 12d and 13d). It should be noted that, although the power changes in the network under examination are not big, assuming that the proposed solution would be implemented in the majority of transformer controllers in HV/MV substations, the total value of reduced power can be significant from the point of view of stable network operation. An undesirable effect resulting from the reduction of the HV/MV substation's preset voltage value is a voltage drop in load nodes (Figures 12e and 13e—the voltage variations are determined in the node of the lowest voltage).

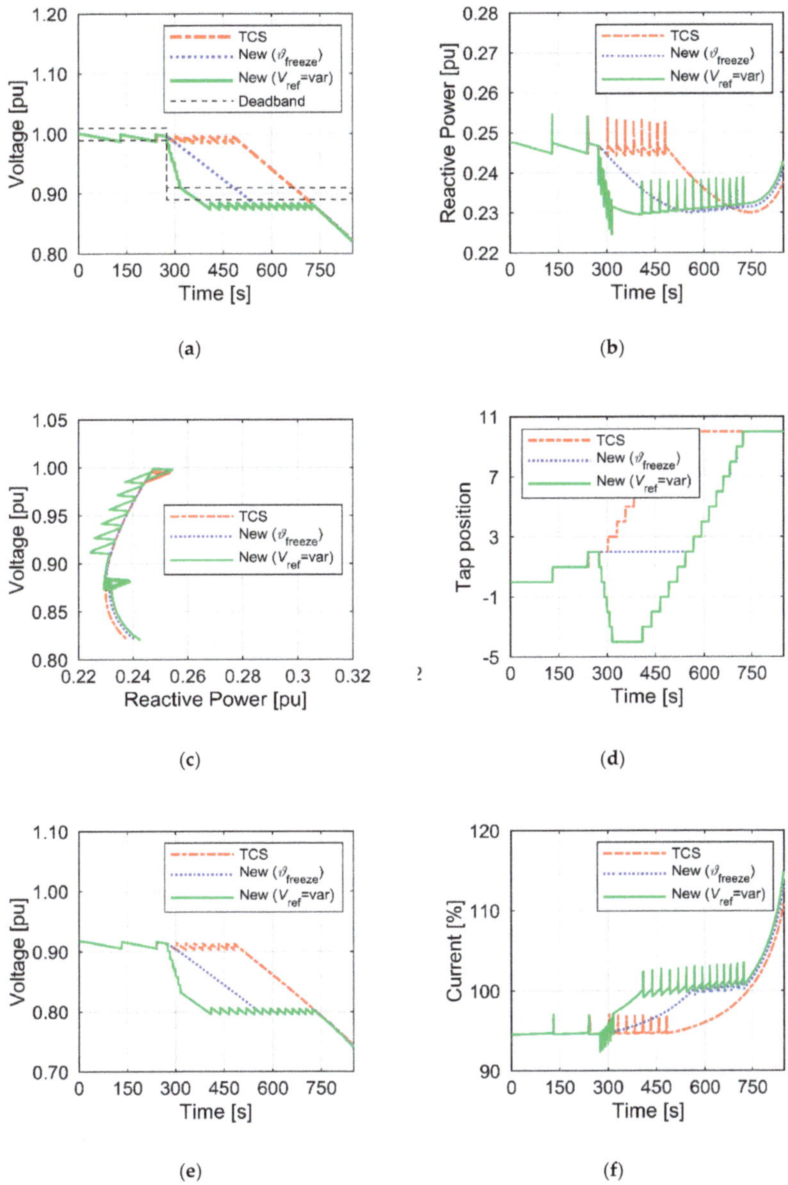

Figure 12. Verification of the effectiveness of the proposed algorithm (voltage drop on the HV side—2.5 kV/min): (**a**) voltage on MV bus of the HV/MV substation; (**b**) reactive power of the transformer; (**c**) characteristics $V_{MV} = f(Q_{MV})$; (**d**) tap position; (**e**) voltage of loads; (**f**) I_{RMS} of the asynchronous motor (expressed as a percentage of the rated current).

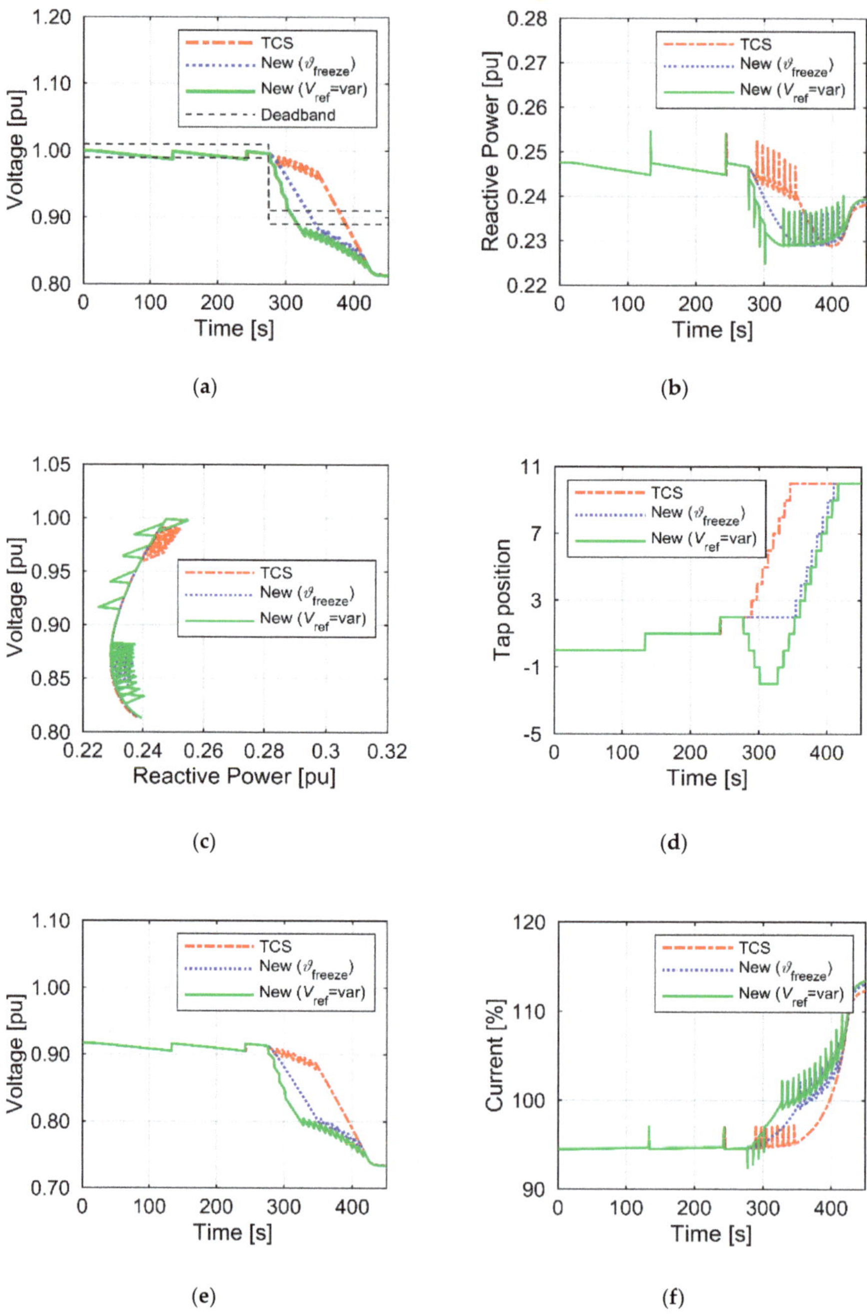

Figure 13. Verification of the effectiveness of the proposed algorithm (voltage drop on the HV side—10 kV/min): (**a**) voltage on MV bus of the HV/MV substation; (**b**) reactive power of the transformer; (**c**) characteristics V_{MV} = f(Q_{MV}); (**d**) tap position; (**e**) voltage of loads; (**f**) I_{RMS} of the asynchronous motor (expressed as a percentage of the rated current).

In the network under analysis, the attained values do not exceed the threshold values typical of undervoltage protection settings. A consequence of voltage reduction in load nodes is an increase in the induction machines' current. However, the attained current magnitudes will not cause a fast shutdown of the equipment. The applied overcurrent protections make use of time-dependent characteristics with a large trip delay set value, in the order of several minutes. In the case under analysis, the maximum overload value did not exceed 2% for the case illustrated in Figure 12f and 10%, as shown in Figure 13f.

With the correct settings of the protection devices, the possible risk of disconnection of the devices connected to the MV grid is small. Please note that the issues discussed in this paper concern the regulation of power transformers supplying the distribution network in conditions of voltage collapse. That's why the incidental disconnection of single devices, in a situation of saving the network from a catastrophic failure, ceases to be relevant.

6.4. Tests on the Physical Model

To make an additional verification of the operational effectiveness and to explore the possibility of implementing the proposed solution in a real facility, tests were carried out in the laboratory using a physical model to simulate the operation of the HV/MV substation. A simplified equivalent circuit diagram is shown in Figure 14. The object of control was the transformer of a functional unit, LOAD1. The disturbance was a change in the voltage V_1 (Bus 4) caused by a linear variation in the reactive power uptake of receivers LOAD2 and LOAD3. The test system was power supplied from a 15 kV network of the Gdańsk University of Technology (GUT) via a power transformer of S_{nT} = 630 kVA.

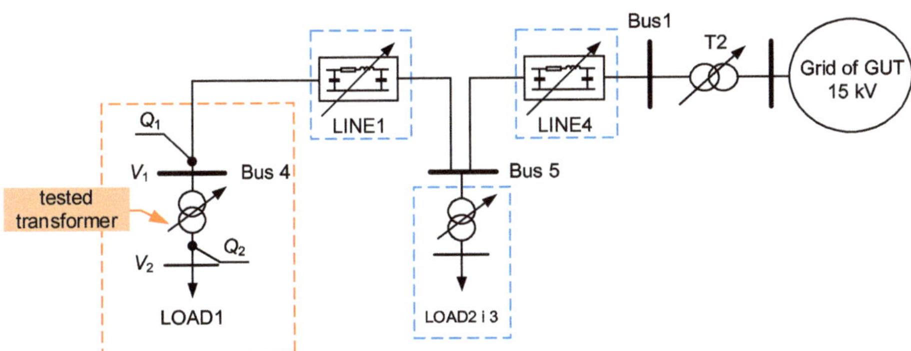

Figure 14. Simplified network diagram used in laboratory tests.

Three algorithm variants, identical to the ones used in the simulation studies, were implemented in the transformer controller. Three cases of average initial voltage decrease rate were emulated, namely:

- Variant A—much smaller than the limiting value determined from Relationship (6) for the data $\Delta v = 0.2\%$, $t_\Sigma = 8$ s; however, causing the activation of the author's algorithm;
- Variant B—close (±5%) to the limiting value determined from Relationship (6);
- Variant C—greater by approx. 10% than the limiting value determined from Relationship (6);
- Variant D—approx. 30 times greater than the limiting value determined for the previous variants, $\Delta v = 0.2\%$, $t_\Sigma = 300$ ms.

In view of the fact that the transformed tested was equipped with a power electronic tap changer, tests were carried out to demonstrate its regulation capabilities, as well as to show the versatility of the proposed solution. In variant D, an assumption was made that switchovers were done without delay in all of the algorithms examined. The preliminary tests showed that a stable tap changing process was attained at a rate of switching between adjacent taps in the range of (300–400) ms.

The obtained results are shown, respectively, in Figure 15—variant A, Figure 16—variant B, Figure 17—variant C, and Figure 18—variant D.

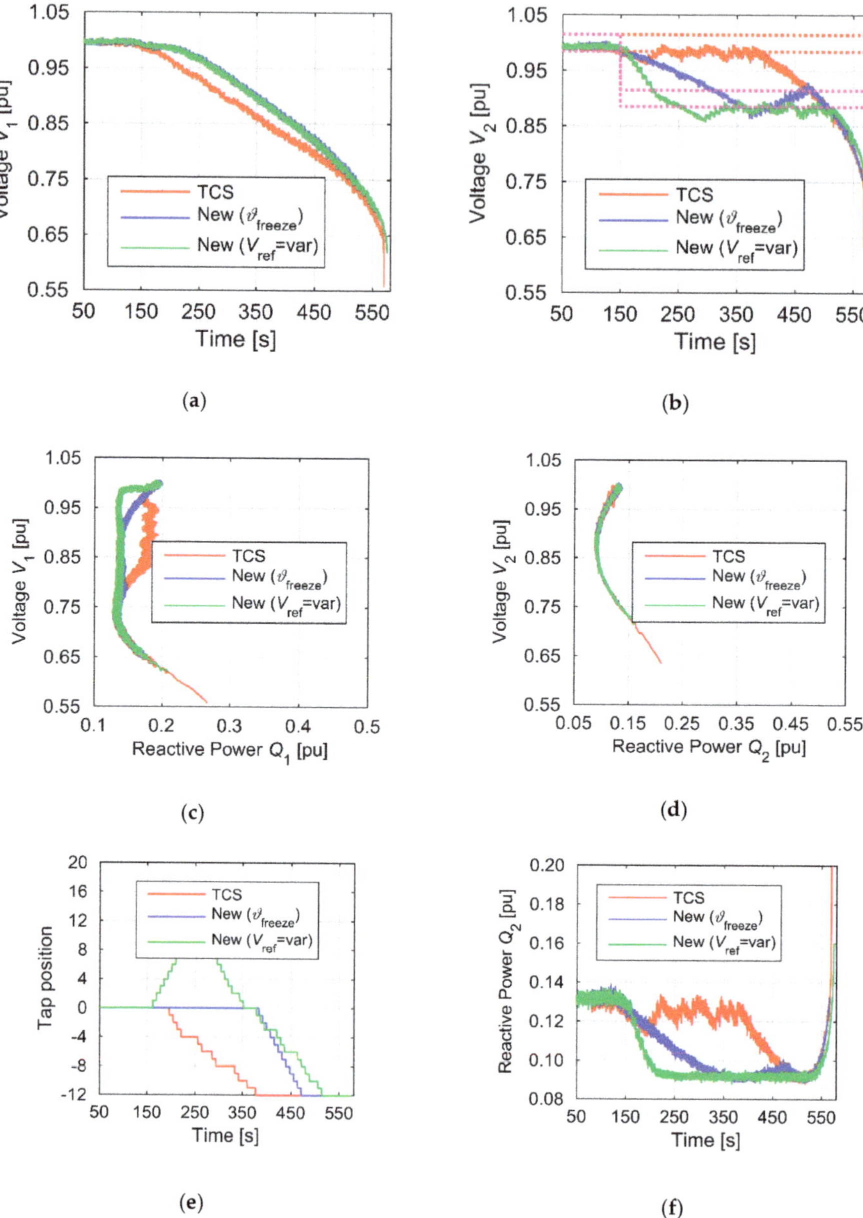

Figure 15. Verification of the effectiveness of the proposed algorithm, tests on the physical model—variant A: (**a**) voltage V_1; (**b**) voltage V_2; (**c**) characteristics V_1 = f(Q_1); (**d**) characteristics V_2 = f(Q_2); (**e**) tap position; (**f**) reactive power Q_2.

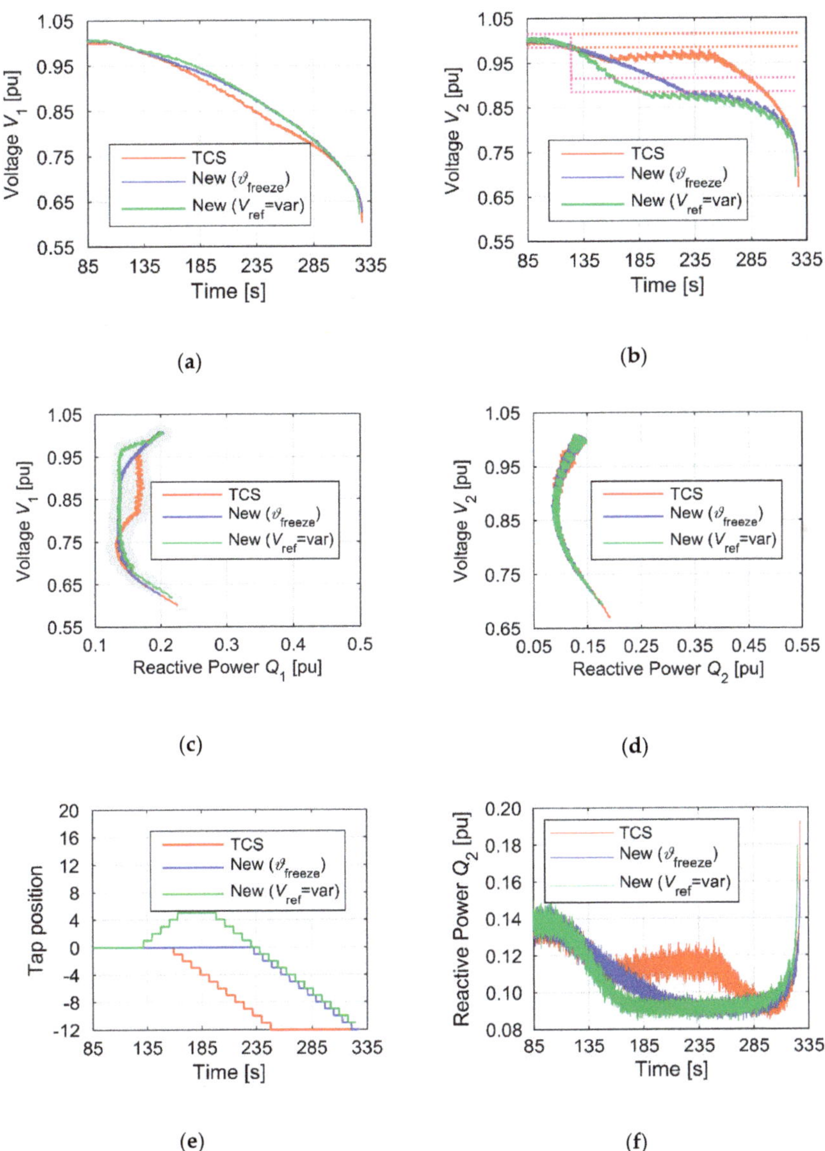

Figure 16. Verification of the effectiveness of the proposed algorithm, tests on the physical model—variant B: (**a**) voltage V_1; (**b**) voltage V_2; (**c**) characteristics V_1 = f(Q_1); (**d**) characteristics V_2 = f(Q_2); (**e**) tap position; (**f**) reactive power Q_2.

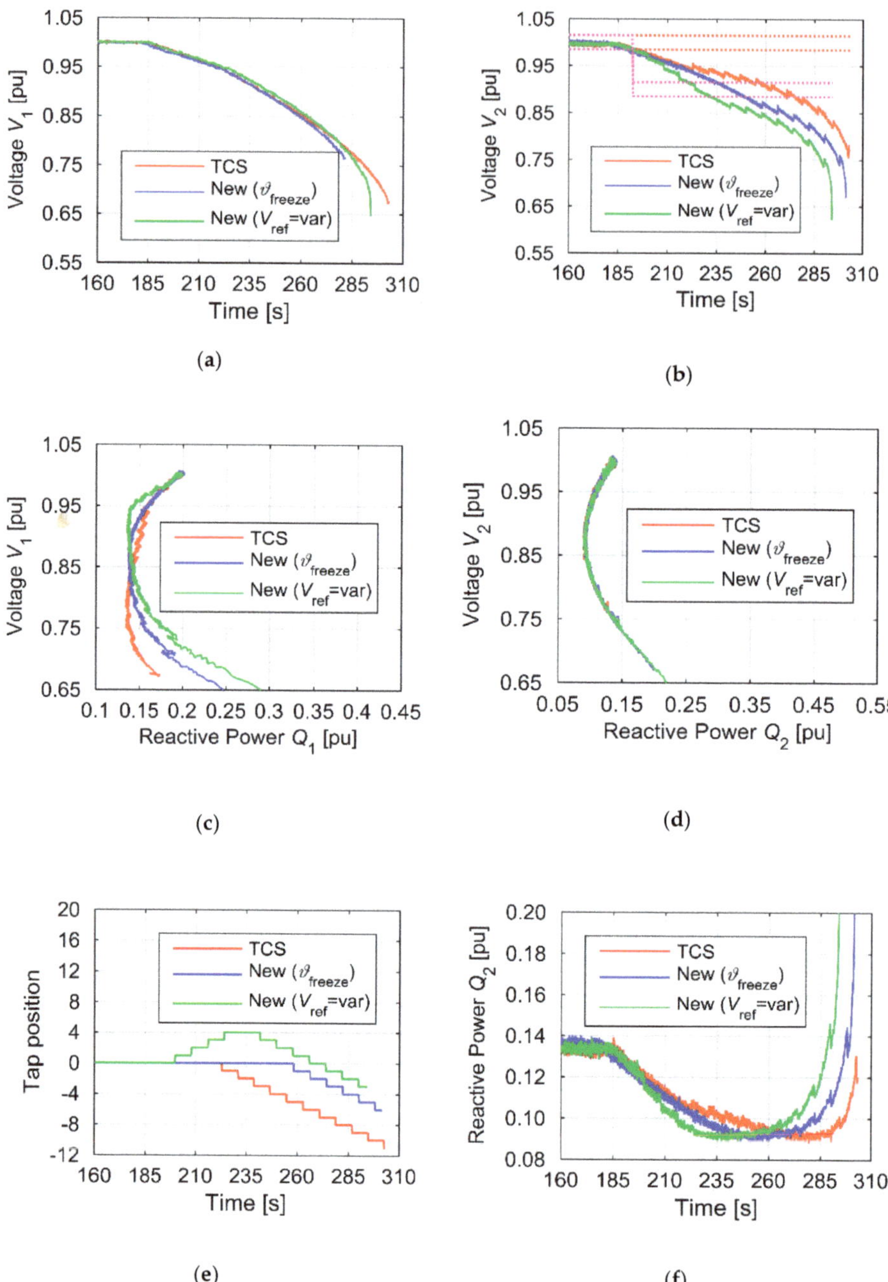

Figure 17. Verification of the effectiveness of the proposed algorithm, tests on the physical model—variant C: (**a**) voltage V_1; (**b**) voltage V_2; (**c**) characteristics V_1 = f(Q_1); (**d**) characteristics V_2 = f(Q_2); (**e**) tap position; (**f**) reactive power Q_2.

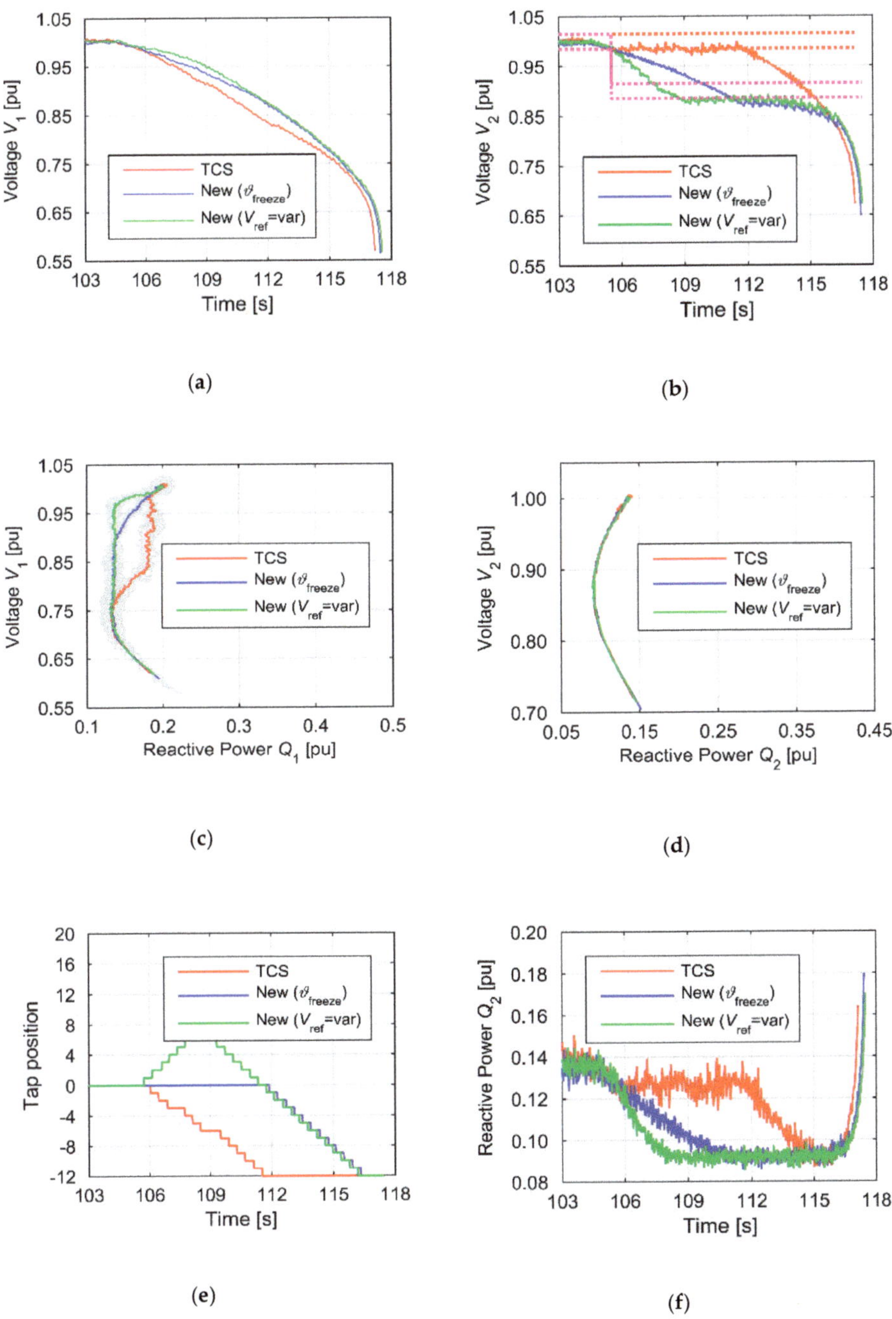

Figure 18. Verification of the effectiveness of the proposed algorithm, tests on the physical model—variant D: (**a**) voltage V_1; (**b**) voltage V_2; (**c**) characteristics V_1 = f(Q_1); (**d**) characteristics V_2 = f(Q_2); (**e**) tap position; (**f**) reactive power Q_2.

The positive effect of the application of the proposed solution is best visible in the situation, where the voltage decrease rate is significantly smaller than the rate of the voltage ratio regulation process (Figure 15). With a typical HV/MV substation transformer voltage regulation algorithm (here called TCS—Transformer Control System), a stable preset value on the secondary transformer side is maintained (Figure 15b, the voltage V_2). From the point of view of the quality of the supply of consumers, this is a desirable action. Unfortunately, in a situation of power deficit in the system, maintaining this control criterion results in a decrease in the limit of voltage stability on the primary transformer side (Figure 15c). Much more advantageous effects are achieved by using the proposed solution (Figure 8). The activation of one of the algorithm's elements results in a change in the preset voltage value, which causes an abrupt increase in deviation, resulting in a change in voltage ratio towards voltage reduction. Because, in the initial phase of the disturbance under examination, the $\mathrm{d}Q/\mathrm{d}V$ coefficient is positive, a fast reduction of reactive power flowing through the transformer follows (Figure 15f). As a consequence, an increase in the reserve of voltage stability in the supply network is obtained (Figure 15c).

The reduction of the reactive power input also results in a decreased drop of the voltage V_1 (Figure 15a). An added positive effect of the preset value adjustment is a change in tap switching direction (Figure 15e). As a result, the present regulation capabilities increase, which results in a longer time of maintaining reduced power. So, positive effects are observed for cases in which the voltage decrease rate is lower than the limiting value described by Relationship (6).

The confirmation is provided also by graphs shown in Figure 16. In a situation, where the voltage decrease rate exceeds the limiting value (Figure 17), a positive effect of reducing the preset value to the value of $V_{ref} = U_{ext.} + \varepsilon$ is only observed in the first voltage decrease phase, when $V \leq V_{ext}$ (Figure 17c). Due to the high voltage decrease rate, the voltage drop is not compensated for by the change of the voltage ratio, which leads to a dramatic increase in reactive power uptake by the HV/MV substation. Because, as a result of changing the critical preset voltage value, the critical voltage value has been attained sooner, the voltage collapse will also come up sooner (Figure 17a,b).

The emulated change rate of the supply voltage V_1 in variant D was, more or less, 30 times greater compared to the remaining variants. As the regulation process rate, resulting from the use of the power electronic tap changer, was almost 20 times greater, the results obtained by applying the proposed solution are equally positive as those obtained from the system with the emulated power electronic tap changer (with the respectively lower voltage decrease rate). Just like for variants A and B, an increase in the voltage stability reserve (Figure 18c), a reduction in the decrease rate of the supply voltage V_1 (Figure 18a), and an increase in the present regulation range (Figure 18f) were obtained.

The effects of the operation of the "New (ϑ_{freeze})" algorithm taken for reference should be regarded as moderate for each of the A-D cases under consideration. With the exception of the case, where the voltage decrease rate exceeds the transformer's regulation capacities, the performance of the reference algorithm is always poorer than the proposed author's algorithm and, at the same time, better than the algorithm currently in use. As the effects of the operation of the reference algorithm are noticeably lower compared to the author's algorithm, the solution shown in Figure 8 should be taken as recommended.

7. Conclusions

The test results presented in this paper have confirmed both the effectiveness and versatility of the operation of the proposed algorithm. The change of the preset voltage value, made in the appropriate moment of the test disturbance, reduced the reactive power input at very low voltages. In a real electric power system, in the situation of an existing reactive power deficit resulting in a voltage drop, such a result of regulation would be very desirable

One of the reasons for the reactive power deficit in the HV network is the simultaneous increase in the demand for reactive power in many HV/MV substations in a given area. This deficit is manifested by a significant increase in the speed of voltage drop in the HV network and its extremely low level. As described in Section 6.1, both of these factors activate the algorithm proposed in the article. Should the proposed solution be implemented in the majority of HV/MV substations, a significant increase in the level of its positive influence on the HV grid could be expected. Simultaneous reduction of reactive power consumption in many HV/MV substations may significantly reduce the risk of voltage collapse in the HV supply network.

The authors are currently conducting multi-variant simulation studies with the use of the real HV grid model, which will prove the correctness of the above statement. The research results will be published in the next paper.

To sum up, the author's algorithm is effective and owing to its simplicity, it has a high potential of practical application, as confirmed by the tests using a real transformer with a capability to change the voltage ratio under loading. The implementation of the algorithm makes it possible to increase the reserve of the voltage stability of the node, in which it is applied, thus mitigating the risk of a voltage collapse occurring.

Author Contributions: Conceptualization, R.M.; methodology, R.M.; software, R.M.; validation, R.M.; formal analysis, R.M., M.I. and P.M.; investigation, R.M.; resources, R.M.; writing—original draft preparation, R.M.; writing—review and editing, M.I. and P.M.; supervision, M.I. and P.M. All authors have read and agreed to the published version of the manuscript.

Funding: The funding sponsors had no role in the design of the study, in the collection, analysis, or interpretation of the data, in the writing of the manuscript, or in the decision to publish the results. This research received no external funding. The APC was funded by the Gdańsk University of Technology.

Conflicts of Interest: The authors declare no conflict of interest.

References

1. Alsafran, A.S.; Daniels, M.W. Consensus control for reactive power sharing using an adaptive virtual impedance approach. *Energies* **2020**, *13*, 2026. [CrossRef]
2. Bignucolo, F.; Caldon, R.; Prandoni, V. Radial MV networks voltage regulation with distribution management system coordinated controller. *Electr. Power Syst. Res.* **2008**, *78*, 634–645. [CrossRef]
3. Jauch, E.T. Possible effects of smart grid functions on LTC transformers. In Proceedings of the 2010 IEEE Rural Electric Power Conference (REPC), Orlando, FL, USA, 16–19 May 2010; Volume 22, p. B1-B1-10.
4. Kulmala, A.; Repo, S.; Bletterie, B. Avoiding adverse interactions between transformer tap changer control and local reactive power control of distributed generators. In Proceedings of the 2016 IEEE PES Innovative Smart Grid Technologies Conference Europe (ISGT-Europe), Ljubljana, Slovenia, 9–12 October 2016. [CrossRef]
5. Ravindra, H.; Faruque, M.O.; Schoder, K.; Steurer, M.; McLaren, P.; Meeker, R. Dynamic interactions between distribution network voltage regulators for large and distributed PV plants. In Proceedings of the PES T&D 2012, Orlando, FL, USA, 7–10 May 2012; pp. 1–8. [CrossRef]
6. Go, S.-I.; Yun, S.Y.; Ahn, S.J.; Kim, H.W.; Choi, J.H. Heuristic coordinated voltage control schemes in distribution network with distributed generations. *Energies* **2020**, *13*, 2849. [CrossRef]
7. Chiandone, M.; Campaner, R.; Bosich, D.; Sulligoi, G. A coordinated voltage and reactive power control architecture for large PV power plants. *Energies* **2020**, *13*, 2441. [CrossRef]
8. Tshivhase, N.; Hasan, A.N.; Shongwe, T. Proposed fuzzy logic system for voltage regulation and power factor improvement in power systems with high infiltration of distributed generation. *Energies* **2020**, *13*, 4241. [CrossRef]
9. Summers, A.; Johnson, J.; Darbali-Zamora, R.; Hansen, C.; Anandan, J.; Showalter, C. A comparison of DER voltage regulation technologies using real-time simulations. *Energies* **2020**, *13*, 3562. [CrossRef]
10. Pudjianto, D.; Ramsay, C.; Strbac, G. Virtual power plant and system integration of distributed energy resources. *IET Renew. Power Gener.* **2007**, *1*, 10–16. [CrossRef]
11. Kim, S.-B.; Song, S.H. A hybrid reactive power control method of distributed generation to mitigate voltage rise in low-voltage grid. *Energies* **2020**, *13*, 2078. [CrossRef]
12. Murphy, C.; Keane, A. Optimised voltage control for distributed generation. In Proceedings of the 2015 IEEE Eindhoven PowerTech, Eindhoven, The Netherlands, 29 June–2 July 2015; pp. 1–6. [CrossRef]
13. CPR-D Collapse Prediction Relay. 2014. Available online: https://www.a-eberle.de/sites/default/files/media/ba_cpr_d_d.pdf (accessed on 22 September 2020).

14. Hofbeck, M.; Sybel, T.; Fette, M.; Winzenick, I. Measurement of the dynamical status of highvoltage-networks with CPR-D. In Proceedings of the South African Power System Protection Conference, Johannesburg, South Africa, 8 November 2008.
15. Hofbeck, M.; Mayer, L.; Sybel, T.; Fette, M.; Werther, B.; Winzenick, I. Collapse Prediction Relay CPR-D—Theory and applications—Implementation into SCADA-systems to support security stability assessment. In Proceedings of the South African Power System Protection Conference, Johannesburg, South Africa, 8 November 2008.
16. Kunkel, P.; Kuznetsov, Y.A. Elements of Applied Bifurcation Theory. New York etc., Springer-Verlag 1995. XV, 515 pp., 232 figs., DM 98.-ISBN 0-387-94418-4 (Applied Mathematical Sciences 112). *J. Appl. Math. Mech./Zeitschrift Angew. Math. Mech.* **1997**, *77*, 392. [CrossRef]
17. Srivastava, K.N. Elimination of dynamic bifurcation and chaos in power systems using facts devices. *IEEE Trans. Circuits Syst. I Fundam. Theory Appl.* **1998**, *45*, 72–78. [CrossRef]
18. Caizares, C.A. On bifurcations, voltage collapse and load modeling. *IEEE Trans. Power Syst.* **1995**, *10*, 512–522. [CrossRef]
19. Małkowski, R.; Szczerba, Z. Sposób Regulacji Transformatorów Zasilających sieć Rozdzielczą. Urząd Patentowy Rzeczpospolitej Polskiej, PL 217312, Warszawa 2014. EN: Method for Regulating Transformers Supplying the Distribution Network. Polish Patent Office No PL 217312, Warsaw, Poland, 2 January 2012.
20. Machowski, J.; Lubośny, Z.; Białek, J.W.; Bumby, J. *Power System Dynamics. Stability and Control*; John Wiley & Sons Ltd.: Hoboken, NJ, USA, 2020; ISBN 9781119526346.
21. Kundur, P.S. Power system dynamics and stability. In *Power System Stability and Control*, 3rd ed.; CRC Press: Boca Raton, FL, USA, 2017; ISBN 9781439883211.
22. Grigsby, L. *Eletcric Power Generation, Transmission, and Distribution*; CRC Press: Boca Raton, FL, USA, 2012; ISBN 9781439856376.
23. Larsson, M. Coordinated Voltage Control in Electric Power Systems. Ph.D. Thesis, Lund Institute of Technology, Lund, Sweden, 2000. ISBN 91-88934-17-9.
24. Vishnu, M.; Kumar, T.K.S. An improved solution for reactive power dispatch problem using diversity-enhanced particle swarm optimization. *Energies* **2020**, *13*, 2862. [CrossRef]
25. Machowski, J.; Kacejko, P.; Robak, S.; Miller, P.; Wancerz, M. Simplified angle and voltage stability criteria for power system planning based on the short-circuit power. *Int. Trans. Electr. Energy Syst.* **2015**, *25*, 3096–3108. [CrossRef]
26. Cresswell, C.; Djokić, S. Representation of directly connected and drive-controlled induction motors. Part 2: Three-phase load models. In Proceedings of the 2008 International Conference on Electrical Machines, ICEM'08, Wuhan, China, 17–20 October 2008.
27. Cresswell, C.; Djokić, S. Representation of directly connected and drive-controlled induction motors. Part 1: Single-phase load models. In Proceedings of the 2008 International Conference on Electrical Machines, ICEM'08, Wuhan, China, 17–20 October 2008.
28. Cheng, Z.; Ren-Mu, H. Analysis and comparison of two kinds of composite load model. In Proceedings of the 3th International Conference on Deregulation and Restructuring and Power Technologies (DRPT 2008), Nanjing, China, 6–9 April 2008; pp. 1226–1230. [CrossRef]
29. Gaikwad, A.M.; Bravo, R.J.; Kosterev, D.; Yang, S.; Maitra, A.; Pourbeik, P.; Agrawal, B.; Yinger, R.; Brooks, D. Results of residential air conditioner testing in WECC. In Proceedings of the 2008 IEEE Power and Energy Society General Meeting—Conversion and Delivery of Electrical Energy in the 21st Century, Pittsburgh, PA, USA, 20–24 July 2008; Volume 5, pp. 1–9.
30. Price, W.W.; Gasper, S.G.; Nwankpa, C.O.; Bradish, R.W.; Chiang, H.D.; Concordia, C.; Staron, J.V.; Taylor, C.W.; Vaahedi, E.; Wu, G. Bibliography on load models for power flow and dynamic performance simulation. *IEEE Trans. Power Syst.* **1995**, *10*, 523–538. [CrossRef]
31. IEEE Standard Standard load models for power flow and dynamic performance simulation. *IEEE Trans. Power Syst.* **1995**, *10*, 1302–1313. [CrossRef]
32. Korunovic, L.M.; Milanovic, J.V.; Djokic, S.Z.; Yamashita, K.; Villanueva, S.M.; Sterpu, S. Recommended parameter values and ranges of most frequently used static load models. *IEEE Trans. Power Syst.* **2018**, *33*, 5923–5934. [CrossRef]
33. Wang, Q.; Wang, H.; Yang, H.; Zhao, R. Load characteristics-oriented control strategy for air conditioners during voltage sags. In Proceedings of the 2017 20th International Conference on Electrical Machines and Systems (ICEMS), Sydney, NSW, Australia, 11–14 August 2017. [CrossRef]

34. Zhang, Q.; Guo, Q.; Yu, Y. Research on the load characteristics of inverter and constant speed air conditioner and the influence on distribution network. In Proceedings of the 2016 China International Conference on Electricity Distribution (CICED), Xi'an, China, 10–13 August 2016. [CrossRef]
35. Kowalak, R.; Małkowski, R.; Szczerba, Z.; Zajczyk, R. *Automatyka Systemowa a Bezpieczeństwo Energetyczne Kraju. TOM 3 Węzły Sieci Przesyłowej i Rozdzielcej*; Wydawnictwo Politechniki Gdańskiej: Gdańsk, Poland, 2013; ISBN 978-7348-467-2.
36. IEA. *The Future of Cooling: Opportunities for Energy-Efficient Air Conditioning*; IEA: Paris, France, 2018. [CrossRef]
37. *Renewables 2018—Global Status Report a Comprehensive Annual Overview of the State of Renewable Energy Advancing the Global Renewable Energy Transition—Highlights of the REN21 Renewables 2018 Global Status Report in Perspective*; REN21: Paris, France, 2018; ISBN 978-3-9818911-3-3.
38. Li, D.; Chen, X.; Modeling, A. A Control Strategy for Static Voltage Stability Based on Air Conditioner Load Regulation. In Proceedings of the 2017 4th International Conference on Systems and Informatics (ICSAI), Hangzhou, China, 11–13 November 2017; Volume 2, pp. 288–293.
39. Tomiyama, K.; Daniel, J.P.; Ihara, S. Modeling air conditioner load for power system studies. *IEEE Trans. Power Syst.* **1998**, *13*, 414–421. [CrossRef]
40. Wu, B.E.I.; Zhang, Y.A.N.; Chen, M. The effects of air conditioner load on voltage stability of urban power system. In Proceedings of the 6th WSEAS International Conference on Power Systems, Lisbon, Portugal, 22–24 September 2006; pp. 40–45.
41. Abe, S.; Fukunaga, Y.; Isono, A.; Kondo, B. Power system voltage stability. *IEEE Trans. Power Appar. Syst.* **1982**, *10*, 3830–3840. [CrossRef]
42. Van Cutsem, T.; Vournas, C. *Voltage Stability of Electric Power Systems*; Springer: Cham, Switzerland, 2008; ISBN 9780387755359.
43. Gao, B.; Morison, G.K.; Kundur, P. Voltage stability evaluation using modal analysis. *IEEE Trans. Power Syst.* **1992**, *7*, 1529–1542. [CrossRef]
44. Rizk, F.A.M.; Trinhhigh, G.N. *High Voltage Engineering*; CRC Press: Boca Raton, FL, USA, 2014; ISBN 9781466513778.
45. Ismail, N.A.M.; Zin, A.A.M.; Khairuddin, A.; Khokhar, S. A comparison of voltage stability indices. In Proceedings of the 2014 IEEE 8th International Power Engineering and Optimization Conference, PEOCO 2014, Langkawi, Malaysia, 24–25 March 2014.
46. Davies, M. The impact of transformer tap-changer operation on voltage collapse. In Proceedings of the 2013 Australasian Universities Power Engineering Conference (AUPEC), Hobart, TAS, Australia, 29 September–3 October 2013; pp. 1–4. [CrossRef]

Publisher's Note: MDPI stays neutral with regard to jurisdictional claims in published maps and institutional affiliations.

Article

Solar Radiation Estimation Using Data Mining Techniques for Remote Areas—A Case Study in Ethiopia

Bizuayehu Abebe Worke [1], Hans Bludszuweit [2] and José A. Domínguez-Navarro [1,*]

[1] Electrical Engineering Department, EINA, University of Zaragoza, 50018 Zaragoza, Spain; buzeabebe@gmail.com

[2] CIRCE Foundation, 50018 Zaragoza, Spain; hbludszuweit@fcirce.es

* Correspondence: jadona@unizar.es

Received: 29 August 2020; Accepted: 26 October 2020; Published: 2 November 2020

Abstract: High quality of solar radiation data is essential for solar resource assessment. For remote areas this is a challenge, as often only satellite data with low spatial resolution are available. This paper presents an interpolation method based on topographic data in digital elevation model format to improve the resolution of solar radiation maps. The refinement is performed with a data mining method based on first-order Sugeno type Adaptive Neuro-Fuzzy Inference System. The training set contains topographic characteristics such as terrain aspect, slope and elevation which may influence the solar radiation distribution. An efficient sampling method is proposed to obtain representative training sets from digital elevation model data. The proposed geographic information system based approach makes this method reproducible and adaptable for any region. A case study is presented on the remote Amhara region in North Shewa, Ethiopia. Results are shown for interpolation of solar radiation data from 10 km × 10 km to a resolution of 1 km × 1 km and are validated with data from the PVGIS and SWERA projects.

Keywords: solar radiation modeling; GIS; interpolation; digital elevation model; data mining; ANFIS

---

1. Introduction

Solar energy yield is related to the quantity of radiation received at a specific geographical location which in turn depends on a number of environmental factors. There are other factors such as temperature, but still, estimation of solar radiation is fundamental requisite for siting of photovoltaic and solar thermal installations. These estimations are calculated from radiation data obtained in meteorological stations or on-site measurements but satellite data is gaining more and more importance. Good estimates are obtained combining long-term satellite-based time series with a short-term measuring campaign of at least one year. However, there are many countries with an insufficient network of meteorological stations and on-site measurements are costly, especially in remote areas. This is a major obstacle for reliable estimates of solar resources for policy makers and regional planning in these countries.

Accurate modeling of solar radiation is a difficult job due to the high number of atmospheric parameters and their spatio-temporal variations. It received special attention over the last three decades due to the rise of solar applications worldwide. Three main groups of models can be identified from the literature:

- Classical models with physical or empirical methods, based on astrophysical, atmosphere and geometrical measurements,
- Geographic information system (GIS) based models which usually combine ground measurements with physical models and
- Image processing models based on data obtained from satellite observations with ground measurement validation.

1.1. Classical Solar Resource Data Modeling Approaches

These classical methods commonly apply indirect approaches and empirical relationships like Angstrom [1], Glover and McCulloch [2], Paulescu [3], and Almorox [4]. Improved methods have been described by Perez [5], and Kamali [6]. Models which in addition use ground measurements of global radiation and its components are described in Kumar [7], and Monteiro [8]. Munkhammar [9] uses a statistical model based on copula to represent the dependence among solar radiation data at several locations. Currently huge progress has been achieved in this field. Several approaches exist to estimate solar radiation at a global scale for any application from simple to in-depth. In any case, ground measurements are the fundamental basis of all classical methods. Chelbi [10] and Robaa [11] estimate the solar radiation in developing countries with different models in countries where it is difficult to obtain measurement data as Tunisia and Egypt.

1.2. Solar Resource Modeling Based on GIS

GIS-based solar assessment has become more and more attractive as modern desktop computers are able to manage large amounts of data and many powerful GIS software tools are available. Rich [12] and Dubayah [13] aimed to estimate the diffuse irradiance of the sky in the absence of clouds with the sky image obtained from a point and assuming isotropic conditions (equivalent to view shed operation in GIS). Kumar [7] and Gueymard [14] presented algorithms to estimate radiation over a large area under clear sky conditions, they use digital elevation and latitude data to evaluate radiation changes in different aspects, slopes and positions of adjacent surfaces. Voivontas [15] developed a solar model using GIS that provides some tools to deal with spatial and temporal differences in solar radiation and demand. Monteiro [16] developed a model for calculating solar radiation maps under real-sky conditions using adaptive triangular meshes, specifically focusing on focusing on accurately defining the terrain surface and the generated shadows. The model can be used as a local server to interface with GIS tools. Piedallu [17] presented GIS-based programs to calculate solar radiation with and without clouds. Shortwave radiation components are calculated considering three sets of parameters—atmospheric attenuation, topographic parameters from the digital elevation model (DEM), and geometric relationship between the earth's surface and the sun, Charabi [18] analyses the influence of the sun in the DEM during a time period using ArcGIS tools. These methods combine ground measurements with physical models. Measurement data availability depends on location and the algorithms are simplifications of physical processes.

1.3. Solar Resource Modeling Based on Satellite Data

Currently, several institutions have implemented their own solar data bases, based on satellite measurement data. Data grids provided is based on cells with a size between 10 km^2 and 100 km^2. Data values are given as average over the area of each cell. Its purpose is to fill the gap where ground measurements are missing, although spatial resolution is low.

An outstanding example of a worldwide solar data base is the SWERA project [19]. It provides data freely available to the public and is the result of a cooperation of renowned organizations such as DLR, NASA, NREL, DTU and UNEP.

The input consisting of high resolution direct normal irradiation (DNI) and global horizontal (GHI) is checked for data quality and repaired by DLR and SUNY methodologies, and has a coarse resolution of 10 km × 10 km.

In the data set the primary component of information for the estimation of irradiation is a digital imagery for cloud detection. The cloud index of 10 km × 10 km resolution is extracted from half-hourly data from Meteosat satellites. The second part of information holds the physical atmospheric data set (including Aerosol Optical Thickness, Total Ozone, Transmission of Raleigh Atmosphere and Mixed Gases, water vapor etc.). The final data set is compiled using satellite imagery and ground measurements provided by the project countries. Furthermore, for the solar radiation data processing data sets from previous processes are combined into geo-referenced maps and site specific hourly time series of GHI and DNI using the methodologies of DLR for Direct Radiation and SUNY for Global Radiation.

Another example of free-access database is PVGIS of the European Joint Research Center (JRC). It contains solar resource data for Europe, Africa, and South-West Asia [20].

Other databases provide specific regional data, such as the Renewable Resource Data Center (RReDC) from NREL for the USA [21], Natural Resources Canada [22], Australian Bureau of Meteorology [23].

Finally, there are of course commercial products such as Vaisala [24] or SolarGIS [25].

Therefore, in regions where no ground measurements are available, several institutions have implemented their own climatological data modeling systems, based on satellite data and integrated on GIS platforms [7,8,13,15,26]. Huld [27] recently developed PVMAPS a set of computational tools and climate data for GRASS GIS to calculate solar radiation on large areas.

1.4. Proposed Resource Modeling with Data Mining

Satellite data is the solution for regions where no ground measurements are available. The main drawback is the coarse spatial resolution. Here is where this work is proposing a refining method based on data mining techniques. The method is explained in general and illustrated with a study case.

The presented method aims to provide a low-cost tool which can be applied even by entities with very limited budget, in order to assess solar development options for remote regions. It is considered especially useful if no reliable data is available from meteorological stations.

The paper is structured in the following way—in the introduction the motivation of the present work is presented. Section 2 describes the proposed geo-statistical data mining methodology. The method is illustrated in Section 3 with a case study. Within this section, aspects such as definition of Adaptive Neuro-Fuzzy Inference System (ANFIS) training parameters, sampling process for training data, and finally results are shown. A summary and concluding remarks are given in Section 4.

2. Geo-Statistical Data Mining Methodology

2.1. General Considerations

As seen before, ground measurements are the fundamental basis for reliable solar resource estimation. However, especially for remote and poor regions, the initial cost of this approach is too high. Furthermore, the conventional approach extrapolating known solar radiation data from reference sites is not feasible. Large distances between meteorological stations render extrapolations too uncertain to be useful.

From existing models such as PVGIS, irradiation data is available with coarse resolution of 10 km × 10 km for example. At the same time, topographic data from high resolution DEM is available for almost any location [20]. Consequently, the objective of this work is to develop a method to increase the spatial resolution of available solar radiation maps combining coarse irradiation data (GHI) with high-resolution topographic data.

The method is based on the hypothesis that atmospheric parameters such as cloud index, water vapor, aerosol and ozone are already incorporated in the initial low-resolution solar irradiation data. In addition, it is assumed that solar irradiation is largely influenced by the local terrain.

Some authors introduce the terrain conditions to improve the calculation of the solar radiation received in a certain area [10,18,28,29]. On flat terrain with clear-sky, solar radiation is almost the same over relatively large areas [7]. In hilly and mountainous terrains, altitude and slope distribution has a greater effect on local climate [30]. Surface radiation can change a lot depending on the frequency and thickness of the clouds [31]. As a result, terrain parameters such as elevation are related to solar radiation because they have a direct impact on cloud covers.

Therefore, the present methodology combines topographic data with geo-referenced GHI values. Geo-statistical information is captured from GHI data available in lower resolution (e.g., cell size of 10 km $\times$ 10 km) using ANFIS to estimate data with higher resolution (e.g., 1 km $\times$ 1 km).

The ANFIS is trained with a reduced representative training set, which is obtained based on statistical analysis of the original data. With a show case, it is demonstrated that the obtained ANFIS model is valid for the whole region under study.

This novel approach constitutes a fundamental difference with regard to the above mentioned models as it represents an indirect assessment. It is not a complete model in itself, but it is a tool to enhance existing model output considering topographical information.

2.2. Data Mining Using ANFIS

Artificial intelligence techniques, such as the well known heuristic method ANFIS have been used successfully in different renewable energy applications [32–35].

The basic steps of the proposed geo-statistical data mining methodology are:

- Definition of training parameters for data mining model,
- Representative data sampling for training parameters,
- Training of data mining model,
- Estimation of high-resolution radiation data using the trained data mining model,
- Validation of results.

In summary, the relationship of spatial variation of global solar radiation (available in low resolution) and terrain parameters (available in high resolution) is captured in a data mining model. This model is later applied to generate a solar radiation map with considerably higher resolution.

The steps mentioned above are illustrated in the next section with a case study from a remote region in Ethiopia.

3. Case Study: Amhara Region Ethiopia

In order to illustrate the proposed method, a study case is presented for the Amhara region in North Shewa, Ethiopia. As input, publicly available long-term data of GHI is used from SWERA project [19] having a coarse resolution of 10 km $\times$ 10 km (100 km^2 cell size). Further, a high-resolution DEM was obtained from GeoCommunity™ website [36]. In Figure 1 geo-referenced maps of elevation (above) and irradiation (below) are shown for the studied Amhara region.

In the following sections, the case study is developed following step by step the proposed method. At the end, the result is a refined map of GHI with a resolution of 1 km $\times$ 1 km.

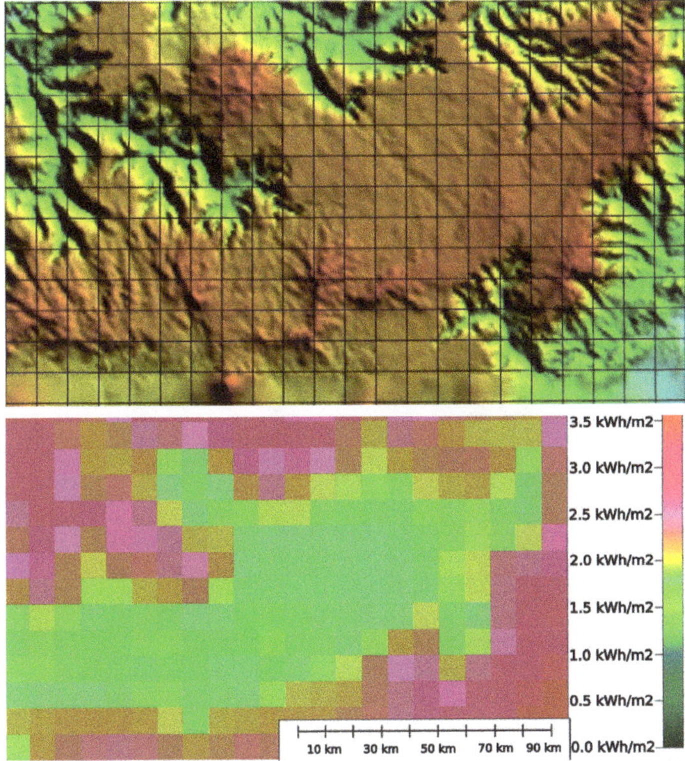

Figure 1. High-resolution elevation map (above) and low-resolution daily solar radiation map with 100 km^2 cell size (below) of Amhara region, North Shewa (Ethiopia).

3.1. Definition of ANFIS Training Parameters

The topographical terrain parameters used to describe the study region are terrain elevation, slope and terrain aspect, as proposed in Reference [7].

It may be mentioned here that the terminology used here is based on GIS standards. Translated to commonly used terms in solar energy assessment, aspect would be azimuth (γ) and slope would be tilt (β). Nevertheless, it is preferred here to maintain the GIS terminology, as these parameters are referred exclusively to the terrain and not to the solar energy capturing system.

In addition, values of standard deviation (STD) of elevation and slope are added to the training set. Tests (not shown here) have shown that by including these additional parameters the stability of model outputs can be improved. For example, estimated negative radiation values are avoided.

Using GIS Global Mapper platform raster maps of 1 km $\times$ 1 km resolution are obtained directly from DEM data [36] for elevation and slope and their corresponding standard deviation for each cell.

A continuous raster of the terrain aspect was created in order to facilitate data sampling (see Section 3.2. This was attained using the built-in shader function for slope direction of the GIS platform of Global Mapper software [37]. In order to obtain useful data, the default grayscale has been adapted. As a simplification of more complex natural phenomena, it is assumed that east and west orientation have the same effect on radiation. Therefore, the scale is defined ranging from 0 (black = north) to 255 (white = south). West and east are equally represented by the value 127 (gray). As a result, a circular scaling is obtained. From the GIS data an average aspect value is calculated for each square kilometer of the region extension. The resulting map is shown in Figure 2.

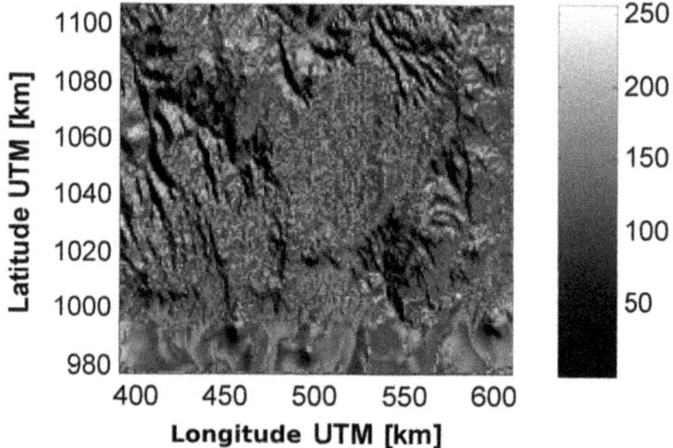

Figure 2. Full grayscale bitmap of aspect of the region terrain.

As a result five parameters compose the input for the training session: terrain elevation, slope (inclination), aspect (orientation) and STD of elevation and slope. The output is solar radiation (see Table 1).

Table 1. Ranges of Adaptive Neuro-Fuzzy Inference System (ANFIS) training data (matrix of 219 columns and 130 rows).

Data	Min.	Max.
Latitude	8°47′29″ N	9°57′29″ N
Longitude	38°0′0″ E	40°0′0″ E
Elevation [m]	500	3800
Aspect [°]	0	360
Slope [°]	0	90
Elevation_STD [m]	0	200
Slope_STD [°]	0	8
Radiation [kWh/(m^{2}d)]	0	3.99

3.2. Representative Data Sampling for ANFIS Training

Solar radiation data of the region under study is only available with a resolution of 10 km × 10 km. It is given in a matrix of 13 rows and 22 columns, covering a surface of 130 km × 220 km. On the other hand, terrain data (elevation, slope, aspect) are available with a resolution of 1 km × 1 km. In order to obtain a representative training sample which establishes a relationship between terrain data and solar radiation, an adequate sampling method is required.

Therefore, in a first step, the resolution of solar data is increased to 1 km × 1 km, such as the terrain layers (see Figure 3). Notice that in this step, only the number of data points is increased and no information is added. This step is merely for convenience for the following steps, as all four matrices M (one for solar radiation and three for terrain properties) have the same size and each cell $X_{i,j}$ is geo-referenced to the same location.

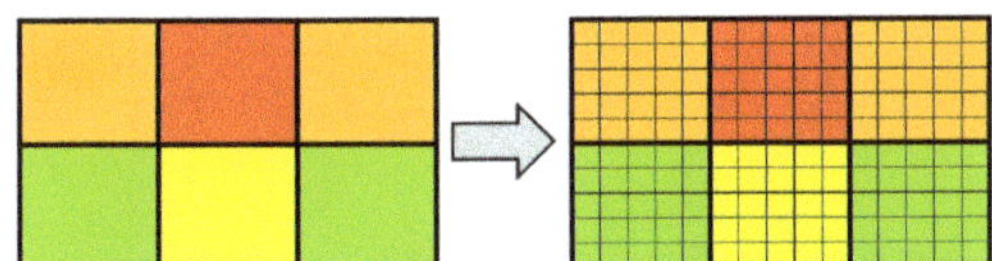

Figure 3. Illustration of the increase of solar radiation data resolution from 10 km × 10 km to 1 km × 1 km.

In a second step a simple sampling method is applied, which makes sure that all solar radiation data is included and a representative selection of terrain data is obtained. This sampling method consists in extracting every tenth row of each matrix M. Hence, the new dataset contains exactly 10% of the original data. In the presented show case, 28,470 values are reduced to 2847 values for every training parameter.

The same procedure is repeated for all five matrices and a training set of five reduced matrices is obtained. With this procedure, ten different training sets can be generated, depending on the row where the selection is started.

It can be expected that any of these sets are suitable as training set. In order to verify if the obtained reduced datasets are representative for the original data and to select the best training set, two additional steps are proposed here. Notice that the representativeness for solar radiation data is given automatically, as exactly one line (ten identical values) is extracted from each cell of 10 km × 10 km. First, histograms are computed for all reduced matrices of terrain data. An example is given in Figure 4, where the histograms of the ten possible training sets for terrain elevation, slope, aspect as well as for the STD of elevation and aspect are shown.

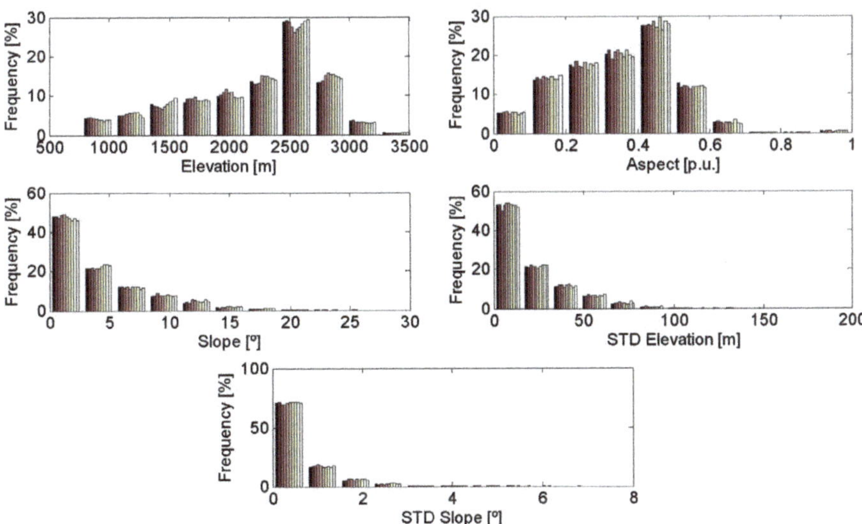

Figure 4. Histograms of the ten possible training sets of terrain parameters from top to bottom: elevation, slope, aspect, STD of elevation, STD of Slope.

In order to evaluate the representativeness of the extracted training sets, in a second step the root mean square error ($RMSE$) is computed for each histogram compared to the histogram of the whole dataset.

$$RMSE_i = \sqrt{\overline{(f_i - f)^2}}, \tag{1}$$

where f_j and f are vectors of 10 values, representing the frequency in the 10 bins of the histogram.

The results for all five terrain variables are shown in Figure 5. The numbering of the histograms is related to the first line i of extraction. In addition, the average of all 5 curves is shown. From this average, it can be derived that the most representative training sets would be those with index $i = 1, 2$ and 9. Nevertheless, it can also be concluded that any training set is suitable, as all sets have distributions which are very similar to each other, with deviations of less than 1%.

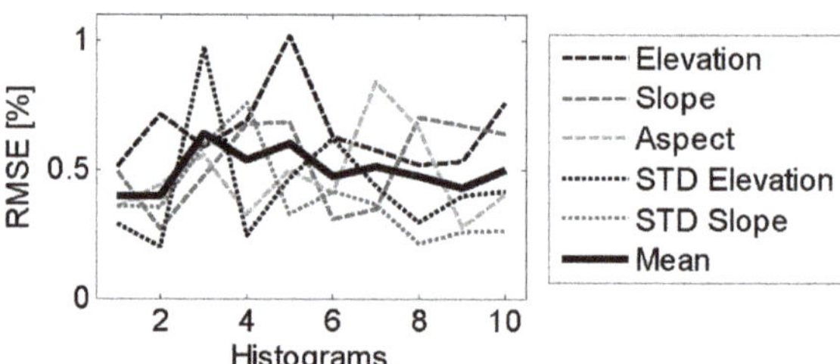

Figure 5. RMSE of the histograms of the ten extracted possible training sets from terrain aspect data.

In order to illustrate the obtained training data set, in Figure 6, a cross-correlation plot of available data and the extracted training data for radiation and elevation is shown. The strong correlation of these two parameters can be observed easily.

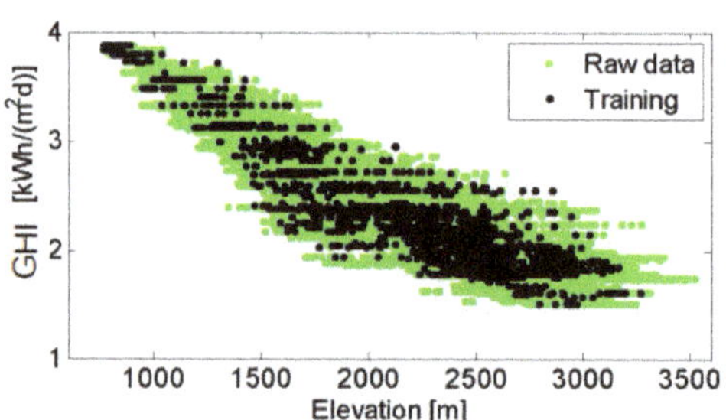

Figure 6. Cross-correlation plot of available (raw) data and the extracted training data for radiation and elevation.

3.3. Training of Neuro-Fuzzy Network

The data mining model has been implemented with Matlab ANFIS. The sub-clustering architecture in combination with hybrid optimization was chosen, because this combination offered the fastest and best approximation of the solar radiation estimation algorithm.

The training procedure used for the selected data is using Matlab built-in ANFIS editor, where a common modeling technique is implemented. After appropriate data sampling (as explained in Section 3.2), the obtained training set can be imported in text format into the ANFIS editor. For effective ANFIS training, the structure of the neuro-fuzzy network should be as simple as possible and be able to capture the desired information. In a first step, the numbers of membership functions are assigned arbitrarily. Then, a hybrid optimization method optimises the ANFIS structure and parameters until a correct reproduction of the training data is achieved. The number of epochs selected was 80–300 and the training population size is 2144 data sets with 5 to 8 alternative runs. At the end of this learning process, it was possible to tailor a membership function such that adequate inference and rule surfaces were obtained. For example, the Fuzzy Inference System (FIS) rule surface in Figure 7 shows that radiation values remain defined for all possible combinations of elevation and slope.

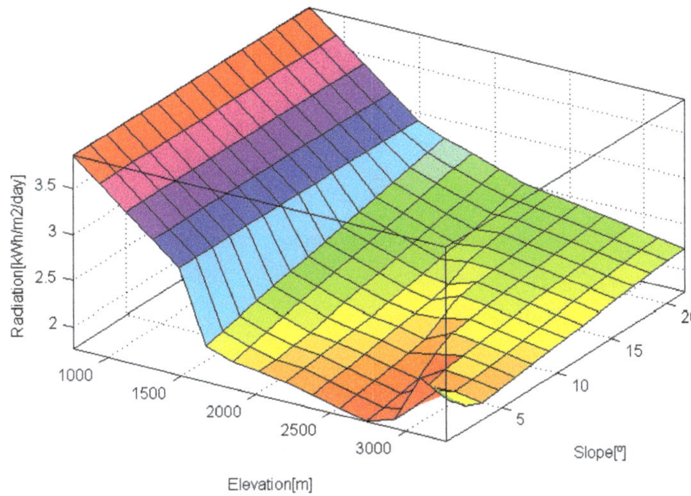

Figure 7. Fuzzy Inference System (FIS) Rule surface of elevation and slope for the estimation of radiation.

3.4. Estimation of High-Resolution Radiation Data

With the trained ANFIS model, solar radiation is estimated for the entire region under study with a resolution of 1 km × 1 km. For convenience, the result is exported to a text file with grid format (x, y, radiation). This file can be imported and represented as another layer for the geo-referenced region in any GIS tool.

In Figure 8 the result of the ANFIS model is compared with original coarse resolution data from the study region of Northern Ethiopia. Both maps are generated with the raster extrapolation tool of ArcGIS software.

Comparing the two maps, the refined resolution of solar radiation data in the lower map is evident. Especially the strong correlation with elevation data can be noticed.

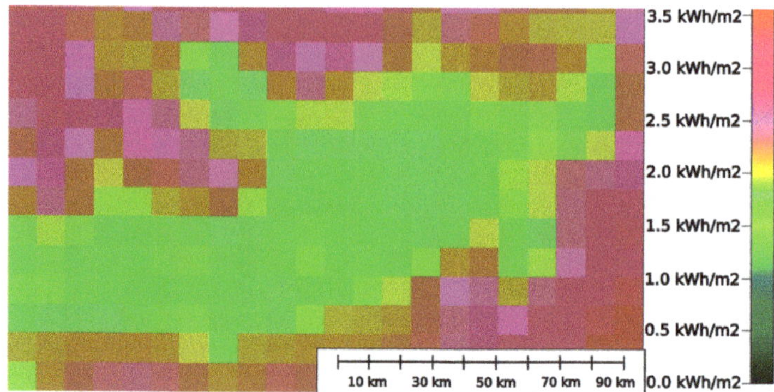

Figure 8. *Cont.*

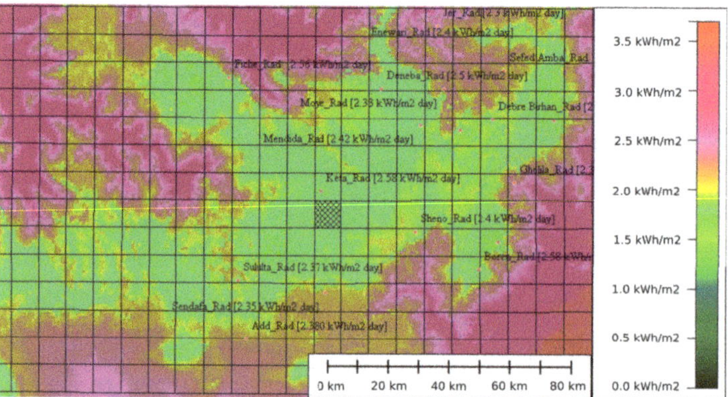

Figure 8. Comparison of coarse daily solar radiation data (**top**) and refined data of 1 km × 1 km (**bottom**), applying ANFIS estimation model.

3.5. Validation of Results

In order to give an idea of the validity of the estimation, two validation steps are presented in this section. First, the consistency of training data and estimation results is evaluated. In a second step, results are compared with two benchmark data sources (PVGIS and SWERA).

The consistency of the ANFIS output with original data is demonstrated representing the data in several different plots.

In Figure 9, the effect of the modification of solar radiation data can be observed. Raw data (coarse radiation) is compared with ANFIS estimation for every 1 km × 1 km grid cell. The graph represents a sequential scan of the map from the upper left corner down to the lower right corner. On the horizontal axis all the pixels of the map are represented in order. Training data is represented at the same position as raw data. Therefore, training data (black dots) is separated into 13 almost vertical lines which represent the 13 rows of the coarse radiation data set. This way it can be seen how three-dimensional data has been represented in a two-dimensional plot. It can be observed how minimum and maximum radiation limits are widely respected and much more intermediate values are generated. In addition, it can be seen how few training data values of solar radiation were employed.

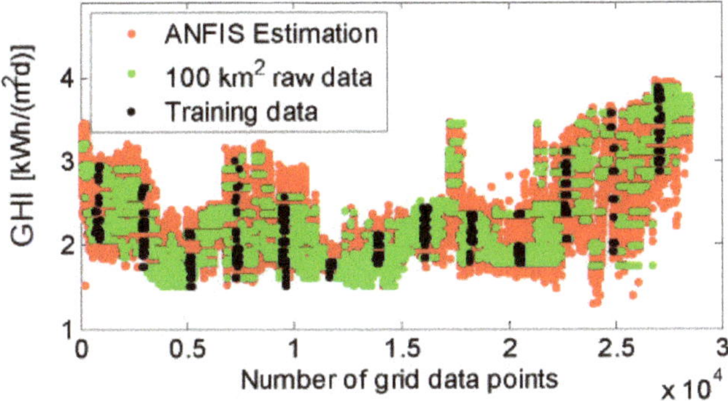

Figure 9. ANFIS radiation extrapolation compared to average (raw) and training data.

In Figure 10 a correlation plot is shown of original (10 km × 10 km) against estimated refined data (1 km × 1 km) of GHI for the training set (black) and the final estimate of the entire region (red). Representativeness of training data becomes evident as it covers the complete range of results. In addition, the final estimate shows almost no outliers, which demonstrates the stability of the model.

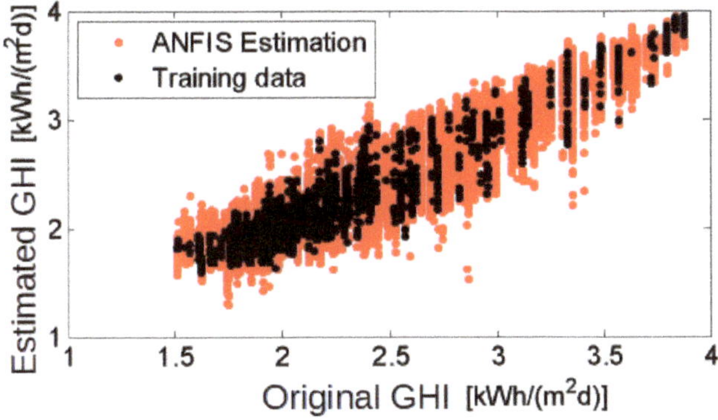

Figure 10. Correlation between 100 km^2 radiation data and 1 km^2 resolution estimated radiation.

Following the same procedure, in Figure 11 three different radiation data sets are represented against the terrain elevation—Original GHI with coarse resolution, refined GHI from the training set and from the final estimate of the entire region. Again, the consistency of the model output becomes visible. Also, a remarkable correlation pattern can be observed between elevation and GHI. This pattern is captured by the ANFIS model and as a result estimated data are much more correlated with terrain elevation than original low-resolution input data.

It must be pointed out here that the observed pattern is a specific feature of the region under study. This by itself is a valuable result of the model, as it indicates an interesting subject, which is worth to be further investigated. Additional studies may determine if this pattern also can be found in regions with different climates and elevation patterns.

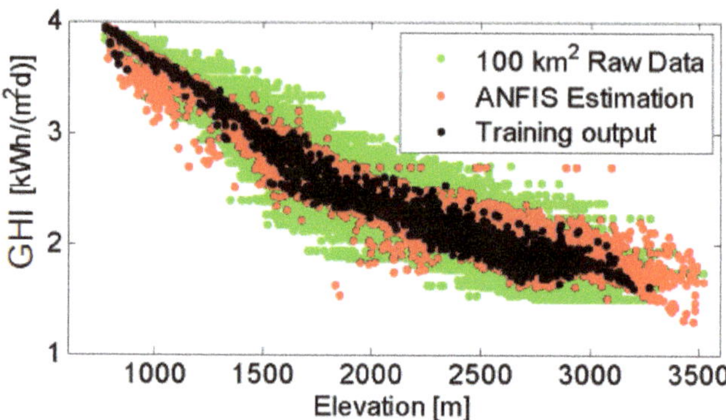

Figure 11. Comparison of raw radiation data versus ANFIS training and final ANFIS estimation as a function of terrain elevation.

The presented correlation plots give plausible evidence that ANFIS estimation has been able to capture the behaviour of the input data to be assessed and produces coherent results. The next step is a validation with available GHI data for some specific locations within the region under study.

The best way to validate the model would be with random samples of long-term on-site measurements. But as these measurements are not available, the validation is carried out with existing data from PVGIS and SWERA as benchmarks. PVGIS was chosed as it is a standard freely available online database and SWERA in order to illustrate the modification which the ANFIS model made at certain locations compared to the original data.

In Table 2, values of GHI are shown for some selected locations within the studied region of Ethiopia (sorted in alphabetical order). ANFIS estimation is compared with original data from SWERA (mean value of 10 × 10 km cell) and PVGIS.

The difference of ANFIS output compared to PVGIS (Average difference: 45% is slightly higher than the difference to SWERA (Average difference: 18%). This is mainly due to the fact that SWERA data was used to train the ANFIS model.

It is worth noticing that for some locations a large difference of GHI can be observed when data from SWERA is compared to PVGIS. The average deviation is with 45% in a similar range as the deviations of the ANFIS results compared to both models (18% and 45% respectively).

It is evident that only on-site measurements can determine with more clarity the degree of accuracy the proposed model is providing.

Table 2. Comparison of ANFIS model output with global horizontal (GHI) obtained from SWERA and PVGIS for selected locations.

No	Location	Latitude	Longitude	Elev.	Glob. Rad. [kWh/m^2 day]			Difference between:		
					ANFIS	SWERA	PVGIS	ANFIS SWERA	ANFIS PVGIS	SWERA PVGIS
1	Addis Ababa	9°1′14″	38°44′35″	2372	2.60	2.10	2.38	0.50	0.22	0.28
2	Dar Neba	9°45′28″	39°11′17″	2675	1.89	1.90	2.50	0.01	0.61	0.60
3	Debre Birhan	9°40′16″	39°32′3″	2777	1.83	1.91	2.41	0.08	0.58	0.50
4	Eneware	9°53′15″	39°8′19″	2428	2.50	2.27	2.40	0.23	0.10	0.13
5	Fiche	9°47′24″	38°43′36″	2852	1.83	2.15	2.56	0.32	0.73	0.41
6	Mendida	9°38′38″	39°18′32″	2701	1.83	1.84	2.42	0.01	0.59	0.58
7	Metehara	8°54′4″	39°54′59″	1091	3.56	3.59	2.55	0.03	1.01	1.04
8	Sululta	9°11′19″	38°45′35″	2594	1.99	1.79	2.37	0.20	0.38	0.58
9	Ejere	8°49′11″	39°15′54″	2319	2.10	2.50	2.55	0.40	0.45	0.05
10	Boren	9°13′56″	39°29′5″	3199	1.53	1.86	2.58	0.33	1.05	0.72
11	Ghelila	9°28′53″	39°35′21″	1928	2.34	2.35	2.34	0.01	0.00	0.01
12	Moye	9°45′28″	39°23′29″	2506	2.25	2.00	2.33	0.25	0.08	0.33
13	Jer	9°45′28″	39°22′10″	1711	2.53	2.30	2.30	0.23	0.23	0.00
14	Debre Zeit	8°45′24″	38°58′56″	1916	2.35	2.59	2.43	0.24	0.08	0.16
15	Sheno	9°19′58″	39°17′46″	2857	1.81	1.82	2.40	0.01	0.59	0.58
16	Sendafa	9°9′7″	39°1′23″	2557	1.91	1.90	2.35	0.01	0.44	0.45
17	Fentale Mount	8°58′21″	39°54′12″	978	2.49	3.41	2.51	0.92	0.02	0.90
18	Angolala	9°38′8″	39°25′54″	2826	1.84	1.88	2.37	0.04	0.53	0.49
19	Megezez	9°18′0″	39°33′1″	3292	1.79	1.79	2.52	0.00	0.73	0.73
20	Abawebe Ager	9°40′2″	39°20′58″	3259	1.74	1.75	2.47	0.01	0.73	0.72
21	Keta	9°27′07″	39°0′0″	2640	1.84	1.77	2.39	0.07	0.55	0.62
22	Sefed Amba	9°28′26″	39°20′28″	2975	2.23	2.28	2.38	0.05	0.15	0.10
			Average		2.13	2.17	2.43	0.18	0.45	0.45

4. Conclusions

The comparison of ANFIS estimation results against SWERA, and PVGIS values for specific validation sites shows a reasonable approximation to both benchmark validation data sets. Both data sets are produced using remote sensed data in combination with NWP and validated by local meteorological stations. Seeing the correlation plots of the parameters it could be demonstrated that the ANFIS technique produces a robust model which follows very well the patterns of the training data. On the other hand it can be noted that with the proposed methodology, good results could be achieved in-spite of limited input data. Finally, it can be said that this methodology is creative and unique in its approach on GIS-based solar assessment as well its application to the case study in Ethiopia.

The main contributions of this work are:

- A GIS-based data mining methodology using Sugeno ANFIS is proposed to interpolate global solar irradiation.
- A clear and reproducible statistical method is proposed to create a reduced and representative training set.
- The proposed model may be easily adapted for any geographical location with limited solar data records.

The presented case study is a simplified training procedure for ANFIS estimation of radiation, even though this procedure is reproducible at any scale and able to contain more parameters in order to improve its performance.

Author Contributions: Conceptualization, B.A.W. and J.A.D.-N.; methodology, B.A.W.; software, B.A.W.; validation, B.A.W. and H.B.; writing—original draft preparation, B.A.W. and H.B.; writing—review and editing, J.A.D.-N.; supervision, J.A.D.-N. All authors have read and agree to the published version of the manuscript.

Funding: This research received no external funding.

Conflicts of Interest: The authors declare no conflict of interest.

Abbreviations

The following abbreviations are used in this manuscript:

ANFIS	Adaptive Neuro Fuzzy Inference System
DEM	Digital Elevation Model
FIS	Fuzzy Inference System
GHI	Global Horizontal Irradiation
GIS	Geographic Information System
PVGIS	Photovoltaic Geographical Information System
RMSE	Root Mean Square Error
STD	Standard Deviation
SWERA	Solar Wind Energy Resource Assessment

References

1. Ångström, A.; Angstrom, A. On Radiation and Climate. *Geogr. Ann.* **1925**, *7*, 122, [CrossRef]
2. Glover, J.; McCulloch, J.S.G. The empirical relation between solar radiation and hours of sunshine. *Q. J. R. Meteorol. Soc.* **1958**, *84*, 172–175, [CrossRef]
3. Paulescu, M.; Stefu, N.; Calinoiu, D.; Paulescu, E.; Pop, N.; Boata, R.; Mares, O. Ångström–Prescott equation: Physical basis, empirical models and sensitivity analysis. *Renew. Sustain. Energy Rev.* **2016**, *62*, 495–506, [CrossRef]

4. Almorox, J.; Arnaldo, J.; Bailek, N.; Martí, P. Adjustment of the Angstrom-Prescott equation from Campbell-Stokes and Kipp-Zonen sunshine measures at different timescales in Spain. *Renew. Energy* **2020**, *154*, 337–350, [CrossRef]
5. Perez, R.; Ineichen, P.; Seals, R.; Michalsky, J.; Stewart, R. Modeling daylight availability and irradiance components from direct and global irradiance. *Sol. Energy* **1990**, *44*, 271–289, [CrossRef]
6. Kamali, G.A.; Moradi, I.; Khalili, A. Estimating solar radiation on tilted surfaces with various orientations: A study case in Karaj (Iran). *Theor. Appl. Climatol.* **2006**, *84*, 235–241, [CrossRef]
7. Kumar, L.; Skidmore, A.K.; Knowles, E. Modelling topographic variation in solar radiation in a GIS environment. *Int. J. Geogr. Inf. Sci.* **1997**, *11*, 475–497, [CrossRef]
8. Monteiro, C.; Saraiva, J.; Miranda, V. Evaluation of electrification alternatives in developing countries-the SOLARGIS tool. In Proceedings of the MELECON '98 9th Mediterranean Electrotechnical Conference, Tel-Aviv, Israel, 18–20 May 1998; Volume 2, pp. 1037–1041.
9. Munkhammar, J.; Widén, J.; Hinkelman, L. A copula method for simulating correlated instantaneous solar irradiance in spatial networks. *Sol. Energy* **2017**, *143*, 10–21, [CrossRef]
10. Chelbi, M.; Gagnon, Y.; Waewsak, J. Solar radiation mapping using sunshine duration-based models and interpolation techniques: Application to Tunisia. *Energy Convers. Manag.* **2015**, *101*, 203–215, [CrossRef]
11. Robaa, S. Validation of the existing models for estimating global solar radiation over Egypt. *Energy Convers. Manag.* **2009**, *50*, 184–193, [CrossRef]
12. Rich, P.; Dubayah, R.; Hetrick, W.; Saving, S. *Using Viewshed Models to Calculate Intercepted Solar Radiation: Applications in Ecology;* American Society for Photogrammetry and Remote Sensing Technical Papers; American Society of Photogrammetry and Remote Sensing: Bethesda, MD, USA, 1994; pp. 524–529.
13. Dubayah, R.; Loechel, S. Modeling Topographic Solar Radiation Using GOES Data. *J. Appl. Meteorol.* **1997**, *36*, 141–154, [CrossRef]
14. Gueymard, C.A. Clear-sky irradiance predictions for solar resource mapping and large-scale applications: Improved validation methodology and detailed performance analysis of 18 broadband radiative models. *Sol. Energy* **2012**, *86*, 2145–2169, [CrossRef]
15. Voivontas, D.; Tsiligiridis, G.; Assimacopoulos, D. Solar potential for water heating explored by GIS. *Sol. Energy* **1998**, *62*, 419–427, [CrossRef]
16. Montero, G.; Escobar, J.; RodrÃguez, E.; Montenegro, R. Solar radiation and shadow modelling with adaptive triangular meshes. *Sol. Energy* **2009**, *83*, 998–1012, [CrossRef]
17. Piedallu, C.; Gégout, J.C. Efficient assessment of topographic solar radiation to improve plant distribution models. *Agric. For. Meteorol.* **2008**, *148*, 1696–1706, [CrossRef]
18. Charabi, Y.; Gastli, A. GIS assessment of large CSP plant in Duqum, Oman. *Renew. Sustain. Energy Rev.* **2010**, *14*, 835–841, [CrossRef]
19. Schillings, C.; Meyer, R.; Trieb, F. *Solar and Wind Energy Resource Assessment (SWERA)*; DLR report to UNEP; 2004. Available online: http://www.en.openei.org/wiki/SWERA/Data (accessed on 24 June 2020).
20. JRC. PVGIS. 2012. Available online: http://re.jrc.ec.europa.eu/pvgis/ (accessed on 24 June 2020).
21. NREL. NREL, National Renewable Energy Laboratory. 2018. Available online: http://www.nrel.gov/rredc/ (accessed on 24 June 2020).
22. NCR. NCR, Natural Resources Canada. 2018. Available online: https://www.nrcan.gc.ca/18366 (accessed on 24 June 2020).
23. BOM. Australian Government, Bureau of Meteorology. 2018. Available online: http://www.bom.gov.au/climate/maps/ (accessed on 24 June 2020).
24. VAISALA. VAISALA. 2018. Available online: https://www.vaisala.com/en/wind-and-solar-online-tools (accessed on 24 June 2020).
25. SolarGIS. SolarGIS. 2018. Available online: http://solargis.info/ (accessed on 24 June 2020).
26. Martínez-Durbán, M.; Zarzalejo, L.; Bosch, J.; Rosiek, S.; Polo, J.; Batlles, F. Estimation of global daily irradiation in complex topography zones using digital elevation models and meteosat images: Comparison of the results. *Energy Convers. Manag.* **2009**, *50*, 2233–2238, [CrossRef]

27. Huld, T. PVMAPS: Software tools and data for the estimation of solar radiation and photovoltaic module performance over large geographical areas. *Sol. Energy* **2017**, *142*, 171–181, [CrossRef]
28. Ertekin, C.; Yaldiz, O. Comparison of some existing models for estimating global solar radiation for Antalya (Turkey). *Energy Convers. Manag.* **2000**, *41*, 311–330, [CrossRef]
29. Duzen, H.; Aydin, H. Sunshine-based estimation of global solar radiation on horizontal surface at Lake Van region (Turkey). *Energy Convers. Manag.* **2012**, *58*, 35–46, [CrossRef]
30. Oumbe, A.; Wald, L. A parameterisation of vertical profile of solar irradiance for correcting solar fluxes for changes in terrain elevation. In Proceedings of the Earth Observation and Water Cycle Science Conference, Frascati, Italy, 18–20 November 2010; p. S05.
31. Sen, Z. *Solar Energy Fundamentals and Modeling Techniques*; Springer: Berlin/Heidelberg, Germany, 2008; p. 276.
32. Mellit, A.; Kalogirou, S. Neuro-Fuzzy Based Modeling for Photovoltaic Power Supply System. In Proceedings of the 2006 IEEE International Power and Energy Conference, Putra Jaya, Malaysia, 28–29 November 2006; pp. 88–93.
33. Mohammadi, K.; Shamshirband, S.; Kamsin, A.; Lai, P.; Mansor, Z. Identifying the most significant input parameters for predicting global solar radiation using an ANFIS selection procedure. *Renew. Sustain. Energy Rev.* **2016**, *63*, 423–434, [CrossRef]
34. Güçlü, Y.S.; Öner Yeleğen, M.; İsmail D.; Şişman, E. Solar irradiation estimations and comparisons by ANFIS, Angström–Prescott and dependency models. *Sol. Energy* **2014**, *109*, 118–124, [CrossRef]
35. Olatomiwa, L.; Mekhilef, S.; Shamshirband, S.; Petković, D. Adaptive neuro-fuzzy approach for solar radiation prediction in Nigeria. *Renew. Sustain. Energy Rev.* **2015**, *51*, 1784–1791, [CrossRef]
36. OpenDEM. OpenDEM home page, 2018. Available online: https://opendem.info/link_geodata.html (accessed on 24 June 2020).
37. Marble, B. *Global Mapper Software*; 2018. Available online: http://www.bluemarblegeo.com/products/global-mapper.php (accessed on 24 June 2020).

Publisher's Note: MDPI stays neutral with regard to jurisdictional claims in published maps and institutional affiliations.

Article

Optimization of Spatial Configuration of Multistrand Cable Lines

Artur Cywiński [1], Krzysztof Chwastek [2,*], Dariusz Kusiak [2] and Paweł Jabłoński [2]

1 Omega Projekt s.j., ul. Topolowa 1, 43-100 Tychy, Poland; artur.cywinski@omega-projekt.pl
2 Faculty of Electrical Engineering, Częstochowa University of Technology, Al. Armii Krajowej 17, 42-201 Częstochowa, Poland; dariusz.kusiak@pcz.pl (D.K.); pawel.jablonski@pcz.pl (P.J.)
* Correspondence: krzysztof.chwastek@gmail.com or krzysztof.chwastek@pcz.pl

Received: 27 September 2020; Accepted: 11 November 2020; Published: 13 November 2020

Abstract: Skin and proximity effects have a considerable impact on current distribution in multistrand cable lines. Under unfavorable heat exchange conditions, some strands may be subject to excessive overheating, which may lead to serious malfunctions or even fires of the installation. The paper proposes a new criterion for a quick choice of spatial configurations, for which the effect might be minimized. A comprehensive analysis of literature cases is provided, including the recommendations of the U.S. National Code and the Canadian standard.

Keywords: multistrand cable lines; ampacity; skin and proximity effects; symmetry

---

1. Introduction

Delivery of electrical energy often requires cables with high current-carrying capacity, that is, ampacity. In some situations it is necessary to use multistrand cables connected in parallel. Since 1957, the publication year of the seminal paper by Neher and McGrath [1], much attention was paid by the engineering community to the problems related to computations of current distribution in such systems. Murgatroyd developed a method to compute total proximity loss per unit length in multistrand bunch conductors of arbitrary shape [2]. Dawson et al. carried out a simplified analysis of nonuniform current sharing based on a coupled circuit model [3]. An analysis of their results, pertaining to DC systems, allowed us to formulate a useful criterion for selecting the most promising spatial configurations of cables after a generalization to three-phase systems. Petty presented a matrix algebra method for solving current distribution among phase conductors [4]. Ghandakly et al. provided current sizing information for bundled cables in high-current applications in the form of ampacity tables [5]. Sellers and Black proposed several improvements to the Neher–McGrath model: a new parameter accounting for unequal heating among the cables, some improved relationships for thermal resistance for the fluid layer existing in pipe-type cables and cables installed in ducts, as well as a modified model for thermal resistance of concrete duct banks with nonsquare cross-sections [6]. Du and Burnett developed a general prediction method for current distribution and examined its usefulness on data concerning supply of Hong Kong office building installations [7]. The textbook by Anders [8] is a comprehensive source of information on steady-state and transient ampacity calculations for electric power cables, in particular for the cases not covered in IEC 60287 standard (derived from the Neher–McGrath paper). Special attention was paid to such problems as derating considerations for cables crossing thermally unfavorable regions and for deeply buried cables. The thermal effect on cable ampacity in multistrand cable systems has also been considered in several recent papers published in MDPI Energies [9,10].

Generally speaking, current distribution in multistrand cable systems is uneven due to several factors, such as skin and proximity effects related to strand geometry [3,5], the conditions of heat exchange with the neighborhood [8–10], soil inhomogeneity, moisture, installation depth (if the cables are buried), harmonic spectrum of currents flowing through the cables (cf. e.g., [11]), and the presence of buried metal objects, to mention but a few. Analytical treatment of all these factors is hardly possible [12], therefore numerical methods, in particular the Finite Element and the Finite Volume methods, have found wide use for engineering purposes.

At the same time, it should be stated that the problem is not merely a theoretical one; as pointed out by Čiegis et al. [13], many existing power lines are overdimensioned by up to 60% in terms of transmitted power, which means additional costs and waste of deficit core material (e.g., copper). Desmet et al. [14] estimate that cable losses reach 5% of the total power consumption. On the other hand, the infrastructure grid has to be redesigned constantly in order to adapt to new conditions resulting from installation of new distributed generating capacities like wind farms and so forth. The wide use of power electronics devices both in industry and at households results in the presence of highly distorted current spectra. All these challenges have to be faced by the designers of cable supply systems. Their negligence may lead to serious malfunctions [11,15] or even fires of cable installations [9]. Such a situation has occurred during professional activities of the first author of the present paper. Despite considerable efforts aimed at a correct choice of cables supplying the main low-voltage switchboard in a car hood factory, after putting the installation into operation, a fire resulting from overloading one phase of the supply system has destroyed completely the cable installation (Figure 1). The considered system was designed in accordance with the guidelines included in the Polish standard PN-IEC 60364-5-523:2001 (derived from translation from IEC 60364-5 part 52 International Standard) and was supposed to withstand a 10% safety margin, however, as pointed out previously, the system experienced a serious fault soon after being put into operation. This fact has inspired the first author to carry out additional studies on current distribution of multibundle cable systems. It was found during a "postmortem" analysis that in the considered system, considerable differences between phase current values were present due to the proximity and skin effects [16].

Figure 1. A destroyed cable supply system [16].

The aim of the present paper is to propose a simple approach to identify the most promising configurations of low-voltage multibundle cable systems, for which the skin and proximity effects are minimized. The original concept of the paper is to avail of some routines available in MATLAB Graph and Network Toolbox [17] in order to facilitate the description of interactions between the cables. It should be remarked that in this manuscript, a simplified model for current distribution is considered. It assumes that current distribution is a function of spatial geometry only. Therefore, the use of procedures related to graph theory is somewhat limited, that is, we use nonoriented graphs as a smart, easily scalable method to separate the description of cable geometry in real-life conditions from the abstract layer that specifies interactions between graph nodes (represented by the cables in the bundle). In this context, we believe the use of adjacency function allows one to simplify the notation and to avoid mistakes. Moreover, we think that the use of notation borrowed from the graph theory allows one to simplify the analysis in those cases when the direct application of barycenter criterion (described in detail in Section 2.3 of the present paper) is not possible (e.g., due to constraints introduced by real-life geometry of the ducts, limited space for cable location, etc.). Then the description of mutual interactions between the strands in the form of a graph may be incorporated as a fragment of some relevant optimization algorithm applied (e.g., to a circuit-based model similar to the one considered by Lee [18]). The full potential of graph theory, which incorporates powerful, highly optimized algorithms for computation of maximal flow between graph nodes (e.g., the Ford–Fulkerson algorithm), is still to be explored in a subsequent publication.

The paper is structured as follows. Section 2 describes the developed laboratory stand. The results of measurements carried out using a single-phase excitation for several cable configurations are presented. A simplified method to determine current distribution using the distance criterion is proposed. Section 3 includes a brief literature review concerning methods of computation of current distribution in multibundle cable systems. Subsequently, a generalization of the results is presented, indicating that in the cases considered as optimal ones, some spatial symmetry patterns may be found. A practical conclusion is the formulation of a criterion, which allows one to choose preliminary configurations to be examined in subsequent steps during an in-depth FEM-based analysis. Section 4 provides a verification of the proposed criterion using the freeware FEMM software for some simple cases.

The ultimate goal of the paper is to provide the designers of cable systems a really simple criterion, which can be easily implemented in a spreadsheet or verified "by hand" computations, allowing them to choose the most promising spatial configurations for subsequent analysis using coupled electromagnetic-thermal FEM calculations.

2. Materials and Methods

2.1. Experimental Determination of Current Asymmetry for Single-Phase Systems

A laboratory stand was developed for the experimental examination of several spatial configurations of single-phase cable systems. The photograph of the stand is shown in Figure 2, whereas in Figure 3, the considered configurations are depicted. Five of them included the strands made of YAKXS 1 × 70 cables, whereas the remaining two were made using YAKXS 1 × 240 cables. All considered configurations were excited from a single phase. For each setup, three values of excitation current were preset, that is, 200 A, 300 A, and the maximal value, possible to be obtained for the transformer used.

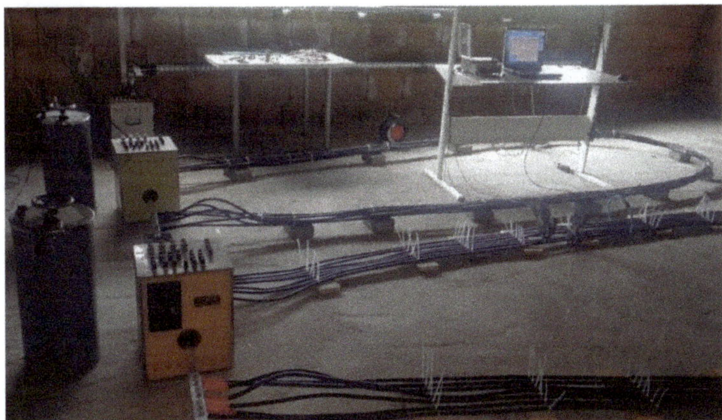

Figure 2. A photograph of the laboratory setup for examination of current distribution in the flat setup.

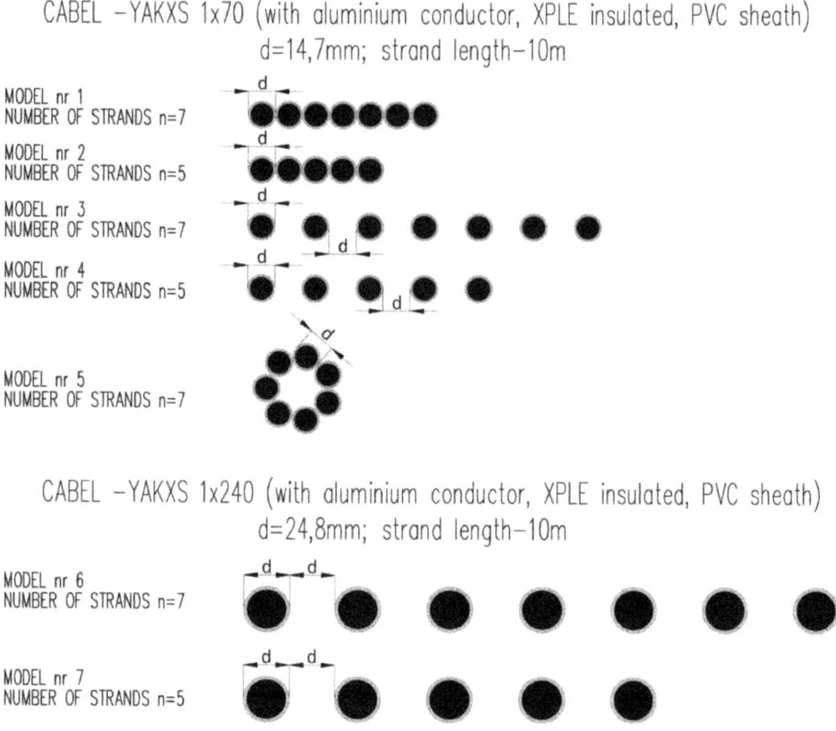

Figure 3. The considered cable configurations.

Ahead of the current measurements, the measurements of resistances of individual strands were carried out. The difference between extreme values was smaller than 1%. The measurements were made using the MMR-620 device from Sonel. The temperature of each strand before the measurements was the same and remained practically constant during the measurements (taking into account relatively small current magnitudes and short measurement time). Temperature recording for each strand was made individually for each phase. A control measurement of the ambient temperature was also carried out. Considering very small values of cable resistances (4.42 mΩ for YAKXS 1 × 70 cables, 1.21 mΩ for YAKXS 1 × 240 cables), it was crucial to prepare the cable endings properly and to minimize the values of contact resistances. The screw connections were applied and the cable endings were pressed from both sides using aluminum bars. Additionally, between the cable ending and the bar, the silver-based contact paste MG Chemicals 8463 was applied. After the connections were made, the measurements of contact resistance were carried out using the MMR-620 device. The average value of resistance for the 70 mm^2 cable ending was 27.6 μΩ, for the 240 mm^2 cable ending—15.4 μΩ. The deviation between the measured extreme values did not exceed 5%.

In order to compare the obtained measurement results for each spatial configuration, the ratio of currents in the extremely loaded strands (the asymmetry coefficient) was determined, as in Equation (1):

$$C_{AS} = \frac{I_{max}}{I_{min}}, \tag{1}$$

where I_{max} is RMS current value in the most loaded strand and I_{min} is RMS current value in the least loaded strand.

Table 1 presents the measurement results for the maximal attainable excitation current value, whereas Table 2 includes the values of asymmetry coefficient for different models, both for the maximal excitation current value and for the preset value 200 A. (Notation: S1—strand 1, etc., M1—model 1, etc.)

Table 1. Current of individual strands—summary for the maximal attainable excitation current.

	YAKY 1 × 70					YAKY 1 × 240	
	M1	M2	M3	M4	M5	M6	M7
S1	116 A	138 A	94 A	104 A	66 A	128 A	148 A
S2	104 A	128 A	84 A	94 A	66 A	106 A	126 A
S3	98 A	126 A	76 A	86 A	66 A	76 A	86 A
S4	92 A	128 A	70 A	92 A	66 A	67 A	122 A
S5	98 A	140 A	72 A	104 A	66 A	76 A	146 A
S6	102 A		84 A		68 A	108 A	
S7	116 A		106 A		66 A	128 A	
EXCITATION	726 A	660 A	586 A	480 A	465 A	689 A	628 A

Table 2. The values of asymmetry coefficient for two values of excitation current.

	PHYSICAL MODEL						
	M1	M2	M3	M4	M5	M6	M7
C_{AS}	1.26	1.11	1.51	1.20	1.00	1.91	1.69
C_{AS} for 200 A	1.28	1.13	1.50	1.13	1.00	1.50	1.50

The analysis of the obtained measurement results indicates that the current distribution in parallel lines is highly dependent on spatial configuration. More uneven distributions are obtained for bigger cable cross-sections. Strand separation, as well as the increase of their amount in the flat configuration, leads to the increase of asymmetry coefficient. For the considered physical models, the most uniform current distribution was obtained for model No. 5.

2.2. Distance-Based Criterion

Lee considered a cable configuration technique for the balance of current distribution in parallel three-phase cables in [18]. The method requires the computation of self- and mutual impedances of the cables, which are geometry-dependent. Figure 4 depicts schematically the couplings between individual strands.

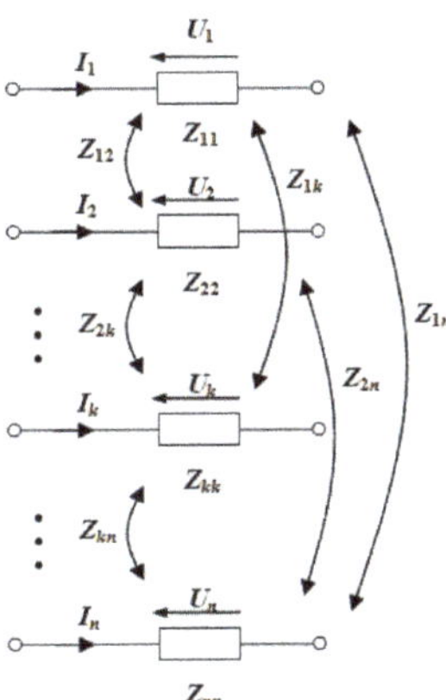

Figure 4. The couplings between individual strands (taken from [19]).

The problem with the method is that the resulting impedance matrix is a full matrix containing imaginary numbers (the elements on the main diagonal are complex numbers, since they include the strand resistances). Moreover, the computed values are relatively small in most practical cases, which causes some numerical problems with the precise determination of the inverse matrix needed in the subsequent computation step. This has led us to the formulation of a simplified method to determine current distribution, relying only on mutual distances between the strands. We have assumed that the current in the individual strand may be approximately proportional to the sum of distances of the strand to its neighbors. We have used the Graph and Network MATLAB toolbox [17] in order to introduce an abstract, geometry-independent layer, which facilitates the computations. An exemplary graph corresponding to model 2 or 4 is depicted in Figure 5a. The strand centers are represented as graph nodes and the distances between them may be written as appropriate weights for the edges (Figure 5b).

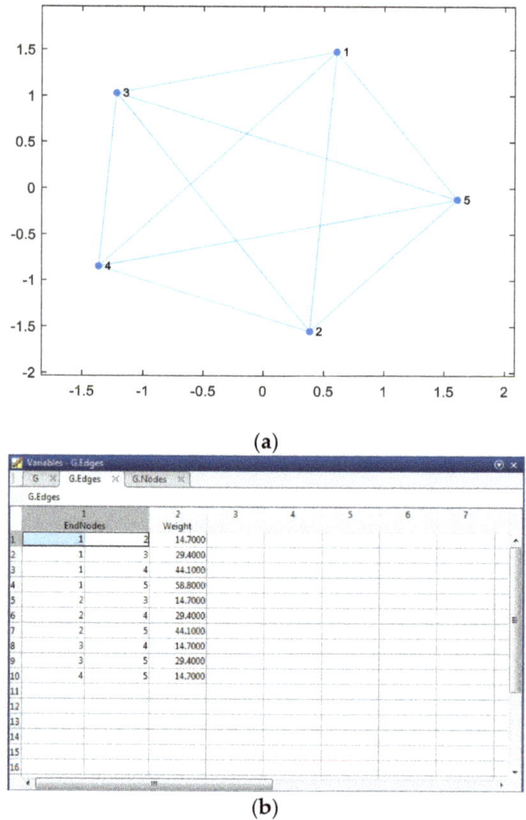

	EndNodes		Weight
1	1	2	14.7000
2	1	3	29.4000
3	1	4	44.1000
4	1	5	58.8000
5	2	3	14.7000
6	2	4	29.4000
7	2	5	44.1000
8	3	4	14.7000
9	3	5	29.4000
10	4	5	14.7000

(b)

Figure 5. (**a**) An exemplary graph corresponding to model 2 or 4. (**b**) An exemplary list of edge connections and distances (in mm) between graph nodes.

The number of all graph edges in the considered case is $\begin{pmatrix} 5 \\ 2 \end{pmatrix} = 10$ (the appropriate MATLAB command is nchoosek (5, 2)). Let us notice that in this case, the size of adjacency matrix is moderate (i.e., 10×10), but in practice for fewer cables per bundle in three-phase systems, it becomes large and the procedure of writing down the connection list becomes error-prone.

Some MATLAB code snapshots are given below:

```
points = [points_x points_y]; % node coordinates
distances = squareform(pdist(points));
G = graph(distances); G.Nodes.x=points(:,1); G.Nodes.y=points(:,2);
A = full(adjacency(G, G.Edges.Dist));
Currents = zeros(length(points_x),1);
for i = 1:length(Currents) Currents(i)=sum(A(i,:)); end
Currents = Currents./sum(sum(A));
```

A comparison of the values of computed and experimentally determined currents for the preset excitation current 200 A is given in Table 3. The discrepancies between the corresponding values do not exceed 21%.

Table 3. The percentage error values for the maximal attainable current values.

Configuration	Percentage Error for Current						
	S1	S2	S3	S4	S5	S6	S7
M1	17.2	2.0	−10.7	−14.3	−10.7	2.0	17.2
M2	19.0	−7.9	−21.0	−10.3	16.3		
M3	−16.9	−0.3	−10.5	−10.3	−5.5	−0.3	3.7
M4	19.0	−7.9	−21.0	−10.3	16.3		
M5	0	0	0	0	0	−3.0	0
M6	0.9	−7.2	5.2	10.2	5.2	−8.9	0.9
M7	6.0	−12.8	9.5	−9.9	7.5		

An interesting case is when the cables are placed on a circle radius (model 5), when the uniform current distribution was obtained. The simplified computational model also indicates that such spatial configuration is close to optimum. The results are the premise for further examination of cases, where the cables are placed in the nodes of equilateral polygons.

2.3. Barycenter-Based Criterion

Let us notice that in this case, the size of adjacency matrix is moderate (i.e., 10 × 10), but in practice for several cables per bundle in three-phase systems, it becomes large and the procedure of writing down the connection list becomes error-prone. When analyzing current distributions in cables with multiple strands, an important factor to be accounted for seems to be the symmetry. Intuitively, the arrangements with certain symmetry lead to even current distribution between the strands. To a certain extent, the symmetry can be taken into account by analyzing the barycenters (gravity centers) of strand groups. Therefore, we propose another simplified criterion for preliminary selection of the most promising arrangements of strands. It is based on minimizing distances between barycenters of strands belonging to individual phases. If (x_k, y_k) are coordinates of the center of strands consisting of phase A ($k = 1, 2, 3, \ldots n$), then the barycenter of phase A equals $x_A = (\sum_{k=1}^{n} x_k)/n$ and $y_A = (\sum_{k=1}^{n} y_k)/n$. Analogous formulas are used for phases B and C. Then we build a triangle of vertices (x_A, y_A), (x_B, y_B), (x_C, x_C) and calculate its area as follows:

$$S = \frac{1}{2} det \begin{bmatrix} x_A & y_A & 1 \\ x_B & y_B & 1 \\ x_C & y_C & 1 \end{bmatrix} \quad (2)$$

By analyzing several examples offered by literature in the subsequent section, we are convinced that the arrangements with as small S as possible are good candidates to keep a high degree of symmetry for current distribution. This criterion has to be modified when the phase barycenters are collinear, because then the triangle area is zero. Without loss of generality, we can assume then that

$Y_A = Y_B = Y_C$. In such a case, it is convenient to introduce discrepancy between phase barycenters as follows:

$$D = \max(X_A, X_B, X_C) - \min(X_A, X_B, X_C) \tag{3}$$

The analysis of several examples in the subsequent section indicates that "good" arrangements have D as low as possible.

3. Relevant Case Studies from Literature

In the paper [3], Dawson and Jain have analyzed chosen methods to balance current distribution in DC circuits (hot wire + earth). The authors have suggested the following methods to obtain more uniform current distributions:

- adding series resistance (not recommended, due to increased losses)
- adding series reactance (enhances the self-inductance/mutual inductance ratio. In this way, the effect of asymmetries in mutual inductance values is diminished. The method is not recommended for circuits operating at higher frequencies).
- transposition either of cables with the same polarity (Figure 6a) or of each pair of cables (Figure 6b)
- separation of cables with complementary polarizations or of groups of cable pairs (Figure 7). Red circle—"hot" (phase) wire, blue circle—return wire.

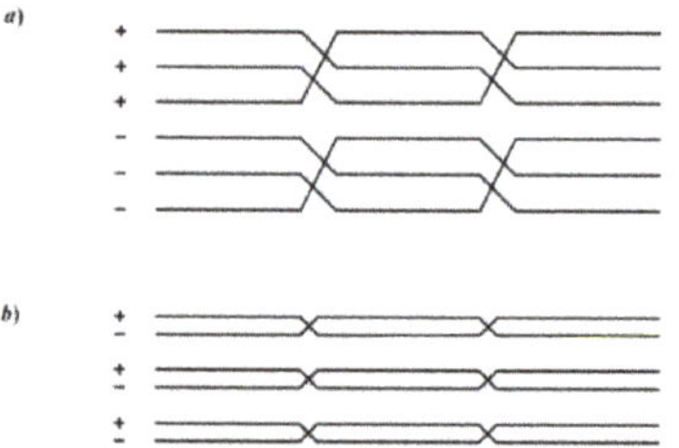

Figure 6. Strand transposition. Own work, based on [3].

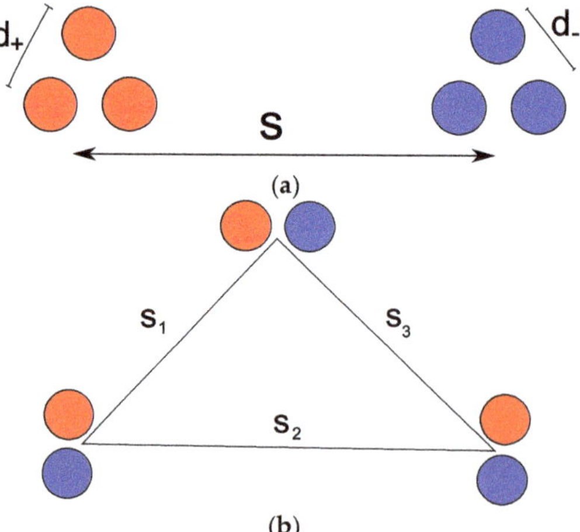

Figure 7. Separation: (**a**) of cables with complementary polarizations, (**b**) of groups of cable pairs. Own work, based on [3].

The authors have also pointed out some examples of optimal spatial configurations, providing uniform current sharing. Some examples are listed below in Figure 8:

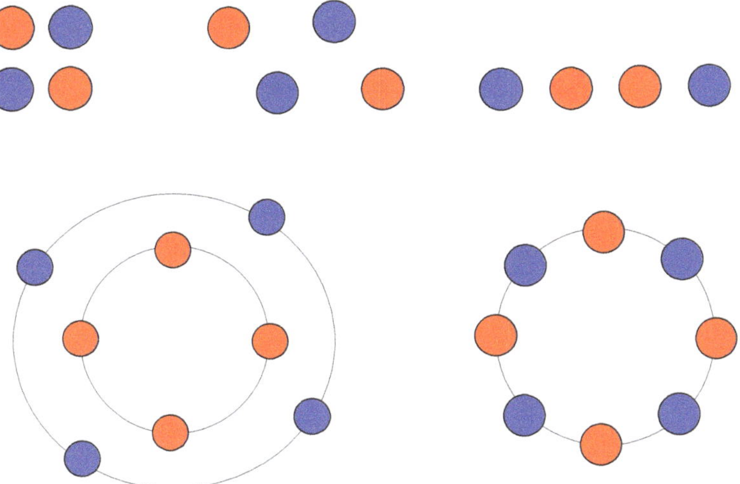

Figure 8. Spatial configurations considered by Dawson and Jain as optimal ones. Own work, based on [3].

Let us notice that all above-given configurations exhibit a common feature, namely the barycenters of supply and return wires approximately overlap. It is evident that there exists a certain symmetry of cable placement in each considered case. At the same time, highly asymmetric layouts (referred to as asymmetry of the first (Figure 9) and the second (Figure 10) kind) are not recommended.

Figure 9. Asymmetric configuration of the first kind. $I_1 = I_3 > I_2$. Own work, based on [3].

Figure 10. Asymmetric configuration of the second kind. $I_2 > I_1$. Own work, based on [3].

Lee has analyzed optimal configurations for three-phase three- and four-wire systems [18]. An exemplary optimal configuration for the flat layout is listed below in Figure 11. Colors brown, black, and red denote successive phases.

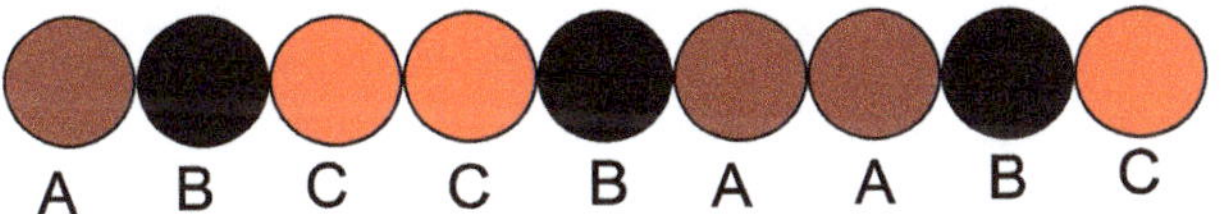

Figure 11. Optimal flat configuration suggested in [18].

If we assume that the center of the leftmost wire has the coordinate $x = 0$, the distance between the centers of successive wires is equal to unity, and the phase currents have the same amplitudes, then the gravity center for the phase L1 (A) is found at the point $\frac{1}{3}(0+5+6) \cong 3.67$, for the phase L2 (B) at the point $\frac{1}{3}(1+4+7) = 4$, and for the phase L3 (C) at the point $\frac{1}{3}(2+3+8) \cong 4.33$. The discrepancy between the barycenter coordinates does not exceed 17%.

For the "two shelves" layout, the optimal spatial configuration is suggested as in Figure 12.

Figure 12. Optimal configuration "two shelves with offset" suggested in [18].

If, similar to the previous case, we assume that the distance between the cable centers is equal to unity and we set the coordinate of the center of the leftmost cable on the lower shelf as (0,0), then it is trivial to compute (e.g., the coordinates of the center of the leftmost cable on the upper shelf as ($\sqrt{3}/2$, 1)). The barycenter of the phase L1 is located approximately in the point (1.956, 0.333), for the phase L2 in the point (2.244, 0.667), for the phase L3 in the point (2.577, 0.333). The whole area of interest is the area of the rectangle with dimensions 5 × 2. The area of the triangle determined by the barycenters of individual phases may be computed using the well-known dependence given by Equation (1). In the considered case, S = 0.0503, which may be referred to as the area of the rectangle ($5{\cdot}2 = 10$), means the triangle area is only 0.5%. Let us notice the vertical symmetry axis passing through the center of the black wire on the lower shelf. The horizontal symmetry could not be fulfilled because of the odd number of considered wires.

Another recommended layout for the nine-cable setup is shown below in Figure 13.

Figure 13. Optimal configuration "two shelves" suggested in [18].

The coordinates of the barycenter for the phase L1 (A) are $x_A = \frac{1}{3}(0+2+3) \cong 1.667$, $y_A = \frac{1}{3}(0+1+0) \cong 0.333$, for the phase L2 (B) $x_B = \frac{1}{3}(0+2+3) \cong 1.667$, $y_B = \frac{1}{3}(1+0+1) \cong 0.667$, and for the phase L3 (C) $x_C = \frac{1}{3}(1+1+4) = 2$, $y_C = \frac{1}{3}(0+1+0) \cong 0.333$. The area determined by the barycenters of individual phases is slightly higher than in the previous case, S = 0.0556, however, it remains at an acceptable level. It should be remarked that this configuration is slightly worse than the previous one, which can be confirmed with the results obtained by Lee himself (cf. the values of indicators δ_T and δ_M listed in Tables 2 and 3 in the paper [12]).

In the case of three-phase systems with neutral wire, the recommended layouts are depicted below in Figures 14–18:

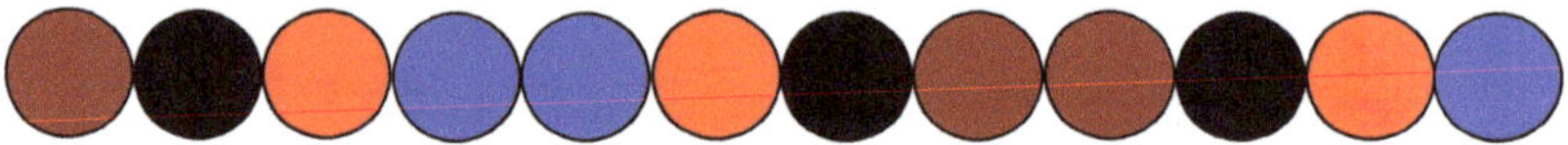

Figure 14. Optimal flat configuration (with neutral wires) suggested in [18].

Figure 15. Optimal configuration "two shelves with offset" (with neutral wires) suggested in [18].

Figure 16. Optimal configuration "two shelves without offset" (with neutral wires) suggested in [18].

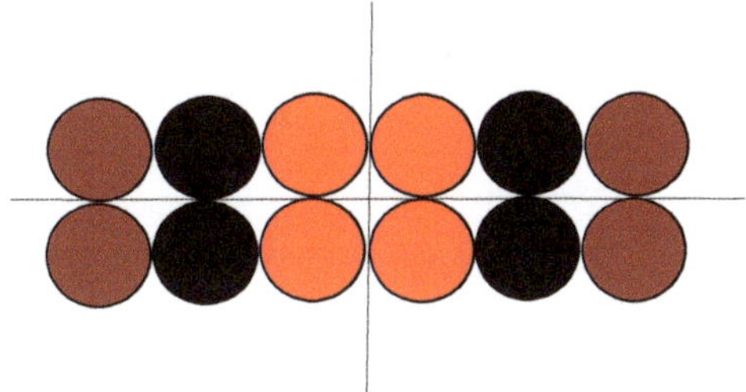

Figure 17. Fully optimal configuration "two shelves without offset" suggested in [18].

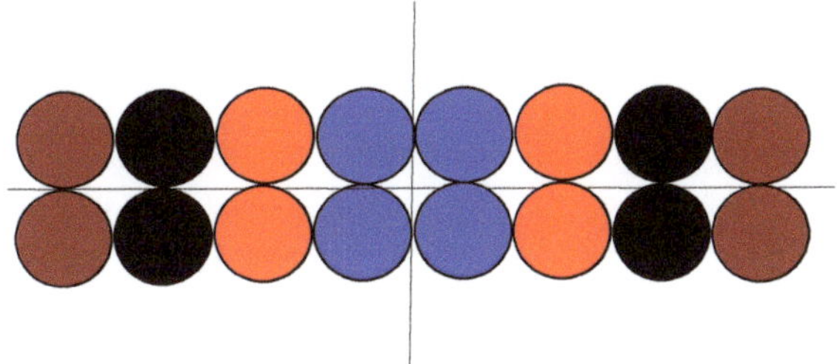

Figure 18. Fully optimal configuration "two shelves without offset" (with neutral wires) suggested in [18].

- Three strands per phase

(a) The flat system

$$x_A = \frac{1}{3}(0+7+8) = 5,\ x_B = \frac{1}{3}(1+6+9) \cong 5.333,\ x_C = \frac{1}{3}(2+5+10) \cong 5.667.$$

$$x_N = \frac{1}{3}(3+4+11) = 6.$$

The maximal discrepancy between the gravity centers for individual phases is equal to 0.667, which, referring to the shelf width (11), is only 6.06%. The gravity center for the neutral wire may be a little distant from the other ones, because under operating conditions (under standard assumptions concerning symmetries of supply and load), the current flowing through the neutral strands takes negligible values with respect to the phase currents.

(b) the layout "two shelves with offset"

$$x_A = \frac{1}{3}\left(0+4+3+\sqrt{3}/2\right) \cong 2.622,\ y_A \cong 0.333,\ x_B = \frac{1}{3}\left(\sqrt{3}/2+3+4+\sqrt{3}/2\right) \cong 2.911,$$

$$y_A \cong 0,667, x_C = \frac{1}{3}\left(1+5+\sqrt{3}/2+2\right) \cong 2.955,\ y_C \cong 0.333,$$

$$x_N = \frac{1}{3}\left(\sqrt{3}/2+1+2+\sqrt{3}/2+5\right) \cong 3.244,\ y_N \cong 0.667.$$

The area determined by the barycenters for individual phases is equal to 0.0073, which, referring to the area of the region with dimensions $5 + \sqrt{3}/2 \times 1$, is only 0.12%.

(c) the layout "two shelves without offset"

$$x_A \cong 2.333,\ y_A \cong 0.333,\ x_B \cong 2.333,\ y_B \cong 0.667,\ x_C \cong 2.667,\ y_A \cong 0.333,$$

$$x_N \cong 2.667,\ y_N \cong 0.667$$

The area determined by the gravity centers for individual phases is equal to 0.0558, which means 1.12% of the whole area for admissible solutions.

- Fours strands per phase

The layout symmetry is evident. For the odd number of strands, it was impossible to achieve an optimal placement for two shelves, but in this case, it is feasible. The presented layouts provide a full load symmetry for the strand wires.

- Recommendations concerning spatial layouts in the North American and the Canadian standards

Wu [20] has compiled a summary of configurations recommended by the U.S. and the Canadian standards for 2 ... 6 strands per phase. We have chosen for the analysis those cases, which are most often used in the practice. Analyzing the cases depicted in Figures 19–24, we can draw the following conclusions:

- the U.S. standards include flat configurations; such solutions are not used in Canada;
- the lumped triangular layout is used both in the U.S. and in Canada;
- the layout with strands lumped within a group (three phases close to each other, either in the flat or in the triangular configuration) may be perceived as three-phase equivalents of the recommended DC layouts, discussed previously;
- interesting conclusions may be arrived at when one analyzes the results illustrating the current nonuniformity for the lumped and dispersed triangular layouts (Table X in [20]). Setting the cables apart is used in practice in order to improve the heat exchange conditions. From the point of geometry, both layouts are equivalent, however, Wu has noticed that, paradoxically, the geometry with dispersed cables is worse, since the indicator I_{max}/I_{min} takes a higher value (1.59 for the dispersed cables, 1.31 for the lumped cables). It is remarkable that phase C features the highest current asymmetry in both cases, which is most probably due to an additional effect related to phase rotation sequence;
- for six strands per phase, there are some layouts recommended by the U.S. standards, which are not in use in Canada (the cables dispersed in a nonuniform manner, placed on three shelves for wire diameter in the range of 500–1000 mm^2, on two shelves for wire diameter up to 250 mm^2). The first aforementioned configuration features a very balanced current distribution (Tables XI and XII in [20]). The values of nonuniformity indicator I_{max}/I_{min} do not exceed 1.04.

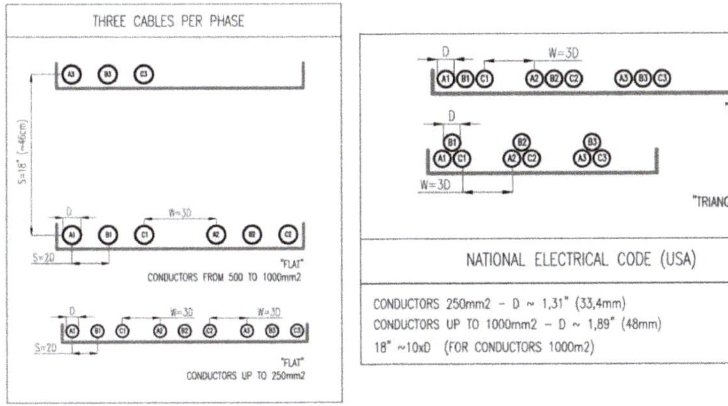

Figure 19. Layouts recommended by the U.S. standard. Own work, based on [20].

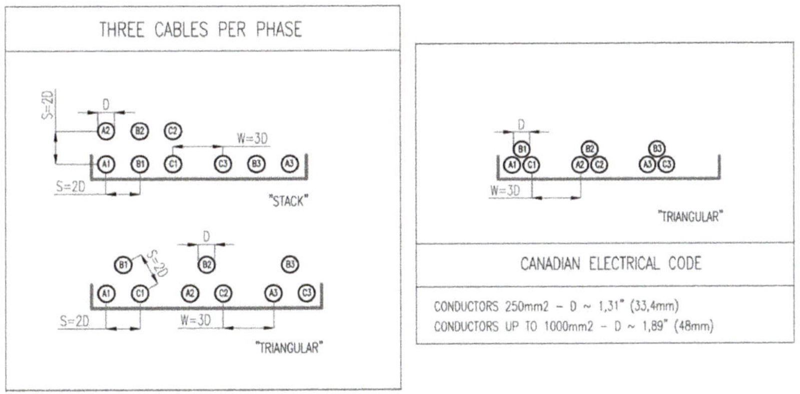

Figure 20. Layouts recommended by the Canadian standard. Own work, based on [20].

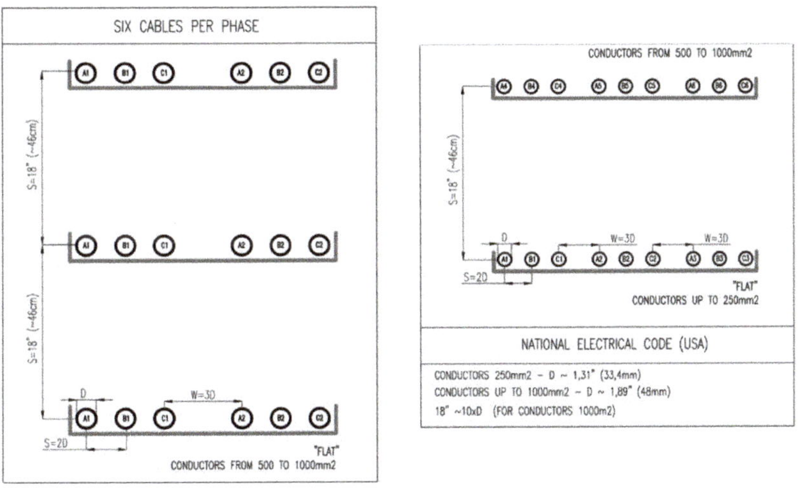

Figure 21. Layouts recommended by the U.S. standard. Own work, based on [20].

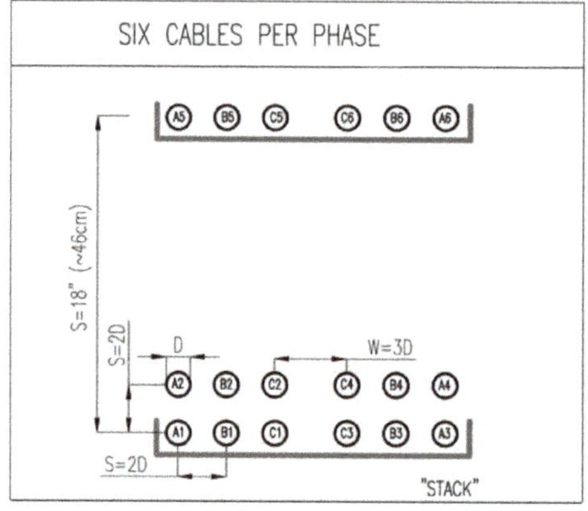

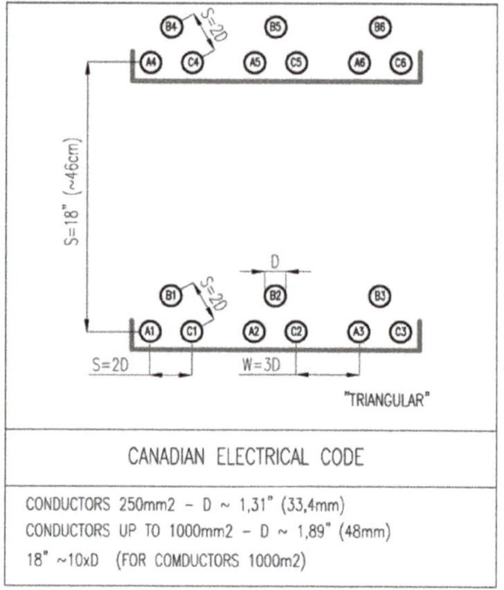

Figure 22. Layouts recommended by the Canadian standard. Own work, based on [20].

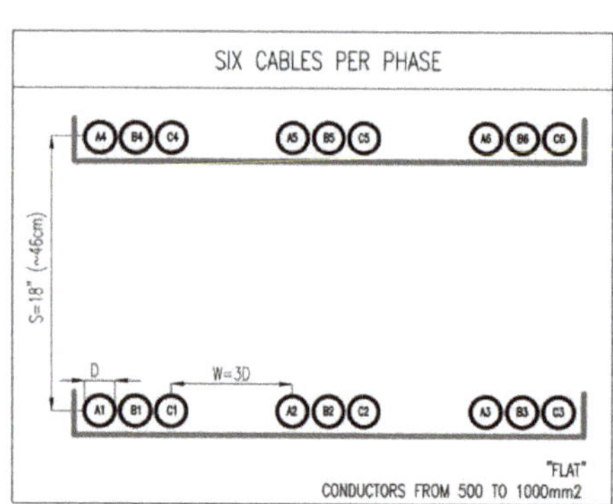

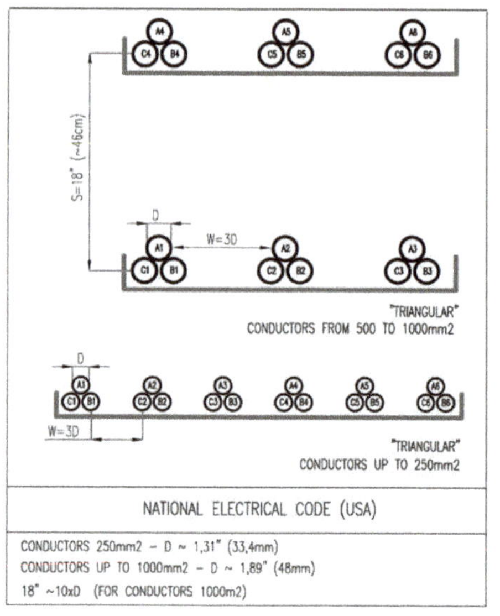

Figure 23. Layouts recommended by the U.S. standard. Own work, based on [20].

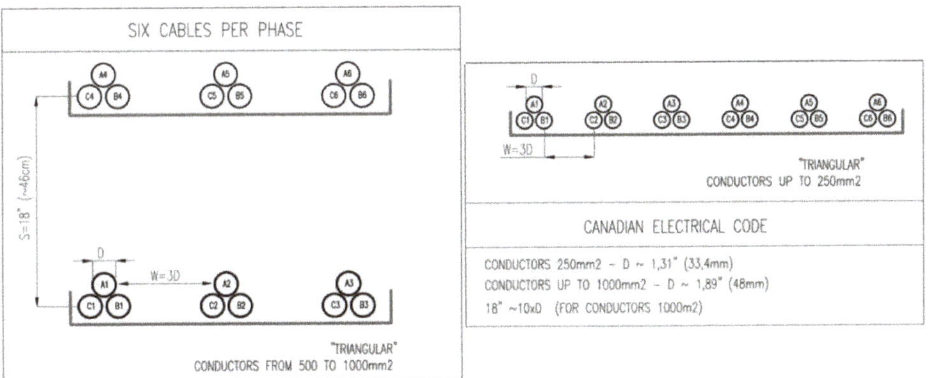

Figure 24. Layouts recommended by the Canadian standard. Own work, based on [20].

For the first aforementioned configuration (depicted in the leftmost part of Figure 21), we have determined the barycenter location for each phase. If we assume that the origin of the coordinate system (point (0,0)) is placed in the middle of the wire A1, and the distance between the shelves (i.e., 18″) is approximately 10 times the wire diameter, then the coordinates of the other wires are: B1(2,0), C1(4,0), A2(10, 0), B2(8,10), C2(6,0), A3(0,10), B3(2,10), C3(6,10), A4(10,10), B4(8,10), C4(6,10), A5(0,20), B5(2,20), C5(4,20), A6(10,20), B6(8,20), C6(6,20).

It is straightforward to compute that the barycenter coordinates for each phase are then the same and they are equal to (5, 10). It means that the proposed criterion for choosing the optimal spatial configuration is fulfilled in this case.

It should be noticed that not all spatial configurations from those depicted in Figures 19–24 result in a fully balanced load of individual strands. A practical conclusion resulting from the analysis of cases listed in [20] is that it is necessary to carry out detailed computations for each scenario, and the entries in the standards should be treated as recommended under certain circumstances.

- Other cases of interest analyzed in the literature

- the case of supply of a textile factory in Greece [21]

Two cable groups consisting of 18 flat-laid, single-conductor cables were laid on a metallic tray in free space, each with 300 mm^2 cross-section, cf. Figure 25. We assume a unit distance between the cable centers, a similar distance assumed between the cables on the internal edges and the compartment bar.

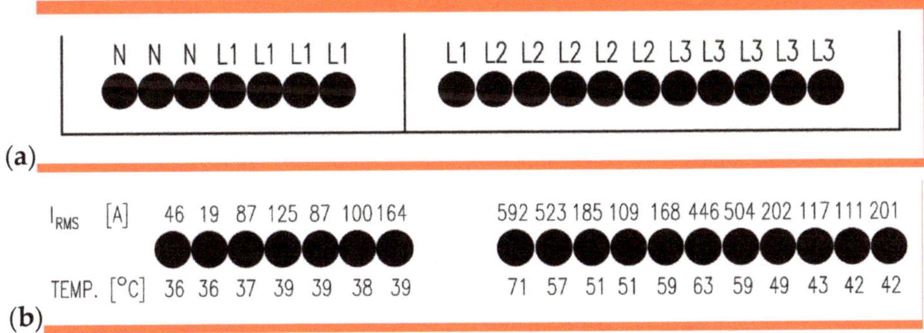

Figure 25. Preliminary cable configuration] (**a**) and the initial current distribution (**b**) [21].

The "x" coordinates of the barycenters for the case depicted in Figure 25 were as follows:

phase A $\frac{1}{5}(3+4+5+6+9) = 5.4$
phase B $\frac{1}{5}(10+11+12+13+14) = 12$
phase C $\frac{1}{5}(15+16+17+18+19) = 17$

For the optimized case (see Figure 26), the corresponding values were, respectively,

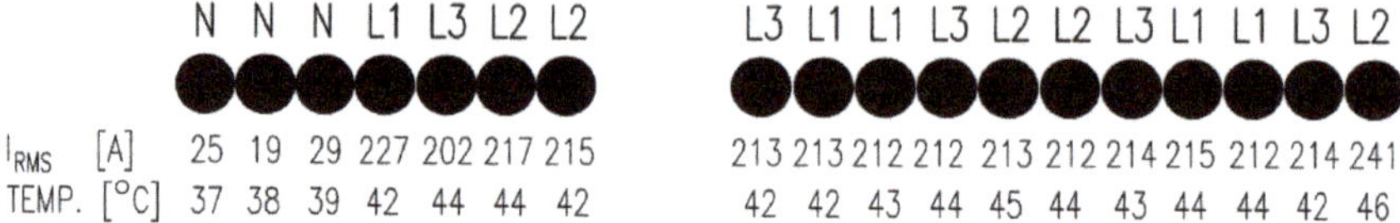

Figure 26. Current and temperature after system layout optimization. Own work, based on [21].

phase A $\frac{1}{5}(3+10+11+16+17) = 11.4$
phase B $\frac{1}{5}(5+6+13+14+19) = 11.4$
phase C $\frac{1}{5}(4+9+12+15+18) = 11.6$

The dispersion between the extreme current values was minimized from 11.6 (which was about 61% of total length considered) down to 0.2 (about 0.61%).

- The case of Teatro Regio in Turin, Italy [15]

The configuration recommended by the authors for the system with six strands per phase plus three neutral wires is depicted in Figure 27.

A A A

BC CB CB

AN AN AN

BC CB BC

Figure 27. Configuration recommended by the authors of [15].

Let us assume that the coordinates of B wire in the leftmost downward edge are (0; 0). Let us assume that the distance between the strand centers in the vertical direction is equal to unity, similarly for the horizontal direction, apart from the upmost row, where the distance is equal to two units, that is, the coordinates of phase A wires are (0; 4), (2; 4), and (4; 4), respectively. The barycenter coordinates are: for phase A (2; 2.5), for B and C phases (2; 5.1), for neutral wires (3; 1). The state close to the optimal one was achieved, and the full symmetry was unreachable due to the presence of neutral wires, which occupied some space. For neutral wires, the barycenter does not have to match the barycenter for phase wires, since under normal operating conditions and supply symmetry, the currents flowing through neutral wires are insignificant.

- Canova et al. have considered the optimal layout of cables using the Vector Immune System algorithm [22]. The case considered was six strands per phase, lack of neutral wire. The configuration indicated by the authors as the optimal one, featuring both the most uniform current partition and the lowest value of magnetic induction, was:
  BCAACBBCA
  ACBBCAACB

If we assume that the coordinates of the center of the leftmost downward A wire were (0;0) and the distance between successive wire centers was equal to three units both for the vertical and the horizontal directions, it is straightforward to compute that the barycenter location for each phase was (12; 1.5). It means that the configuration found by the optimization algorithm is indeed the optimal one and the symmetry state was achieved.

- Lee has considered a similar case (six wires per phase, two shelves), but with neutral wires [18]. Let us assume that the origin of the coordinate system was placed in the center of the leftmost downward wire (phase A) and that the distance was equal to unity.
  ABCNNCBANABC
  ABCNNCBANABC

The barycenter coordinates for phases A and B were equal to (5.67; 0.5), for phase C (6; 0.5), for the neutral wire (5; 0.5). The state close to the optimal one was achieved, and the full symmetry was unreachable due to the presence of neutral wires, which occupied a part of available space. For the neutral wires, the barycenter location does not have to correspond to the barycenter locations for phases since under normal operating conditions and for supply symmetry, the current values flowing though neutral wires are insignificant.

4. Verification of the Proposed Criterion for Optimal Current Distribution Using FEM

Taking into account the computation results from the last section, we can draw a conclusion that the criterion based on barycenter location might be useful for quick selection of potentially optimal spatial configurations, which might be helpful for the designers of cable systems. Additional verification for some simple scenarios is provided in this section, using the freeware FEMM software [23].

Comparing three cases depicted in Figures 28–30, it can be stated that the current uniformity is obtained for the system ABCCBA, for which the phase barycenter locations overlap. For the other two configurations, the current distribution is nonuniform. Table 4 contains the results of FEM-based computations for individual phases. The subscript "1" denotes the leftmost wire, the subscript "2" the rightmost one.

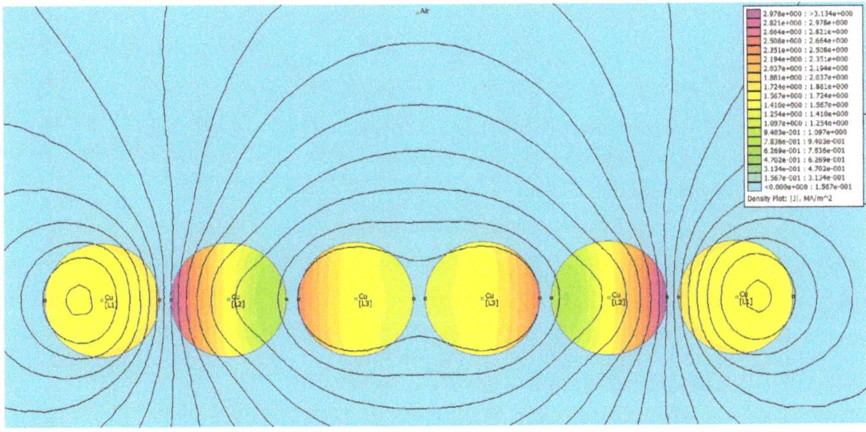

Figure 28. Current densities for the flat ABCCBA configuration.

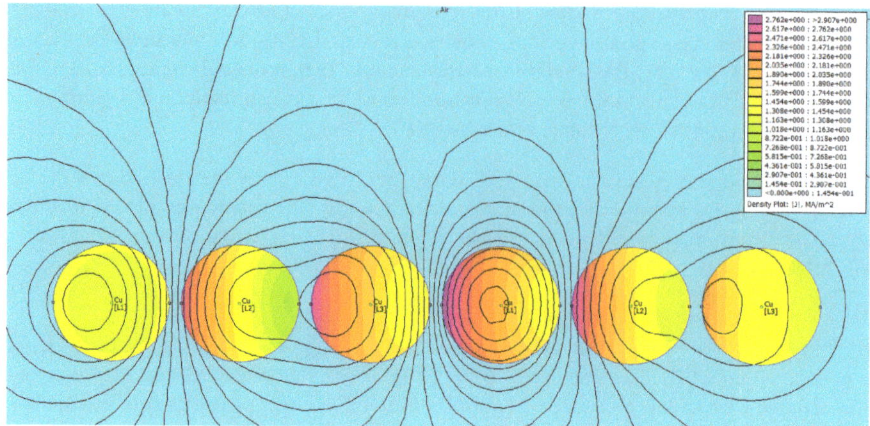

Figure 29. Current densities for the flat ABCABC configuration.

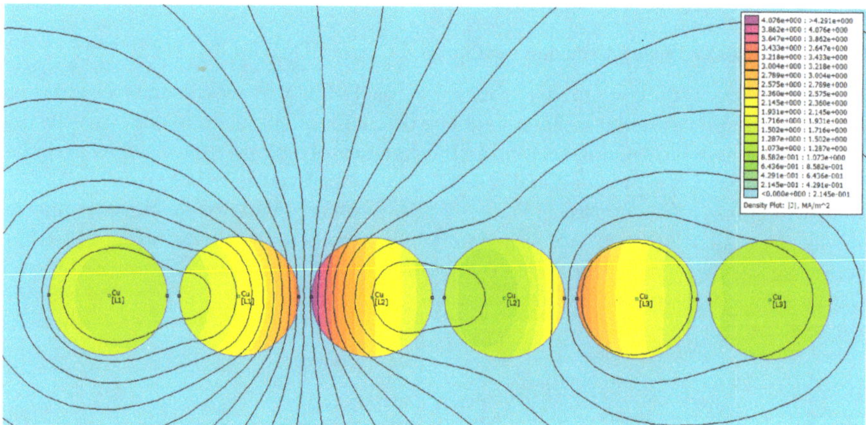

Figure 30. Current densities for the flat AABBCC configuration.

Table 4. The current percentage values in individual phases (flat configurations).

Configuration	Phase A		Phase B		Phase C	
	A_1	A_2	B_1	B_2	C_1	C_2
ABCCBA	50%	50%	50%	50%	50%	50%
ABCABC	38%	62%	48%	53%	57%	46%
AABBCC	46%	61%	78%	37%	64%	36%

Comparing three cases depicted in Figures 31–33, it can be stated that the current uniformity is obtained for the system ABC/CBA, for which the phase barycenter locations overlap. For the other two configurations, the current distribution is nonuniform. Table 5 contains the results of FEM-based load computations for individual phases. The subscript "1" denotes the upper wire, the subscript "2" the lower one.

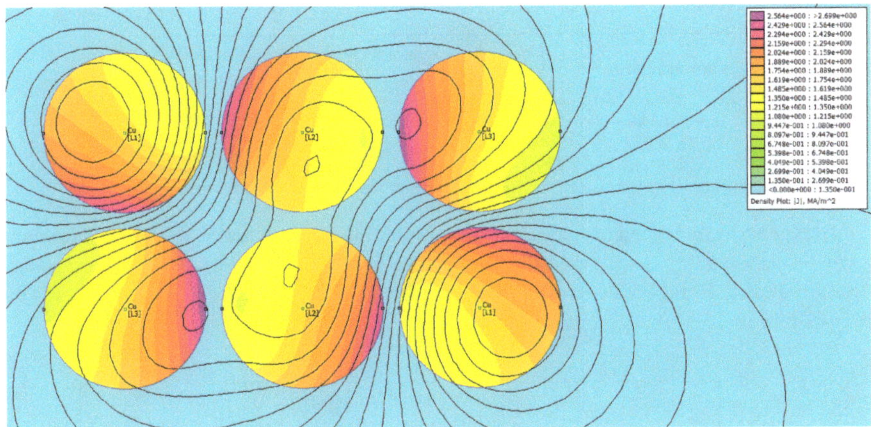

Figure 31. Current densities for the ABC/CBA configuration.

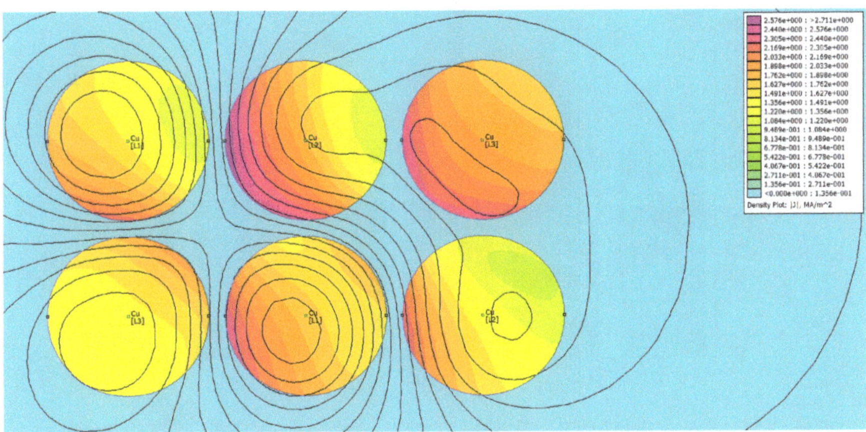

Figure 32. Current densities for the ABC/CAB configuration.

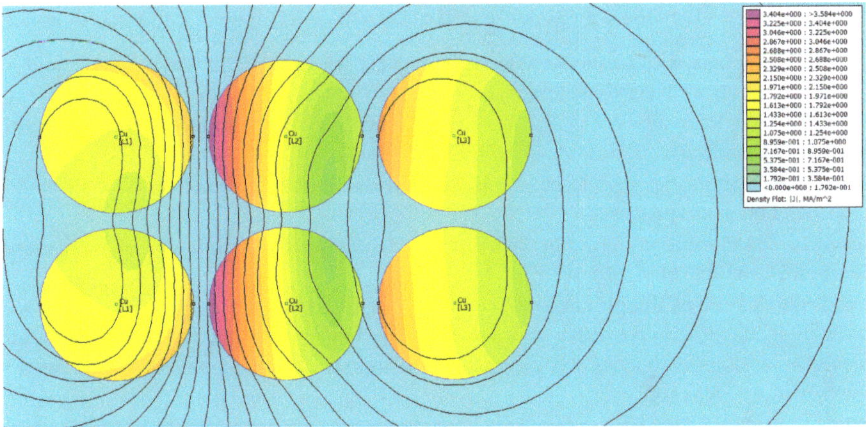

Figure 33. Current densities for the ABC/ABC configuration.

Table 5. The current percentage values in individual phases ("two shelves" configurations).

Configuration	Phase A		Phase B		Phase C	
	A_1	A_2	B_1	B_2	C_1	C_2
ABC/ABC	50%	50%	50%	50%	50%	50%
ABC/CBA	50%	50%	50%	50%	50%	50%
ABC/CAB	47%	54%	57%	43%	60%	46%

If the loss distribution shall be taken into account for individual phases, the best configuration of the six ones considered in this section shall be the layout ABC/CBA. Fractions of total loss dissipated in the wires are depicted graphically in Figure 34. Wire numbering goes from left to right for the flat layouts, for "two shelves" it is given as: wire 1–6.

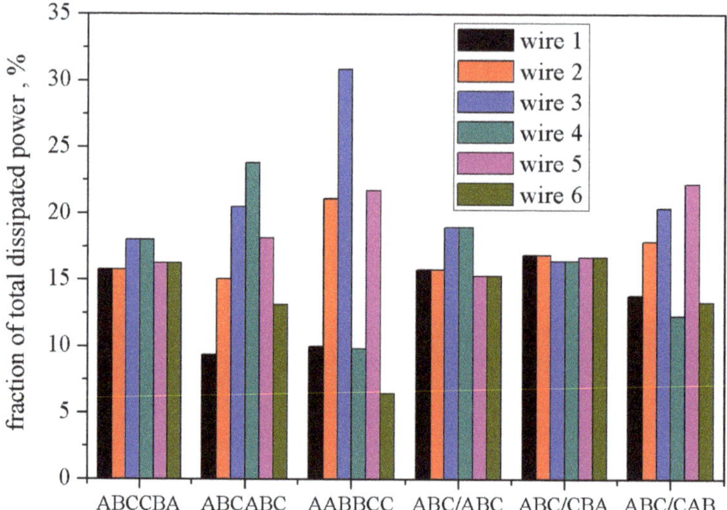

Figure 34. Loss distribution in individual wires for the configurations depicted in Figures 28–33.

5. Conclusions

The paper covers the following issues related to current distribution in multistrand cables:

- for the single-phase excitation case, a description of experimental stand was presented and several cable configurations were considered;
- we suggested to avail of MATLAB routines related to nonoriented graphs, in particular the adjacency function, in order to simplify the analysis;
- a simplified relationship for determination of current distribution based on the distance of the cable to its neighbors was proposed;
- several practical solutions found in the literature, including recommendations of the National U.S. Code and Canadian standard, were analyzed;
- a common feature of all optimal configurations was found, namely the barycenter locations for individual phases should overlap in order to provide uniform current distribution in the strands;
- a FEM-based verification of the proposed criterion was carried out for simple cases using the freeware FEMM software.

The proposed criterion might be useful for the designers of multistrand cable systems in order to reject some less optimal layouts quickly. It should be remarked that a final in-depth analysis of the performance of the multistrand cable systems should rely on FEM computations taking into account coupled electromagnetic-thermal phenomena. This issue shall be the subject of forthcoming research.

Author Contributions: Conceptualization, A.C. and K.C.; methodology, K.C.; software, D.K.; validation, A.C., K.C., and D.K.; formal analysis, P.J.; investigation, A.C.; resources, A.C.; data curation, K.C.; writing—original draft preparation, K.C.; writing—review and editing, A.C. and P.J.; visualization, A.C., K.C., and D.K; supervision, P.J.; project administration, K.C.; funding acquisition, K.C. and P.J. All authors have read and agreed to the published version of the manuscript.

Funding: This research received no external funding.

Conflicts of Interest: The authors declare no conflict of interest.

References

1. Neher, J.; McGrath, M. The calculation of the temperature rise and load capability of cable systems. *AIEE Trans. Part III* **1957**, *76*, 752–757. [CrossRef]
2. Murgatroyd, P.N. Calculation of proximity losses in multistranded conductor bunches. *IEE Proc.* **1989**, *126*, 115–120. [CrossRef]
3. Dawson, F.P.; Jain, P.K. A simplified approach to calculating current distribution in paralleled power buses. *IEEE Trans. Magn.* **1990**, *26*, 971–974. [CrossRef]
4. Petty, K.A. Calculation of current division in parallel single-conductor power cables for generating station applications. *IEEE Trans. Power. Deliv.* **1991**, *6*, 479–483. [CrossRef]
5. Ghandakly, A.A.; Curran, R.L.; Collins, G.B. Ampacity ratings of bundled cables for heavy current applications. *IEEE Trans. Ind. Appl.* **1994**, *30*, 233–238. [CrossRef]
6. Sellers, S.M.; Black, W.Z. Refinements to the Neher-McGrath model for calculating the ampacity of underground cables. *IEEE Trans. Power Del.* **1996**, *11*, 12–30. [CrossRef]
7. Du, Y.; Burnett, J. Current distribution in single-core cables connected in parallel. *IEE Proc.-Gener. Transm. Distrib.* **2001**, *148*, 406–412. [CrossRef]
8. Anders, G.J. *Rating of Electric Power Cables in Unfavorable Thermal Environment*; J. Wiley & Sons, Inc.: Hoboken, NJ, USA, 2005.
9. Xiong, L.; Chen, Y.; Jiao, Y.; Wang, J.; Hu, X. Study on the Effect of Cable Group Laying Mode on Temperature Field Distribution and Cable Ampacity. *Energies* **2019**, *12*, 3397. [CrossRef]
10. Zhu, W.W.; Zhao, Y.F.; Han, Z.Z.; Wang, X.B.; Wang, Y.F.; Liu, G.; Xie, Y.; Zhu, N.X. Thermal Effect of Different Laying Modes on Cross-Linked Polyethylene (XLPE) Insulation and a New Estimation on Cable Ampacity. *Energies* **2019**, *12*, 2994. [CrossRef]
11. Rasoulpoor, M.; Mirzaie, M.; Mirimani, S.M. Calculation of losses and ampacity derating in medium-voltage cables under harmonic load currents using finite element method. *Int. Trans. Electr. Energ. Syst.* **2017**, *27*, e2267. [CrossRef]
12. Jabłoński, P.; Szczegielniak, T.; Kusiak, D.; Piątek, Z. Analytical–Numerical Solution for the Skin and Proximity Effects in Two Parallel Round Conductors. *Energies* **2019**, *12*, 3584. [CrossRef]
13. Čiegis, R.; Jankevičiūtė, G.; Bugajev, A.; Tumanova, N. Numerical simulation of heat transfer in underground electrical cables. In *Progress in Industrial Mathematics at ECMI 2014, Mathematics in Industry 22*; Russo, G., Vincenzo, C., Giuseppe, N., Vittorio, R., Eds.; Spring Nature: Cham, Switzerland, 2014; pp. 1111–1119. [CrossRef]
14. Desmet, J.; Vanalme, G.; Belmans, R.; Van Dommelen, D. Simulation of Losses in LV Cables due to Nonlinear Loads. In Proceedings of the 2008 IEEE Power Electronics Specialists Conference, Rhodes, Greece, 15–19 June 2008; Volumes 1–10, pp. 785–3182683. Available online: http://hdl.handle.net/1854/LU-3182683 (accessed on 1 September 2020).
15. Freschi, F.; Tartaglia, M. Power lines made of many parallel single-core cables: A case study. *IEEE Trans. Ind. Appl.* **2013**, *49*, 1744–1750. [CrossRef]
16. Borowik, L.; Cywiński, A. Current-carrying capacity parallel single-core LV cable. *Przegl. Elektrotechn.* **2016**, *1*, 71–74. [CrossRef]
17. Mathworks. *MATLAB Toolbox for Graph and Network Algorithms*; Mathworks: Natick, MA, USA, 2015.
18. Lee, S.-Y. A cable configuration technique for the balance of current distribution in parallel cables. *J. Marine Sci. Techn.* **2010**, *18*, 290–297.

19. Jabłoński, P.; Kusiak, D.; Szczegielniak, T.; Piątek, Z. Reduction of impedance matrices of power busducts. *Przegl. Elektrotechn.* **2016**, *12*, 49–52. [CrossRef]
20. Wu, A. Single-conductor cables in parallel. *IEEE Trans. Ind. Appl.* **1984**, *20*, 377–395. [CrossRef]
21. Gouramanis, K.; Demoulias, C.; Labridis, D.; Dokopoulos, P. Distribution of non-sinusoidal currents in parallel conductors used in three-phase four-wire networks. *Electr. Power Syst. Res.* **2009**, *79*, 766–780. [CrossRef]
22. Canova, A.; Freschi, F.; Tartaglia, M. Multiobjective optimization of parallel cable layout. *IEEE Trans. Magn.* **2007**, *43*, 3914–3920. [CrossRef]
23. Meeker, D. Available online: http://www.femm.info (accessed on 1 September 2020).

MDPI
St. Alban-Anlage 66
4052 Basel
Switzerland
Tel. +41 61 683 77 34
Fax +41 61 302 89 18
www.mdpi.com

Energies Editorial Office
E-mail: energies@mdpi.com
www.mdpi.com/journal/energies

www.ingramcontent.com/pod-product-compliance
Lightning Source LLC
LaVergne TN
LVHW070104170726
843515LV00007B/1609

9783036507484